THE GRASSLANDS OF
THE UNITED STATES

Other Titles in
ABC-CLIO'S
NATURE AND HUMAN SOCIETIES SERIES

NATURE AND HUMAN SOCIETIES

THE GRASSLANDS OF THE UNITED STATES
An Environmental History

James E. Sherow

A B C 〰 C L I O

Santa Barbara, California • Denver, Colorado • Oxford, England

Library of Congress Cataloging-in-Publication Data
Sherow, James Earl.
  The grasslands of the United States : an environmental history / James E. Sherow.
  p. cm. – (Nature and human societies series)
  Includes bibliographical references and index.
  ISBN 978-1-85109-720-3 (hard copy : alk. paper) – ISBN 978-1-85109-725-8 (ebook)
1. Human ecology–United States. 2. Indians of North America–Ethnobotany. 3.
Grassland ecology–United States. 4. Nature–Effect of human beings on–United
States. 5. United States–Environmental conditions. I. Title.
GF503.S54 2007
333.740973–dc22

                                                                    2007000197

10 09 08 07   1 2 3 4 5 6 7 8 9 10

Production Editor: Kristine Swift
Editorial Assistant: Sara Springer
Production Manager: Don Schmidt
Media Editor: John Withers
Media Resources Coordinator: Ellen Brenna Dougherty
Media Resources Manager: Caroline Price
File Management Coordinator: Paula Gerard

ABC-CLIO, Inc.
130 Cremona Drive, P.O. Box 1911
Santa Barbara, California 93116–1911

This book is also available on the World Wide Web as an ebook. Visit
http://www.abc-clio.com for details.

This book is printed on acid-free paper ∞

Manufactured in the United States of America

CONTENTS

SERIES FOREWORD

L ong ago, only time and the elements shaped the face of the earth, the black abysses of the oceans, and the winds and blue welkin of heaven. As continents floated on the mantle, they collided and threw up mountains or drifted apart and made seas. Volcanoes built mountains out of fiery material from deep within the earth. Mountains and rivers of ice ground and gorged. Winds and waters sculpted and razed. Erosion buffered and salted the seas. The concert of living things created and balanced the gases of the air and moderated the earth's temperature.

The world is very different now. From the moment our ancestors emerged from the southern forests and grasslands to follow the melting glaciers or to cross the seas, all has changed. Today the universal force transforming the earth, the seas, and the air is for the first time a single form of life: we humans. We shape the world, sometimes for our purposes and often by accident. Where forests once towered, fertile fields or barren deserts or crowded cities now lie. Where the sun once warmed the heather, forests now shade the land. We exterminate one creature only to bring another from across the globe to take its place. We pull down mountains and excavate craters and caverns; drain swamps and make lakes; divert, straighten, and stop rivers. From the highest winds to the deepest currents, the world teems with chemical concoctions that only we can brew. Even the very climate warms from our activity.

And as we work our will upon the land, as we grasp the things around us to fashion them into instruments of our survival, our social relations, and our creativity, we find in turn our lives and even our individual and collective destinies shaped and given direction by natural forces, some controlled, some uncontrolled, and some unleashed. What is more, uniquely among the creatures, we come to know and love the places where we live. For us, the world has always abounded with unseen life and manifest meaning. Invisible beings have hidden in springs, in mountains, in groves, in the quiet sky and the thunder of the clouds, in the deep waters. Places of beauty from magnificent mountains to small, winding brooks have captured our imaginations and our affection. We have perceived a mind like our own, but greater, designing, creating, and guiding the universe around us.

The authors of the books in this series endeavor to tell the remarkable epic of the intertwined fates of humanity and the natural world. It is a story only now coming to be fully known. Although traditional historians have told the drama of men and women of the past, for more than three decades now, many historians have added the natural world as a third actor. Environmental history by that name emerged in the 1970s in the United States. Historians quickly took an interest and created a professional society, the American Society for Environmental History, and a professional journal, now called *Environmental History*. U.S. environmental history flourished and attracted foreign scholars. By 1990 the international dimensions were clear; European scholars joined together to create the European Society for Environmental History in 2001, with its journal, *Environment and History*. A Latin American and Caribbean Society for Environmental History should not be far behind. With an abundant and growing literature of world environmental history now available, a true world environmental history can appear.

This series is organized geographically into regions determined as much as possible by environmental and ecological factors, and secondarily by historical and historiographical boundaries. Befitting the vast environmental historical literature on the United States, four volumes tell the stories of the North, the South, the Plains and Mountain West, and the Pacific Coast. Other volumes trace the environmental histories of Canada and Alaska, Latin America and the Caribbean, Northern Europe, the Mediterranean region, sub-Saharan Africa, South Asia, Southeast Asia, East Asia, and Australia and Oceania. Authors from around the globe, experts in the various regions, have written these volumes, almost all of which are the first to convey the complete environmental history of their subjects. Each author has, as much as possible, written the twin stories of the human influence on the land and of the land's manifold influences on its human occupants. Every volume contains a narrative analysis of a region along with a body of reference material. This series constitutes the most complete environmental history of the globe ever assembled, chronicling the astonishing tragedies and triumphs of the human transformation of the earth.

Creating the series, recruiting the authors from around the world, and editing their manuscripts has been an immensely rewarding experience. I cannot thank the authors enough for all of their effort in realizing these volumes. I owe a great debt to Kevin Downing, who first approached me about the series, and Steven Danver at ABC-CLIO, who has shepherded the volumes through delays and crises all the way to publication. Their unfaltering support for and belief in the series were essential to its successful completion.

Mark Stoll
Department of History, Texas Tech University
Lubbock, Texas

ACKNOWLEDGMENTS

As a reference work, this volume owes much to the scholarship of others. I have included a smattering of my own original research which draws upon secondary literature in ecology, geography, agronomy, climatology, sociology, geology, archaeology, and history, and owe a great debt of appreciation to a handful of scholars who have ensured that I have not appeared more ignorant than I am. Brad and Lauren Ritterbush are excellent Great Plains archaeologists and their review of Chapter 1 improved it immensely. Jared Orsi, a rising star among environmental historians, contributed an excellent case study on Zebulon Pike, and read and critiqued the first four chapters. His criticism and suggestions strengthened this work considerably. Series editor Mark Stoll, an astute and highly knowledgeable environmental historian of the grasslands in his own right, read the entire manuscript. His keen discernment contributed to the overall quality of this volume. Steven Danver, acquisitions editor at ABC-CLIO, also read the manuscript in its entirety. His constant encouragement helped give me the strength to do what needed to be done. Alex Mikaberidze, submissions editor at ABC-CLIO, made substantive and valuable recommendations for improvement, and Kristine Swift's keen editorial eye streamlined and strengthened my prose.

Four graduate students assisted with the early secondary research included in this work. Margaret Bickers, Tim Hoheisel, Kent LaCombe, and Chris Vancil helped me find the gaps in my own knowledge of the grasslands. Too often, the wetlands of the grasslands are overlooked as essential components forming these ecosystems, and Margaret's case study, Wetlands of the American Grasslands, provides insights into how important numerous types of wetlands are to functioning grasslands. She further contributed by authoring the entries on the altithermal, Agnes Chase, Cheyenne Bottoms, Charles and Mary Ann Goodnight, Lady Bird Johnson, Playa Lakes, Pleistocene Extinctions, Walter Prescott Webb, and wildflowers.

Undeniably, my deepest gratitude goes to Bonnie Lynn-Sherow, a first-rate environmental historian if I may say so. I am truly fortunate to have her not only as a valuable and esteemed colleague, but one who has joined me in my journey through the grasslands and as my partner in life. I count myself doubly blessed for being able to combine love and scholarship into one relationship. In addition to giving the chapters a sharp, discerning editorial critique, she wrote the entries on George Catlin, *Lone Wolf v. Hitchcock* (1903), and winter wheat. This work has greatly benefited from her reading, and I have profited from her unwavering support, encouragement, and love.

A reader might be tempted to think this volume would be flawless with so much expertise assisting my efforts. But alas, I am all too human, and I bear the responsibility for whatever errors and shortcomings appear in this volume. Hopefully, these will be few, and this work will provide readers with a better understanding of the incredibly rich and textured history of the grasslands that I call home.

Jim Sherow
Kansas State University

INTRODUCTION

oring, flat, monotonous—these are words often used to describe the grass-lands of the United States. Kansas is, in fact, flatter than a pancake. This is the painstaking finding, anyway, of three geographers, two from Texas State University in San Marcos, and one from Arizona State University in Tempe. In 2003, they made two digital relief maps, one of the east to west topo-graphical profile of Kansas, and one of the surface of a "well-cooked" pancake purchased from the International House of Pancakes. When the reliefs of both were placed into the same scale and compared, the pancake had more eleva-tional variation than Kansas (see Fig. 1), or as stated in the case study, the de-gree of flatness in Kansas "might be described, mathematically, as 'damn flat'" (Fonstad, Pugatch, and Vogt 2003). There we have it, scientific proof that people crossing the grasslands find more topographical excitement in their pancakes than they do in crossing Kansas.

So why would anyone want to write an environmental history of this obvi-ously humdrum place? Well, maybe the grassland biomes, life communities, are not so vapid when given deeper consideration. They might provide important insights into the place and role that humans occupy on this tiny sphere of rock and water as it hurtles through infinite time and space. As John Zimmerman, a renowned prairie ornithologist, often reminded people, to "see" a grassland you have to look into it, not at it. The grasslands have been likened to an ocean with undulating waves of grass resembling waves on the surface of a sea (Zim-merman 1990, 1993). And like the ocean, we gain few insights about its depths from whitecaps.

Grasslands may very well be the most "imperiled ecosystem" on Earth (Samson and Knopf 1996). Will we, as humans, lose anything if they pass into extinction? Eugene M. Poirot, a Missouri farmer and environmentalist, believed the vitality of grasslands was important to the very survival of humankind.

> The once great prairies with their fruits and wildlife nourished our nation through its weak infancy. They nourished it again through its reckless and wasteful adolescence. The nation has now reached a maturity which should make it capable of recognizing that the prairie can no longer give that which it does not have, and that as man destroys it, he destroys himself (Poirot 1964).

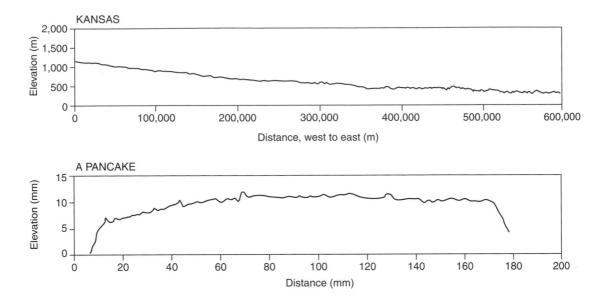

Figure 1 *Illustration showing that Kansas is flatter than a pancake.*

Others, such as Walt Whitman, and more recently William Least Heat-Moon, see the grasslands as America's characteristic landscape (Whitman 1892; Heat-Moon 1991). Indeed, there is much to lose should these insights bear any truth. Moreover, the study of grassland ecology has contributed to many important groundbreaking works. For example, out of Frederick Clements's studies of Nebraska grasslands came the climax theory of ecology, and James Malin's Kansas grasslands accounts were arguably the first environmental histories ever written. These two scholars asked important questions about the ecological relationships people had with their environments, even if they came to dramatically different conclusions. Clements saw people as disruptors of ecosystems that would otherwise maintain themselves in healthy, self-sustaining balance. Malin, on the other hand, had faith in people's abilities to control and enhance their surroundings through science and applied technology. More recently, Frank and Deborah Popper have gained national and international attention by suggesting that farming the Great Plains was a mistake and that Americans should resurrect the shortgrass prairies of old as one immense national park, a "buffalo commons."

Today, archaeological, ecological, climatological, and historical work reveals that grassland evolution is dynamic. The co-evolution of humans and grasslands is becoming clearer. It is not a stretch of the imagination to think one has given rise to the other, and more clearly now than before, humans must be seen as a "keystone species," their fate interlocked within this unfolding ecological drama. If this work does nothing else, it is hoped that it conclusively shows humans as a contributing, modifying species of grassland ecosystems.

Like any single volume, this work falls short in covering every aspect of grassland history. A problem from the outset was the decision as to which grasslands should be depicted. The grasslands of central Canada are omitted as are those of northern Mexico. While coverage of the Southwest and Great Basin grasslands are included, this volume stops short of discussing the Palouse, certainly an important region that will be covered in a volume on the Pacific Northwest. At the last moment, the editors and I agreed to include a case study on the Rocky Mountain West, which undeniably warrants a volume of its own, had resources and time permitted. Not every topic or study germane to the grasslands will be found in this volume, despite the fact that I went many thousands of words beyond what the series editor initially wanted written. However, this work will give the reader a fresh view of the grasslands and new insights about their historical dynamics and complexity. At the same time, it should be remembered that the environmental history of the grasslands remains an incomplete story, one still evolving and full of future discoveries. This work is one stop along that path.

References

Fonstad, Mark, William Pugatch, and Barndon Vogt. 2003. "Kansas Is Flatter Than a Pancake." *Annals of Improbable Research* 9 (May/June): 16–17.

Heat-Moon, William Least. 1991. *PrairyErth (A Deep Map).* Boston: Houghton Mifflin.

Poirot, Eugene M. 1964. *Our Margin of Life.* Raytown, MO: Acres USA.

Samson, Fred, and Fritz L. Knopf, eds. 1996. *Prairie Conservation: Preserving North America's Most Endangered Ecosystem.* Washington, DC: Island Press.

Whitman, Walt. 1892. *Prose Works.* Philadelphia: David McKay.

Zimmerman, John L. 1990. *Cheyenne Bottoms: Wetland in Jeopardy.* Lawrence: University Press of Kansas.

Zimmerman, John L. 1993. "The Birds of Konza: The Avian Ecology of the Tallgrass Prairie." Lawrence: University Press of Kansas.

THE EMERGENCE OF GRASSLAND RELATIONSHIPS PRE-1500 CE

THE ARRIVAL OF THE IMMIGRANTS

Many types of grassland communities have existed on the North American continent over the last 10,000 years. Grassland communities may be divided into four major groupings, or biomes. Generally defined, a biome is an ecological community dominated by a certain kind of plant. These plant communities, from west to east, are the intermountain and desert grasslands, the shortgrass Great Plains, the mixed-grass prairies, and the tallgrass prairies (see map on p. 2) (Brown 1985). The genesis of the grassland biomes, or ecological communities, of North America was the result of a combination of climatic change, geological forces and processes, shifting plant and animal communities, and human management decisions. Over the last two centuries, strangely enough, the forces shaping the relationships that gave rise to these biomes are also, in many ways, the same ones that are destroying them.

Before discussing much else, one word needs some clarification. Ecosystem is a useful term when employed carefully to describe life communities. Yet the term must be used with precision, as reputable scholars have questioned the concept of "ecosystem theory" as an apt descriptor of the ecological relationships into which every living thing, including people, is immersed. Ecosystem theory flourished after World War II and seemed to explain much. When used to define an area influenced and shaped by the interaction of organisms and abiotic forces, it is quite useful (Golley 1993). When ecosystem theory is used to describe a "precisely defined object of a predictive model," it starts becoming less useful. More recent scientific and historical research bear witness to ecological processes that show little resemblance to stable systems without catastrophic events or human interference. Rather, systems now appear as the product of flux and random events, and lacking stable equilibriums. In short, as the honored ecologist Robert V. O'Neill has noted, ecosystem theory as a predictive means "is a product of the human mind's limited ability to understand the complexity of the real world" (O'Neill 2001, 3276).

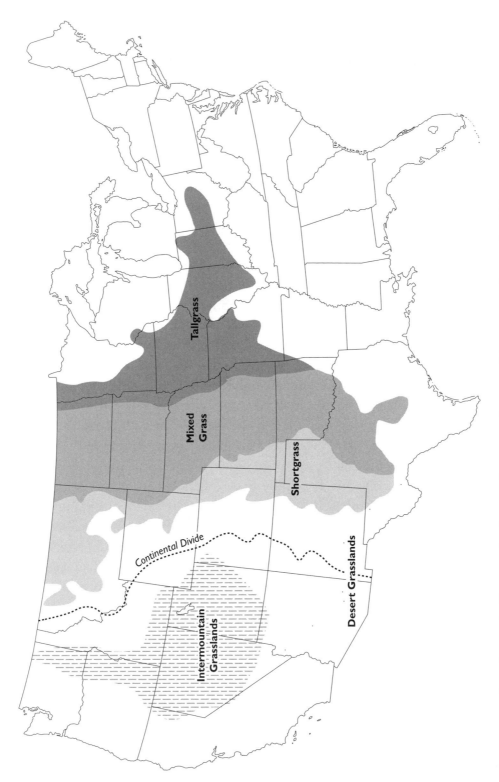

Tallgrass

Mixed Grass

Shortgrass

Continental Divide

Desert Grasslands

Intermountain Grasslands

Approximate boundaries of tallgrass, mixed-grass, shortgrass, desert, and intermountain grassland biomes.

Let us deal with the rise of the grasslands as ecosystems, but ecosystems understood as the relationship of fluctuating biotic and abotic entities and forces through time. Those biomes, or life communities, were formed by the arrival of immigrating plants and animals in the wake of melting glaciers toward the end of the Pleistocene Era. Consider the grasslands between the Rocky Mountains and the deciduous forests to the east. In present-day Kansas, there are only about four or five plant species in the remaining, isolated grasslands that were indigenous to the Pleistocene. All the rest of the grasses and flowering flora came from elsewhere in the years following the retreat of the glaciers (Axelrod 1985). For example, the tallgrasses had their origins in the eastern and southeastern forests of North America (Reichman 1987). In short, all of these grasses share a surprising temporal similarity: they are recent arrivals to the area that in time became the grasslands of North America. Now *recent* is a relative term and must be understood within the context of geological time.

More than tallgrasses made their way out onto the spaces opened up by retreating glaciers. The Red Hills of south central Kansas became a crossroads of migrating plants such as the American elm (*Ulmus americana*), which now grows in the moist areas along the eastern edge of the hills. This tree preceded advancing glaciers, and when they retreated, some populations of elm remained behind in those areas conducive to their health and reproduction. In essence, as one biologist put it, the Red Hills of Kansas stand as a plant refuge, one harboring a great variety of genera (Barrell 1975).

Other immigrants abound throughout the Great Basin and plains. In one striking case, a strong connection exists between Great Basin flora to the west of the cordillera and that found to the east of the line. A daisy, the Rocky Mountain Easter (*townsendia montana*), began a long migration down the west side of the Rocky Mountain Range into the Great Basin during a glaciated episode, one causing harsh cold conditions throughout the mountains, and of increasing rainfall throughout the Great Basin. The plant flourished in this wetter clime and advanced eastward throughout the grasslands along the southern reaches of present-day Arizona, New Mexico, and Texas, and then north into present-day Kansas. It found residency in the Red Hills of Kansas and can be found there today. Yet, it is no longer connected by an uninterrupted community of daisies to its immediate kin that thrive around the contiguous borders of Idaho, Wyoming, and Utah.

Some migrants became isolated in similar movements out of the mountains onto the plains. Today, Ponderosa pines (*Pinus ponderosa*) flourish along highly dissected areas in Nebraska. Also, red cedars (*Juniperus virginiana*) and hackberries (*Celtis occidentalis*) live in isolated stands throughout the roughly eroded areas of Nebraska. Most likely, these trees are remnants of the once vast woodlands covering the plains at the end of the last major glaciation (Wells 1965). When the short and mixed-grasses commenced their movement into the region west of the Rocky Mountains from the desert plains of the southwest, a mere 15,000 to 10,000 years ago, the expanse of the central grasslands was dominated by a "semi-open forest and woodland with scattered grassy patches." Around 14,000 years ago, the flat, shortgrass tableland known as the Llano Estacado supported open woodlands populated by pine and spruce, and as the climate gradually warmed, oak-juniper woodlands may have dominated the scene as little as 10,000 years ago (Axelrod 1985).

Before immigrating grasses could take root in these areas, the soils had to be right for them. Recently, ecologists have reaffirmed the importance of soil characteristics in determining where certain grasses will thrive. Many of the areas that eventually evolved into grasslands first were inland seas or lakes that once gone left behind deep sedimentary deposits. Other soils had their nature shaped by the depositional work of glaciers and winds as each force gathered or unloaded finely eroded rocks across broad sweeps of the North American continent. These beds created the sine qua non for grass communities that took root in them (Hook and Burke 2000).

Often, the story of the grasslands unfolds devoid of the formative relationships between humans and other plants and animals, and this is a mistaken rendering of events. Human beings managed the area and shaped the grasslands ecosystems that evolved throughout the end of the Pleistocene and continued to flourish throughout the Holocene until around 1850 CE. Recognizing the human complement to those life communities is crucial. Not only did humans affect the plant arrangements of the regions, people also played a significant, if hotly debated, role in the megafaunal extinctions, which in turn shaped the animal communities occupying the regions by 1500 CE. Many environmental historians and anthropologists have been asserting this proposition for some time. Presently, ecologists are recognizing the same thing. Again, in his MacArthur Award Lecture, O'Neill identifies human beings and their societies as a "keystone species" in any ecosystem. In his words, "*Homo sapiens* is not an external disturbance, it is a keystone species within the system" (O'Neill 2001, 3281). This insight has great importance given how other ecologists and historians have argued that bison have been the keystone species of the grasslands, or how others might point to prairie dogs fulfilling the same role. More likely, animals and plants both benefited from human fire practices. Grasses adapted exceptionally well to human fire practices as the flames cleared an unobstructed pathway into which grasses migrated. Their root and leaf systems also adjusted to, and thrived within, regularized burning regimens practiced by human beings. Certain herbivores also adapted well in those burned-over grass communities. As animal populations grew, human hunting practices also flourished and continually adapted to changing plant and animal connections. For millennia, these relationships co-evolved and mutually complemented each other within a certain range of climatic fluctuation.

The human part of the life relationships in the grasslands and intermountain regions took several different forms in the thousands of years following the retreat of the last glaciers. Human practices and occupation varied through time as people responded to a host of other factors such as erratic shifts in climate and variations in animal populations. Human migratory patterns and resource exploitation strategies in hunting and riverine agricultural practices were reflected in their material cultures. Human actions, tempered by chaotic climatic conditions, greatly affected animal and plant populations as they, too, attempted to adjust to the weather. During all of these years, humans were the keystone species in shaping the ecological relationships, ones binding them to all living things in the region, and all together, these relationships created the grassland communities of North America.

WHAT IS A GRASS?

A good place to grasp these evolutionary relationships is through an understanding of grasses. Grasses, which number some 10,000 species, along with humans, made a late appearance on Earth. For illustrative purposes, let us round off the age of the Earth at 4.5 billion years. If the age of the Earth is made the equivalent of one solar year, then one day is the equivalent of 12.3 million years. According to reputable scholars, contemporary grass ancestors made their first appearance on the planet around 60 to 70 million years ago, or during the last 5 to 6 days of our "Earth year." The first human ancestors, on the other hand, began appearing around 6 million years ago, or in about the last 12 hours of our Earth year. Beginning in the last 70 seconds of this Earth year, humans and grasses began emerging as dominant species in what would become the grasslands of North America. While arriving late on the scene, human societies and grasses have certainly complemented the expansion of each other. Grasses have nourished human civilizations, such as they have been, and human beings, in returning the favor, have aided and abetted the spread of grassland ecosystems in a few select places on the planet, such as the central and intermountain regions of the North American continent (Caetano-Anollés 2005).

The word *grass* has an interesting history. It is a derivative of an old Aryan word, *ghra-*, which means to grow, and much later some English words like *grain*, *green*, and, of course, *grow* would trace their ancient origins to *ghra-*. The

Romans probably used it to form their word *gramen*, or grass (Dayton 1948). Within this word lies the health, and perhaps sum, of human life. John James Ingalls, a senator from Kansas, observed this connection in an article that he penned and published first in the *Kansas Magazine* in 1872. "Grass feeds the ox: the ox nourishes man: man dies and goes to grass again; and so the tide of life, with everlasting repetition, in continuous circles, moves endlessly on and upward, and in more senses than one, all flesh is grass" (Ingalls 1948, 8). While Senator Ingalls understood, even if in a somewhat simplified manner, the importance of the grasslands to humanity, much about this life community, or biome, remained a mystery to him and the people of his times, and to those who had passed before him.

When Euro-Americans first described grasses throughout North America they usually grouped them into two categories: "warm-season" and "cool-season" grasses. Not surprisingly, warm-season grasses were observed growing late in the spring, thriving throughout mostly dry summers, maturing in the fall, and going dormant throughout the winter. "Cool-season" grasses on the other hand flourished well in wetter, cooler environments. Such places might encompass the northern prairies early in the spring or later in the fall, or the more southeasterly grasslands throughout the winter. In 1968 two biologists, W. J. Downton and E. Tregunna, figured out the different ways in which "warm-season" and "cool-season" grasses combined visible light and carbon dioxide to form carbohydrates, the basic fuels of all biological life. A simple, but crucial, problem confronts grasses: they must take in carbon dioxide while at the same time retaining the water necessary for their lives. Plant leaves regulate the absorption of carbon dioxide, which is indispensable for photosynthesis, through microscopic holes in their leaves called stomata. When a climate is wet there is little need for a leaf to close its stomata to prevent water loss, so it can easily capture as much carbon dioxide as its stomata can inhale. But if the climate is hot and dry, keeping the stomata open can prove exceptionally costly for a plant. Water loss through open stomata will result in dehydration and plants easily die as a result. Shutting off the stomata for a long period results in another problem: the plant has no means to absorb carbon dioxide for photosynthesis. This would be like a human stopping breathing in order to avoid dehydration in a desert. Such a decision would have only brief beneficial results (Redmann and Reekie 1982).

So, how did grasses fix the problem of inhalation without dehydration? Downton and Tregunna observed that warm-season and cool-season grasses differ in how they delivered carbon dioxide to where photosynthesis occurs. Cool-season grasses, which they labeled "C_3 grasses," absorbed and distributed carbon dioxide differently from the warm-season grasses, which they labeled "C_4

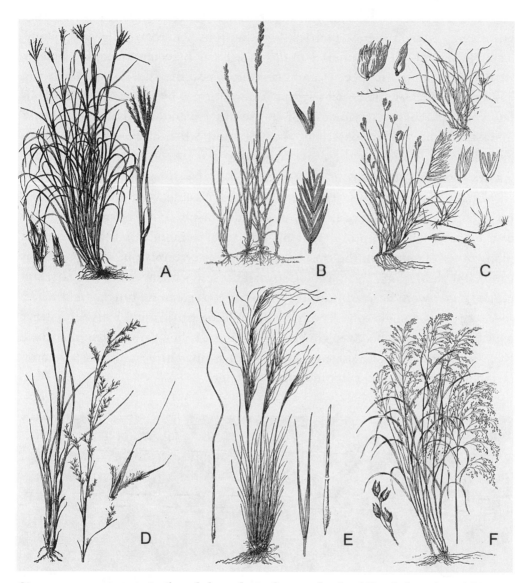

Six common grass species found throughout the grasslands of North America: a) Big Bluestem, b) Western Wheatgrass, c) Buffalo Grass, d) Little Bluestem, e) Needle-and-Thread, f) Switchgrass. (U.S. Department of Agriculture)

grasses." In short, C_4 grasses are more efficient at grabbing carbon dioxide from the atmosphere when they open their stomata, the pores in their leaves. This capacity allows these plants to shut their stomata more often and for longer periods of time than C_3 plants, and allows C_4 grasses to thrive in much drier climes. There is a cost involved to C_4 grasses, however, as they are less efficient at fixing carbon dioxide than are the cool-season grasses when water or temperature are not controlling factors (Reichman 1987). The C_4 plants once thoroughly dominated the landscape of the central grasslands. Throughout the eastern extent, or the tallgrass prairies, Big Bluestem (Andropogon gerardi) flourished; in the mixed-grass prairies to the west Little Bluestem (Andropogon scoparius) was the main species; and everyone knew the shortgrass plains and desert grasslands by the carpet-like appearance formed by the prevalence of Buffalo Grass (Buchloe dactyloides). Of course other plants lived alongside these mainstays, but any half-observant traveler knew when he or she had entered one plant association or the other by simply detecting which of these three grasses was most abundant.

Grasses have formed several symbiotic relationships with other species. An important interdependence is formed with organisms living below the surface of the earth. One group of organisms had an extremely important effect on many other life-forms in the grasslands, yet it was certainly one of the smallest to inhabit the region. Today ecologists call these organisms *mycorrhizae.* Mycorrhiza is a word derived from the Greek *myco*, meaning fungus, and *rhiza*, meaning root—fungus-root. The fungi resembling tiny, round balls with what appears to be a branched web growing out of it are called "arbuscular" or "little tree" mycorrhiza. These organisms average a width of five microns, a micron (symbolized by μ) being one-millionth of a meter.

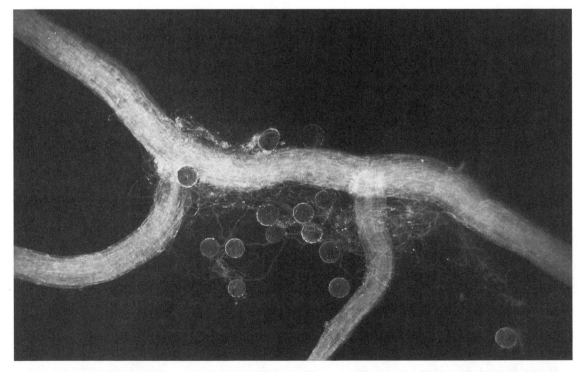

Averaging a mere five microns in diameter, arbuscular mycorrhizal play an important symbiotic role with grasses. (Sara Wright/USDA)

Among tallgrasses as many as twenty species of arbuscular mycorrhiza live in symbiosis with other plants. Healthy C_4 grasses will have their outer root cells thoroughly colonized with these fungi. In fact, tall bluestem or buffalo grass will fail to grow without a healthy bond with the fungi. Phosphorus is a crucial element to the well-being of any C_4 grass, and these fungi effectively transfer this essential element to the root system of the plant, and through the hyphae, or web extensions of the fungi, they transport phosphorous to adjacent plants. Prairie fires have the effect of increasing mycorrhizal populations by stimulating a greater production of spores. Heavy grazing will reduce the diversity of the mycorrhizal species in the soil and diminish the colonization of grass roots. The webs of the fungi, however, react by spreading their strands throughout the soil.

The arbuscular mycorrhizal symbionts of C_4 grasses also live within a narrow range of soil moisture. If the fungi receive too much rain then they will drown, and the grass loses its nutritional value. In the past, Euro-Americans called grasses in this condition "washy" because of the high water content in the plant and its poor nutritional value for grazing livestock. When the soils lose too much water through evaporation, then the fungi will die from dehydration. The grass hosts follow suit and die from two causes: a lack of water for the internal distribution of nutrients and the lack of the nutrients themselves, which the fungi had transported. In the historic grasslands, people struggled with their livestock during those seasons when an El Niño sent precipitation soaring. In this case, plants and animals suffered from more than just floods and nutritionally weakened plants—from a broken relationship of fungi to grasses, and grasses to livestock. All of this happened sight unseen for the humans who had to survive through those days.

There had to be more than a good relationship between plants, microbes, and soils for grasses to flourish in areas between the eastern woodlands and Rocky Mountains, and throughout many parts of the intermountain regions between the Front Range of Colorado and the Sierra Nevada Range to the west. Three other notable variables contributed to the success of grass biomes in these areas. The first is already noted: the plants had evolved by adjusting to the aridity of these lands. The second variable precedes the former. It would be very difficult for grasses to adapt to aridity if the lands were not subhumid to arid in the first place. The third, and perhaps most important variable to affect grasses, was human behavior.

Humans learned to wield one of the most effective agents ever to modify a landscape: fire. Without recurring fires throughout the subhumid to arid regions of North America, plant compositions would have been much different 500 years ago from what they were. There is an odd place in south central Kansas called Flowerpot Mound. Surrounded by grassland, prairie and sand sages blanketed the top. What accounts for this lonely outpost of sage in a sea of grass? The likely answer: fire and grazing animals could never reach the top of the formation. A slight overhang of sandstone caps the top of Flowerpot, and consequently the tableland atop seldom if ever was visited by fire or herbivores. For several thousands of years fire and grazers, on the other hand, shaped the grasslands around it. How different the Great Basin and grasslands would look today if not for the effects of fire and ungulates (Barrell 1975).

THE PLEISTOCENE SCENE

Continuing the metaphor of the Earth year, in the last seventy seconds the area that becomes the grasslands has undergone several dramatic climatic changes. During this brief span, just over a 10,000-year geological interval, the climate of the Earth warmed considerably. Scientists argue over what caused this heating up of the surface. Some think the explanation lies with plate tectonics, or the shifting up and down, and movement apart, of the continental plates. This activity causes some areas to receive greater precipitation and to heat more readily than others. This same geological occurrence could have increased volcanic activity, which in turn would have filled the atmosphere with dust. With enough dust in the air, solar energy reaching the surface of the planet would have been greatly altered and thereby would have affected global temperatures. Some scholars followed the lead of Milutin Milankovitch, a Russian climatologist who suggested that the fluctuations in orbit of the Earth around the Sun, which would vary the amount of solar energy reaching the surface of the planet, produced a cause and effect related to glaciation and glacier retreat (Hays, Imbrie, and Shackleton 1976).

Before this spate of heating began in earnest, let us say around 11,000 to 10,000 years ago, extensive glaciers covered large areas of the northern regions of the North American continent. A succession of oak woodlands and spruce forests dominated what would become the central grasslands. The evidence for this is especially prevalent at such heavily studied sites such as Konza Biological Research Station in the Flint Hills of Kansas. Other places, such as the Llano Estacado, the great, elevated plateau in the Panhandle of Texas, may have been covered in spruce and pine forests. This assertion is questioned by some who believe winds deposited spruce pollen from other regions across this great tableland. Still, the proponents of woodlands picture an emerging grassland ecosystem supporting small horses, giant bison, camels, and mammoths—species long extinct in the same area when the Spanish explorer Francisco Coronado arrived nearly 11,000 years later. Other areas seemed to have been exceptionally dry. A sparse, shortgrass biome dominated the area to the west of the open woodlands. Closer to the northern ice sheet a polar desert held sway. While there is some indication of human occupancy in the region, the aridity of the place rather than the cold probably limited where humans and wildlife could flourish. Most of the water was locked in ice and was largely unavailable for several forms of wildlife and humans.

Large lakes, several hundred square miles in area, covered large portions of what people today call the Great Basin. Temperate woodlands, many paleoclimatologists assert, surrounded these inland water bodies. Rainfall was much higher than today, and the temperatures, not surprisingly, were much lower. A few scholars, however, believe the pollen data are deceptive, and they draw a picture of semi-desert floral spans braided throughout open forests. However one depicts the Pleistocene Great Basin, it certainly was more heavily covered in trees and lakes than it is today. Toward the southern end, where the California Imperial Valley is today, lakes abounded there, too. One particularly large freshwater lake, called Lake Cahuilla, covered not only where the Salton Sea is today, but an immense region surrounding it. Once the glaciers began retreating to the north, the Colorado River flowed with more volume than ever after, and the lower Colorado River Valley had numerous lakes and a meandering river. What eventually becomes the Imperial Valley started out as a vast river delta, one collecting rich sediments carried by the river on its way to the Gulf of California.

Animal populations, too, differed considerably from those encountered by the humans who would occupy these regions some two to four thousand years later. Scholars refer to some of these animals as the megafauna, or large animals. These animals are quite familiar to us. The great mammoths, mastodons, ground sloths, bisons, bears, saber-toothed tigers, and wolves all occupied a place in that rapidly changing, highly volatile world. Yet other species less known, long extirpated or extinct, also populated parts of that incredibly diverse landscape. Fish swam in lakes where some of the driest deserts in North America are today. Tiny horses and large camels ran through the savannah woodlands where farmers harvest prolific wheat crops now.

Scholars usually refer to the first humans who lived during these times as "paleo-Indians." Certain, well-defined stone points are associated with these people, and the first recognized group who once lived in these areas is called "Clovis." They received this name because E. B. Howard and John Cotter, the first archaeologists to investigate their relics, studied the stone points associated with this ancient people near present-day Clovis, New Mexico, in 1932. These finely crafted points, similar to ones eventually discovered throughout the North American continent, had a distinctive design. Those people shaped stone into six-inch-long, two-inch-wide, fluted points. These points, scholars have shown, are closely connected to hunting megafauna. Generally, this culture flourished from around 11,500 to 11,000 years ago.

One of the most controversial topics regarding this culture is how it spread so quickly throughout these areas. Some scholars believe that the intermountain areas were heavily populated before the advent of this particular projectile, and the *idea* associated with this hunting technique radiated throughout the population. Another group of scholars believes people migrated throughout the hemisphere with this technology. If these areas were populated before the idea of the Clovis point, then the peopling of the intermountain areas began far earlier than 11,500 years ago. But if the technology spread with the arrival of people, then perhaps the human occupation of these same areas did not begin in earnest until after 11,500 years ago. To date, Clovis culture is the earliest recognizable human occupancy of what would become the grasslands of North America.

The most contentious issue revolving around these people is their contribution to the disappearance of the megafauna (from the Latin, huge + animal). Years ago, the archaeologist Paul S. Martin put forth the controversial thesis that human beings were primarily responsible for causing the extinction of these animals throughout the grasslands. His thesis goes something like this: paleo-Indians developed such effective hunting techniques that before they realized it they had driven animals such as the woolly mammoth (*Mammuthus primigenius*), American mastodon (*Mammut americanum*), giant ground sloth (*Megatherium americanum*), and giant bison (*Bison antiquus*), among others, into oblivion (Martin and Klein 1984; Grayson and Meltzer 2002).

While some archaeological evidence seems to support Martin's thesis, many other ancient indicators do not. For example, the early Holocene cave camps of the peoples who once lived in the Great Basin have been examined. Their eating habits suggest little, if any, reliance upon extant megafauna even though those animals also roamed and grazed throughout the region when it was rich in lakes and riparian woodlands. Fish, small mammals, and other collectable flora constituted those people's diets. Their eating habits seemed to remain fairly consistent despite the disappearance of the megafauna around them, and they persisted despite a rapidly changing climate. This is one reason that some archaeologists refuse to classify these people as paleo-Indians, as they had shunned any reliance upon eating megafuana. Rather, these scholars prefer to call these same people paleo-archaic, indicating a reliance on hunting and gathering animals and plants that have existed into present times (Blackmar 2001; Beck and Jones 1997).

Still, other evidence suggests a dynamic link to human causation for the Pleistocene extinctions. Archaeologist Thomas Van Devender points out that climatic shifts during this time were not severe enough to cause extinction. Similar climatic shifts had occurred during previous interglacials without resulting in the disappearance of these animals. Moreover, the types of plants that these animals relied upon flourished thousands of years after the great extinctions. At best, he contends, a combination of sudden climatic shifts along with human hunting pressures led to the sudden collapse of those animal populations (Van Devender 1995).

Gary Haynes, an archaeologist, offers an explanation based on some of the most recent paleological reconstructions of Pleistocene climates, animal populations and habits, and Clovis peoples' subsistence strategies. Haynes notes how climatologists generally agree on the broad, fast-warming trends that ended the last large-scale glaciation. Haynes also highlights how water sources formed the most important resource for megamammals, or herbivores weighing over one megagram. He suggests that as the climate throughout North America became drier, grazing resources and water sources became more localized and separated, much like islands in an ocean. This gave Clovis people a distinct advantage in locating these large animals as their ranges became ever more constricted to fewer and fewer locales. In short, climate change led to greater ease in hunting, which led to the extirpation of the megamammals. Haynes does not rule out the possible presence of "hyperdiseases" such as tuberculosis in the mastodont population, or exceptionally active, global volcanic activity, or asteroid or comet impacts that could have resulted in dramatic climatic shifts extreme enough to lead to extinction. Nonetheless, such factors, he contends, are "plausible" but poorly fit existing archaeological and paleoclimatological facts (Haynes 2002).

Evidence of another hunting culture rests atop the archaeological remains of the Clovis peoples. It is likely that they were the direct descendants of the Clovis people, and that they eventually became the foragers of the Archaic period. Scholars refer to the post-Clovis remains as Folsom. They have been designated such for the same reasons the Clovis people received their labeling: George McJunkin, an African-American cowboy, discovered some projectile points of a distinctive style in a gully near the town of Folsom, New Mexico, in 1927. Upon learning of this discovery, anthropologists noted the association of these Folsom points with the bones of now extinct forms of bison, confirming humans in North America at the end of the Pleistocene. For a long time, scholars thought the Folsom culture disappeared around 8000 BCE, but more recent scholarship suggests that in some areas Folsom peoples may have flourished to around 5000 BCE.

Scholars refer to the next distinct group of Indian peoples who followed in the wake of Paleo-Indians as foragers of the Archaic tradition. Two general traits characterized those people: (1) they lacked agriculture, and (2) they manufactured no ceramics. Given these two traits, in some places, such as the desert grasslands throughout the Great Basin and those extending onto the Columbia Plateau, the Archaic tradition persisted well beyond the years of European contact. Archaic peoples lived as skillful hunters throughout the grasslands. Archaeological remains attest to their reliance on bison hunting throughout the central grasslands. These people developed the techniques of bison traps. In the Great Basin grasslands, Archaic peoples developed resourceful hunting and gathering strategies that remained largely intact into the nineteenth century.

Notably, the Archaic peoples began manipulating fire. Setting fire to the grasslands paid great dividends for enhancing grass growth, which in turn attracted and nourished scores of herbivores, which in turn simplified hunting for the humans who had set the fires in the first place. Several early references to Archaic peoples throughout the Great Basin attest to this practice. Present-day Morgan Valley in Utah at one time reflected the aftermath of human fire management prior to Euro-American settlement. In July 1846, Edwin Bryant recorded in his journal luxuriant grass prospering everywhere in the basin with wild rose blooms showing through the stands. Wild currants abounded. The Gosiutes, a people who practiced an Archaic lifestyle, used fire to surround and gather rabbits (Stewart 2002).

Human-managed burnings of the grasslands have an exceptionally deep history extending, most likely, to the late Pleistocene and early Holocene. In what is southern Alberta, Canada, today, human adaptations to climatic changes based upon available water resources, changing plant communities, and hunting strategies are evident in the archaeological record. Rapidly warming climates created sharply isolated valleys within which spruce and entirely different animal populations took hold between 8000 to 6000 BCE. More intriguing is the suggestive evidence that the people living on the fringes of those grasslands were beginning to employ fire management practices in shaping their surroundings (Clark et al. 2001). What becomes clear about Paleo-Indian and Archaic lifestyles is the connections between adaptation strategies and climatic flux. Changes in weather patterns had decided effects on what animals and plants lived where, and if at all. These changes, in turn, altered the technologies people used in extracting or acquiring sustenance, or where they might migrate as they sought places of richer abundance. Reconstructing climatic change throughout the grasslands, although a difficult undertaking, is fundamental for understanding how past occupants lived in this biome.

CLIMATIC CHANGES

No question about it, recurring patterns of warm, dry periods and cooler, wetter years have been a commonplace feature of the grasslands. That said, reconstructing any past climate requires great care. Scholars use several indicators when depicting ancient weather patterns. One common means is the use of proxies. The presence of fossil remains of C_4 or C_3 grasses is commonly used as an indicator of dry and hot, or wet and cool conditions. Yet, soil erosional patterns may mix these indicators and lead to an erroneous interpretation of how wet or dry conditions may have been (Caran 1998). Still, C_4 and C_3 grasses leave indicators in soil layers. Their organic decays deposit different levels of measurable carbon isotopes. Measuring these remains gives an indicator of whether C_4 or C_3 plants once dominated, and respectively, these plants are fine markers of dry or wet climatic conditions. Gradually, scientists are beginning to construct a clearer picture of the shifting plant communities throughout the grasslands in response to changes in precipitation and the occurrence and frequency of fires.

Scholars also use pollen as a proxy indicator of climatic conditions. Again, care is called for when interpreting pollen contents in soil layers. For example, spruce pollen is abundant throughout the Llano Estacado soil layers dating around 8000 BCE (Broutillet and Whetstone 1993). Does the presence of these pollens indicate a spruce-covered tableland or were these pollens lifted from regions far removed, carried by winds over great distances, and deposited atop the Llano Estacado? What does the lack of soil characteristics throughout the tableland common to spruce forests indicate?

Regardless of the difficulty in reconstructing its effects, climate has been the main partner influencing the actions of grassland peoples, and this relationship, more than anything else, has determined the nature of this biome. So, what has been the nature, or the essential reproductive elements, of climate in this region? First, it is important to recognize the mutable tendencies of climate. One good definition of climate is the "composite weather of a region" (Broutillet and Whetstone 1993). So, what are the factors contributing to the past climates of the grasslands? First, the patterns of global air currents are an essential element in shaping all climates on the face of the Earth. So it has always been with the continental United States; three mighty, shifting air masses have intersected over the grasslands and have acted as a mighty force shaping the ever-changing boundaries of plant communities. These alternating currents have varied the climates of these intermountain regions, and life-forms within these areas have variously perished, struggled, or thrived as they adapted, or failed to adapt, to shifting winds.

The first air mass is formed by a large continental air conditioner of snow and ice covering the Arctic that chills the atmosphere above. This heavy air drifts south and the greater the volume the farther south the air travels. The second air mass is formed by the Pacific Ocean warming the atmosphere above it. This accumulation flows to the east as a result of the rotation of the Earth. As it does, the West Coast mountains capture most of its moisture, which results in dry zephyrs caressing the grasslands to the east. The solar heating of the Gulf of Mexico generates the third mass, a moisture-laden, energized air mass that flows northward.

Over these same years the shifting intersections of these major air masses have created generally defined, seasonal patterns (see map on p. 17). For example, in winter a diagonal line of cold polar air masses can stretch from the southern tip of Lake Michigan to the Big Bend of the Rio Grande River. In the late spring, the farthest reach of the warm gulf air masses forms a line from the middle of the Front Range of the Rocky Mountains in Colorado to the southern half of Minnesota, and with a gentle curve, swings eastward along the southern tips of the Great Lakes. A frontal boundary from around Denver, Colorado, eastward angling to the southern tip of Lake Michigan forms a demarcation separating the northern flow of the gulf masses from the southern flow of the polar masses in winter.

What is exceptionally interesting about these lines is the way the boundaries of plant communities seem to follow these same, shifting margins. Plant communities associated with the mixed-grass and shortgrass regions flourish northwest of the southernmost reach of the wintertime polar air masses. To the west of this same line the shortgrass and xeric grasslands have dominated, and the desert grasslands remain below the boundary marking the extent of where northern gulf and southern polar air masses meet in winter. The tallgrass prairies emerged westward out from the eastern woodlands and have generally prospered along the winter line dividing the reaches of the northern polar air masses and the southern gulf air masses. Certainly, a relationship exists between these frontal patterns and the location of plant communities (Hayden 1998).

While these frontal patterns are typical of recent years, this has not always been the case. Periods of global warming and cooling have produced different forces giving rise to the resources used by humans to survive in the grasslands. One of the most important water sources used by contemporary people is the Great Plains Aquifer. Around 5 million years ago, the temperature of the North American continent rose. Over the next 5 million years, the cooling and heating of the earth resulted in massive glaciers covering and then retreating from the Rocky Mountains. During each warming episode, the glacier melts carried with

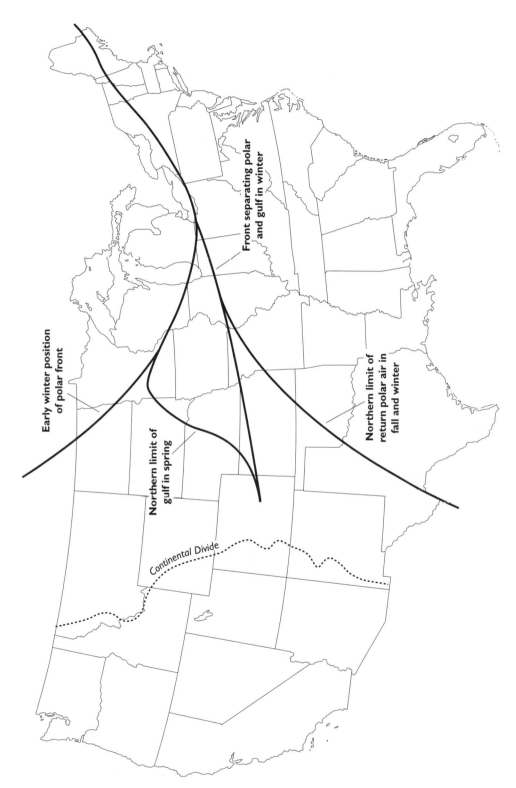

Front separating polar and gulf in winter

Early winter position of polar front

Northern limit of return polar air in fall and winter

Northern limit of gulf in spring

Continental Divide

Approximate frontal boundaries of the major continental air masses.

them huge deposits of sediments. These flows blanketed the land to the east and deposited sediments across large portions of the region. In many places east of the Rocky Mountains, this ebb and flow of mounting and melting glaciers, of relatively dry years followed by times marked by huge rivers fed by glacial thaws, lasted until around 10,000 years ago. Perhaps it is not wise to use the word "lasted," as this geological sequence of events might be unabated. Possibly, humans are simply experiencing a warm lull before a recurrence of another ice age in the distant future. Nonetheless, these sediments have acted as one large continental sponge absorbing the glacial melts streaming eastward away from the Rocky Mountain Front Range. Over hundreds of thousands of years, these water-saturated sediments were covered and sealed underground. Today, this vast underground reservoir of fossil water, the Ogallala Aquifer, underlies much of the shortgrass regions of North America and forms the major water supply for center pivot irrigation farming.

Within these larger patterns of shifting climatic change were strong fluctuations of shorter duration. Scholars have referred to these as El Niños, La Niñas, and Viejo Niños. In the late 1500s, Peruvian fishermen noticed the Pacific Ocean temperatures rising and their catch plummeting. As this phenomenon occurred around Christmastime, they called this event the "Christ Child" or *El Niño*. These fishermen were experiencing the effects of periodic warming and cooling of the Pacific Ocean. This movement of warm ocean water toward the Americas has affected the historical continental climates, and scholars have identified some general climatic characteristics associated with these episodic shifts in oceanic warming and cooling. Increased precipitation is one of the most important implications of El Niños for most of the grasslands.

When the periods of El Niños are examined over the last 100 years, during each one the rainfall in Kansas averaged 136 percent to 176 percent above the average rainfall over the last century during the months of December through March. For Texas the averages increased between 116 percent to 196 percent above the average for the last 100 years. In the Great Basin of Utah, where desert grasslands abound, rainfall averages increased from 100 percent to 122 percent above the average for the last 100 years. When El Niños occur, the eastward flowing, Pacific winds weaken across the grasslands. This allows now feebly contested, warm, moist, gulf air currents to flow northward, and as a consequence, more rainfall descends upon most of the grasslands. An opposite effect to El Niño also occurs, and this is called La Niña. While these episodes occur periodically, these Pacific Oceanic warming and cooling cycles have not been regularly spaced. However, they seem to come in roughly alternating periods of three to seven years with some spacing of years between each event.

To complicate the matter even more, broader cooling and warming events are becoming more apparent. The oceanographer Francisco Chavez has termed these events El Viejo, the old man. These cycles range around twenty years in duration, and occurrences of El Niños and La Niñas are variously influenced by whether they appear during an El Viejo (Chavez 2003). These cycles have certainly been at work over the last four centuries, and there is no reason to doubt their existence over the last 10,000 years. And while one can reasonably assume that these weather patterns have contributed to shaping plant and animal communities throughout the grasslands over these same several millennia, the exact nature of those relationships, at least with current analytical tools, is difficult to describe with any certainty. Still, over the last 500 years, the effects are becoming a little bit clearer.

Even if the details of climatic change and its paleobiogeographic effects are only dimly understood, certain grand climatic patterns do appear evident. The first is often referred to as the Postglacial, or Boreal (8500 to 7000 BCE), followed by the Altithermal or Hypsithermal (7000 to ca. 2500 BCE), and the last, which is ongoing, is called the late Holocene Neoglacial. During the first period, glaciers underwent a general retreat as the climate of the grasslands warmed. It is now becoming clearer that this warming occurred rapidly, in the blink of an eye when considered in the context of geologic time. Detailed studies of the shifting grassland/forest frontiers in North Dakota, Minnesota, and Wisconsin bear testimony to those trends.

By 7000 BCE, a considerably dry and hot period of several thousand years settled over the grasslands. In some parts of the grasslands, the annual temperatures did not start falling until as late as 3,000 years ago. During this time, the grasslands stretched their greatest extent to the east. Scholars call these years the Altithermal (from the Latin for high + temperature), or the Hypsithermal, (from the Greek for high + temperature). These years produced one of the more dramatic landscape changes in the grassland—the formation of majestic sand dunes. The most notable, perhaps, are the Sand Hills in the present-day central portion of Nebraska. Around 6500 BCE, winds began depositing silts and fine grains of quartz. Smaller dune areas formed throughout the present-day Arkansas River Valley of Colorado and Kansas, in the northeastern portions of Colorado, in the intermountain regions of central Wyoming, southern Colorado (where the Great Sand Dunes National Park is today), and northern New Mexico. But it was in present-day Nebraska where the dunes acquired their most spectacular extent, covering over 20,000 square miles. The greatest building of these probably occurred between 6500 to 3500 BCE, with later dune-building activity reappearing between 1000 BCE to 1000 CE. During both of these episodes, this area was a windswept region of shifting sands devoid of vegeta-

Only during the last 1,500 years have the Sand Hills of Nebraska been "stabilized" by a cover of vegetation. (Tom Bean/Corbis)

tion. Only during the last 1,500 years have the dunes been "stabilized" by a cover of vegetation (Swinehart 1989).

Throughout the desert grasslands, the Hypsithermal appears to have been beneficial for the spread of C_4 plants and associated grazers and browsers. Apparently, the summer monsoons of 4,000 years ago were much stronger than they are today throughout desert grasslands. As a result, the archaeological record suggests that during the hypsithermal the grasses attained their greatest spread throughout the area (Van Devender 1995). During the Holocene Neoglacial, the climate again shifted generally to cooler seasonal temperatures and greater rainfall as the years of contact between Europeans and Indian peoples approached. Even during this period, irregular fluctuations of hotter and colder years occurred. In those years, peoples, plants, and animals made various adaptations and adjustments in reaction to those changes in weather; these sometimes mutually benefited each other, and sometimes not.

ARCHAIC TO ANASAZI

One group of grassland peoples changed their lifestyles very little from the time of the Hypsithermal until their first association with European-introduced horses. In the Great Basin desert grasslands, the residents maintained an Archaic way of life, or one based primarily upon hunting, fishing, and collecting. Yet their way of living supported a fairly large population of people, perhaps 100,000 or more. Living in close proximity to water sources, whether streams, rivers, lakes, or wetlands, held the key to their well-being. They located the vast majority of their village sites near these water supplies. Some of these people may have experimented with incipient agriculture. For example, in present-day Utah there is some archaeological evidence of this having taken place briefly between 350 to 1250 CE. Moreover, the archaeological record indicates the presence of regional trade, so these people were not entirely dependent on local resources for their subsistence (Fish and Fish 1994).

To the south of these peoples, however, two other "transegalitarian" societies developed based on agriculture supplemented with hunting and gathering in a diverse environment. Today, scholars call these peoples the Anasazis and the Hohokams. Two concurrent things allowed these people to develop cultures remarkably different from the Archaic tradition and the Plains Woodland and Plains Village traditions throughout the grasslands east of the Rocky Mountains. First, they were some of the first peoples on the North American continent to farm corn, and second, the climate had cooled and become wet enough to support this form of staple crop production (Stuart 2000).

Some time passed before this crop came to resemble its current form. Around 5000 BCE, people began growing some of the first varieties of corn characterized by small cobs, most likely originating in present-day northern Mexico. By 1000 BCE, peoples in the area were planting and harvesting many varieties of corn as well as squash. These people gradually jettisoned the Archaic lifestyles of their neighbors to the north and slowly created more permanent communities. With the addition of beans to their agricultural production, these people had reliable staple crops that could support a burgeoning population around 1000 BCE. For the next 500 years, rainfall averages dropped, and these people learned that agricultural production, if carefully practiced, could produce enough abundance to sustain them through lean years. While some of their kin reverted to Archaic traditions, around 2,000 years ago, farming families grew and became more powerful. These people gradually built the Anasazi and Hohokam cultures and simultaneously absorbed or dominated most of the hunting peoples around them.

The Hohokam, a Pima word meaning "all used up," began settling the desert grasslands in the present-day Santa Cruz, Salt, and Gila River valleys around 800 BCE. They began moving away from an Archaic lifestyle with the farming of corn. Roughly 100 to 1350 CE, perhaps as many as 20,000 to 45,000 people lived in these river valleys. By 1000 CE, they had established irrigation systems, over 120 miles of canals watering somewhere between 30,000 to 60,000 acres of land in the Phoenix basin alone. During this time, the Hohokam overcame arid conditions, less than eleven inches of precipitation annually, through the use of more elaborate irrigation systems and the exploitation of wild plants and animals in the region. On the high slopes of the surrounding mountains, cacti fruits were harvested and small game were hunted. On the lower terraces and mesas approaching the river, mesquite beans were gathered and agave was grown. In the floodplains, where grass and gallery forests flourished, the Hohokam hunted an abundance of game animals and planted crops in the rich, deep, alluvial soils.

Settled in mostly small, scattered communities, the Hohokams prospered until around 1450 or 1500 CE. At this point, even though their culture had endured repeated droughts, their food production system came unglued. One of the more likely explanations for this is offered by environmental historian Michael F. Logan. Just prior to 1500 CE, the Salt, Gila, and Santa Cruz Rivers underwent a period of *arroyoization*, causing a deeply eroded streambed. At this point, the Hohokam could not divert stream flows into their irrigation works, and their agricultural production diminished significantly. This process may have occurred solely as the result of climatic change, or may have been coupled to, or aggravated by, some Hohokam management practices to increase runoffs into the rivers. Two hundred years later, the Santa Cruz River, ever shifting its course, had also refilled with sediments to the point that the Pimas (Akimel O'odham), the likely descendants of the Hohokams, were relocating to the valley and taking up irrigated agriculture again. The timing was poor for them, as it was then that Spaniards also began exploring these desert grasslands. This meeting of cultures would have harsh repercussions for the Pimas.

Concurrent with the rise of the Hohokams, the Anasazis expanded and built a flourishing civilization to the north. Around 1200 CE they had built an elaborate civilization with a locus of power in Chaco Canyon, New Mexico. A series of "great houses," such as Pueblo Bonito, appear to have been gathering places for Anasazi culture and economics. Today, the many abandoned kivas give stark testimony to the importance of religious and civic ceremony. While archaeologists and anthropologists differ in their conclusions, the apparently empty rooms in the great houses may have served as storage bins that held grain supplied by the small farm communities throughout the Four Corners region. Or these rooms may have simply been some sort of monument within the great houses. Archaeological excavations at Chaco Canyon also reveal a once-flourishing exchange of exotic goods from distant lands. Hohokam pottery, pipestone from Alberta, seashells from the California and Texas coasts, macaws and metal works from Mexico were all traded in the great markets such as those that existed at one time in Pueblo Bonito. A superbly engineered road network radiated out from Chaco Canyon and led to small farming outposts, other Anasazi settlements, foreign markets, and some unknown destinations, possibly "spirit paths" (Wicklein 1994). Why then, did such a flourishing, prospering civilization collapse suddenly some 800 years ago?

David Sturat, an anthropologist, offers some intriguing explanations for the Anasazi breakdown and its ramifications. First, a rigid class system seemed to prevail. The residents of the small farm communities experienced harsh working conditions, malnutrition, exceptionally high infant mortality rates of around 45 percent, and storage bin space equal to living space in their dwellings. The few residents of the Chaco Canyon great houses, on the other hand, enjoyed better nutrition, lower infant mortality rates, greater material riches, and storage bin space equal to two-thirds of the space in the cities.

Not only was life more difficult in the small, regional farming communities, but it was also more precarious. Farm practices rapidly depleted the fertility of desert grassland soils, and those farm families who could moved and built new farm communities farther and farther away from the centers of power. Moreover, those left behind lived in weakened ecological relationships to the land, ones that minor droughts, such as experienced in 1100 CE, utterly unraveled. As farm production plummeted, eventually even the vast stores in centers

Anasazi ruins at Pueblo Bonito. (Corel)

such as Pueblo Bonito gave out. An erratic recovery occurred during the 1100s CE, but not enough to sustain the core cities, and people had largely abandoned these, like Pueblo Bonito, by 1200 CE.

Adding to this, perhaps a few decades before, a new group of people living a distinctly Archaic lifestyle emerged upon the scene, the Diné, or commonly called Navajos and Apaches today. These people may have been seeking greater food resources and took advantage of the weakened Anasazis. Most likely, they also learned farming and riverine village life from the Anasazis. Sturat argues that violent, unsettled times followed in the wake of this collapse. Ample evidence indicates Anasazi population displacement, resettlement in fortified dwellings, return to simpler and farther spaced agricultural settlements, and war with other peoples throughout the desert grasslands for the next 200 years. Stuart believes the descendants of the Anasazis learned several important lessons, chiefly, how to create a more egalitarian society based upon more sustainable food procurement that blended irrigated agriculture in river valleys, hunting on the grasslands, and collecting throughout the mountainous foothills of the Southwest. These people, he asserts, became the Pueblo culture of the Rio Grande and Pecos river valleys.

During the 1400s, an emerging economic and ecological relationship grew between the easternmost Pueblo peoples throughout the Rio Grande River Valley with those peoples living in the grasslands to the east. Most likely, the newly arrived Athapascan-speaking peoples, the Diné, and those whom the Spanish would later call the Jumanos began striking up extensive trade relationships with the agricultural people living in their adobe villages to the west (Spielmann et al. 1990). During this same period, settlements such as Pecos or Taos grew as important trade centers. Grassland peoples brought exotic stones, such as obsidian and turquoise, and animal products in exchange for the ceramics and agricultural goods produced in the region surrounding the Pueblean market sites. Demonstrably, prior to European contact, these eastern Pueblean centers showed a growing inventory and diversity of trade goods. This meant rising consumption rates for the residents and visiting traders, and lower costs of trade goods as stores were increased.

The increase of this trade may indeed reflect an effective ecological adaptation strategy by the Pueblos and grassland peoples to the east. Irrigated agriculture and growing populations tied the Pueblo peoples to a limited number of sites suitable for these practices. To sustain their populations, trade with the grasslands people brought an additional source of energy to their cities and created greater systematic stability for their culture. Adding to their energy supplies was important given the unpredictability of climatic fluctuations with the threat of attendant crop failures. Grassland peoples to the east benefited in a

Ruins of Pecos Pueblo, trading center of the Pueblo Indians. (David Muench/Corbis)

similar manner in that this trade extended the range in which they gathered their own resources. In short, the grassland peoples and Puebleans were diversifying and extending the energy flows supplying their relative populations (Bronitsky 1982).

PLAINS WOODLAND AND PLAINS VILLAGE CULTURES

Peoples throughout the grasslands from southern Canada south into present-day Oklahoma generally practiced one of two forms of subsistence. One took the form of small farmsteads whose people hunted and gathered primarily, but who also gardened. The other took the form of large earth lodge villages populated by people who farmed and did seasonal hunting of game such as bison. The Plains Woodlands Culture, sedentary hunters who lived in small settlements and practiced some gardening, generally flourished between 1 and 1000 CE. The Central Plains tradition was similar to the Woodlands culture, and archaeologists identify these people as living in the grasslands from around 1000 to 1500 CE. The Plains Village Culture, or the proto Mandans, Hidatsas, Arikaras, Pawnees, and Wichitas, built large earth lodge villages located in ma-

This William Jackson photograph of a Pawnee earth lodge is a fine depiction of how the Plains Village cultures once housed themselves. (National Archives and Records Administration)

jor river valleys. Gradually, these settlements became fortified, and given these peoples' reliance upon horticulture, they limited themselves to rather fixed sites. The Central Plains Tradition and the Plains Village Culture flourished prior to 1500 CE, but not without significant adaptations and adjustments to changes in climate, and attendant shifts in plant and animal populations.

For example, around 2,800 to 2,000 years ago, in present-day Delaware Canyon, in Oklahoma, only a few people lived there, and those who did practiced an Archaic lifestyle. Paleoclimatic studies suggest a much drier climate than now. In the next 1,000 years, the climate turned much wetter, and many people populated the canyon. They farmed, lived in small villages, hunted deer, and developed wide-flung trade relationships. Around 1300 CE, the climate turned drier, and these people adjusted to it. They intensified their agricultural practices and began hunting bison. Given the ongoing climatic change, bison were attracted to reliable water sources, which placed them in closer proximity to the Plains Village peoples living in the canyon. Also, these same people began locating temporary hunting and gathering stations somewhat removed from their village sites. This subsistence strategy might have been the precursor to the hunting and agricultural rhythms of the Caddos, who are likely the descendants of these people (Ferring 1986). By 1400 CE, Athapascan speaking peoples, the Diné, had successfully inserted themselves into these relationships in competition with the Plains Village peoples.

While Plains Village peoples certainly adjusted their lifestyles to climate change, their own adaptation strategies could prove troublesome regardless of drought or abundant precipitation for stable crop production. In some respects, Caddoan-speaking peoples may have had just as much trouble developing long-lived agricultural practices in their part of the grasslands as had the Anasazi, but without the same self-destructive results for themselves. For example, some scholars believe the abandonment of Plains Village sites and a northward migration of Caddoan-speaking peoples began as the result of drought around 1200 to 1250 CE. Other scholars offer interpretations for these movements. Plains anthropologist Lauren Ritterbush believes the influx of an aggressive people, called the Oneota, displaced the Plains Village peoples (Ritterbush 2002). Another explanation is that the practice of swidden horticulture resulted in these migrations (Blakeslee 1993), while some scholars think timber depletion in the river valleys led to displacement (Griffin 1977).

ON THE EVE OF EUROPEAN CONTACT

Regardless of various migrations by peoples in the Central Grasslands, a sizable population lived there around 1500 CE. Today, reasonable estimates for the population of the region hover around two million people. This is a fantastic number given estimates of Indian peoples throughout North America by scholars prior to the 1980s. For example, Arrell Morgan Gibson's standard textbook, *The American Indian: Prehistory to the Present*, estimated that in 1500 CE, around 1.5 million people, total, lived in what would become the continental United States. Now, that estimate hovers around 20 million or more, with 2 million alone living in the Central Grasslands (Gibson 1980).

Clearly, from the last glaciation to 1500 CE, Indian peoples living in the grasslands adapted to many different ecosystems. They managed grass production with fire. They developed effective skills and technologies for harvesting a wide variety of wild animals. In groups marked by kin, language, village, and cities, they fought, traded, and worked together. They farmed river basins, gathered fruits in the uplands, and hunted bison, elk, and deer across the broad sweeps of tallgrass and mixed-grass prairies, the shortgrass Great Plains, and desert grasslands. A number of populations and cultures developed, changed, and reorganized themselves over countless thousands of years. Some peoples migrated in response to sporadic fluctuations in precipitation while others found efficient means to endure chaotic weather without abandoning their homes. In all of this, proximity to ample water was the key to their well-being. Whether it was the at the end of the cold Pleistocene or during the searing droughts of the 1200s CE, water determined the health and location of game animals and was required for domestic purposes. Riparian ecosystems produced a variety of plants suitable for eating or wearing. Streams and rivers formed the nurturing source of all agriculture, whether those of the Plains Village peoples, the floodwater and later floodplain farming of the Anasazi, or the irrigation systems of the Hohokams.

Variable relationships among humans, plants, animals, and water produced an ever-changing mosaic called the grasslands. Humans, more than any other species or force, acted as the keystone. Prior to 1500 CE, grasslands generally responded well and flourished in their relationships with human fire and hunting practices, but less well at times in response to farming and trade practices. It was these ecological relationships that Europeans first encountered, relationships resulting in lands teeming with bison, elk, deer, cougars, wolves, and prairie dogs providing game for hunters; river valleys populated by peoples practicing all forms of agriculture and living in a variety of communities; arid grasslands effectively harvested and hunted by peoples living near streams and wetlands. At the same time, hidden from European eyes, were the difficulties these peoples experienced maintaining reproducible cultures and ecosystems in these grasslands, regions marked by constant change.

References

Axelrod, Daniel. 1985. "Rise of the Grassland Biome, Central North America." *Botanical Review* 51 (April–June): 163–201.

Barrell, Joseph. 1975. *The Red Hills of Kansas: Crossroads of Plant Migrations.* Rockford, IL: Natural Land Institute.

Beck, C., and G. Jones. 1997. "The Terminal Pleistocene/Early Holocene Archaeology of the Great Basin." *Journal of World Prehistory* 11 (June): 161–236.

Blackmar, Jeannette M. 2001. "Regional Variability in Clovis, Flosom, and Cody Land Use." *Plains Anthropologist* 46, 175:65–94.

Blakeslee, Donald J. 1993. "Modeling the Abandonment of the Central Plains: Radiocarbon Dates and the Origin of the Initial Coalescent." *Plains Anthropologist* 38, 145:199–214.

Bronitsky, Gordon. 1982. "The Southwest and the Plains: Ecology and Economics." *Plains Anthropologist* 27, 95:67–73.

Broutillet, Luc, and David Whetstone. 1993. "Climate and Physiography of North America." In *Flora of North America North of Mexico,* edited by the Flora of North America Editorial Committee. New York: Oxford University Press.

Brown, Lauren. 1985. *Grasslands: A Comprehensive Field Guide, Fully Illustrated with Color Photographs, to the Birds, Wildflowers, Trees, Grasses, Insects, and Other Natural Wonders of North America's Prairies, Fields, and Meadows.* New York: Alfred A. Knopf.

Caetano-Anollés, Gustavo. 2005. "Grass Evolution Inferred from Chromosomal Rearrangements and Geometrical and Statistical Features in RNA Structure." *Journal of Molecular Evolution* 60:635–652.

Caran, S. Christopher. 1998. "Quaternary Paleoenvironmental and Paleoclimatic Reconstruction: A Discussion and Critique, with Examples from the Southern High Plains." *Plains Anthropologist* 43, 164:114–124.

Chavez, Francisco, et al. 2003. "From Anchovies to Sardines and Back: Multidecadal Change in the Pacific Ocean." *Science* 299 (January 10): 217–221.

Clark, James, et al. 2001. "Effects of Holocene Climate Change on the C_4 Grassland/Woodland Boundary in the Northern Plains, USA." *Ecology* 83, 3:620–636.

Dayton, William A. 1948. "The Family Tree of Gramineae." In *Grass: The Yearbook of Agriculture,* edited by Alfred Stefferud. Washington, DC: Government Printing Office, 637–639.

Ferring, C. Reid. 1986. "Late Holocene Cultural Ecology in the Southern Plains: Perspectives from Delaware Canyon, Oklahoma." *Plains Anthropologist* 31 (November): 55–82.

Fish, Suzanne K., and Paul R. Fish. 1994. "Prehistoric Desert Farmers of the Southwest." *Annual Review of Anthropology,* 23:83–108.

Gibson, Arrell Morgan. 1980. *The American Indian: Prehistory to the Present.* Lexington, MA: D. C. Heath and Company.

Golley, Frank Benjamin. 1993. *A History of the Ecosystem Concept in Ecology: More Than the Sum of the Parts.* New Haven: Yale University Press.

Grayson, Donald K., and David J. Meltzer. 2002. "Clovis Hunting and Large Mammal Extinction: A Critical Review of the Evidence." *Journal of World Prehistory* 16 (December): 313–359.

Griffin, David E. 1977. "Timber Procurement and Village Location in the Middle Missouri Subarea." *Plains Anthropologist* 22, 78, pt. 2:177–185.

Hayden, Bruce P. 1998. "Regional Climate and the Distribution of Tallgrass Prairie." In *Grassland Dynamics: Long-Term Ecological Research in Tallgrass Prairie,* edited by Alan K. Knapp, John M. Briggs, David C. Harnett, and Scott L. Collins. New York: Oxford University Press, 19–34.

Haynes, Gary. 2002. *The Early Settlement of North America: The Clovis Era.* New York: Cambridge University Press.

Hays, J. D., John Imbrie, and N. J. Shackleton. 1976. "Variations in the Earth's Orbit: Pacemaker of Its Ice Ages." *Science* 194 (December 10): 1121–1132.

Hook, Paul B., and Ingrid C. Burke. 2000. "Biogeochemistry in a Shortgrass Landscape: Control by Topography, Soil Texture, and Microclimate." *Ecology* 81, 10:2686–2703.

Ingalls, John James. 1948. "In Praise of Blue Grass." In *Grass: The Yearbook of Agriculture.* Washington, DC: Government Printing Office, 6–8.

Logan, Michael F. 1999. "Head-Cuts and Check-Dams: Changing Patterns of Environmental Manipulation by the Hohokam and Spanish in the Santa Cruz River Valley, 200 to 1820." *Environmental History* 4 (July): 403–430.

Martin, Paul S., and R. G. Klein, eds. 1984. *Quaternary Extinctions.* Tucson: University of Arizona Press.

O'Neill, Robert V. 2001. "Is It Time to Bury the Ecosystem Concept? (With Full Military Honors, Of Course!)." *Ecology* 82, 12:3275–3284.

Redmann, R. E., and E. G. Reekie. 1982. "Carbon Balance in Grasses." In *Grasses and Grassland: Systematics and Ecology,* edited by James Estes, Ronald J. Tyrl, and Jere N. Brunken. Norman: University of Oklahoma Press, 195–231.

Reichman, O. J. 1987. *Konza Prairie: A Tallgrass Natural History.* Lawrence: University Press of Kansas.

Ritterbush, Lauren W. 2002. "Drawn by the Bison: Late Prehistoric Native Migration into the Central Plains." *Great Plains Quarterly* 22, 4:259–270.

Spielmann, Katherine A., Margaret J. Schoeninger, and Katherine Moore. 1990. "Plains-Pueblo Interdependence and Human Diet at Pecos Pueblo, New Mexico." *American Antiquity* 55 (October): 745–765.

Stewart, Omer C. 2002. *Forgotten Fires: Native Americans and the Transient Wilderness,* edited by Henry T. Lewis and M. Kate Anderson. Norman: University of Oklahoma Press.

Stuart, David E. 2000. *Anasazi America: Seventeen Centuries on the Road from Center Place.* Albuquerque: University of New Mexico Press.

Swinehart, James B. 1989. "Wind-blown Deposits." In *An Atlas of the Sand Hills,* edited by Ann Bleed and Charles Flowerday. Lincoln: The Conservation and Survey Division of the University of Nebraska, 43–56.

Van Devender, Thomas R. 1995. "Desert Grassland History: Changing Climates, Evolution, Biogeography, and Community Dynamics." In *The Desert Grassland,* edited by M. P. McClaran and T. R. Ban Devender. Tucson: University of Arizona Press, 68–99.

Wells, Philip V. 1965. "Scarp Woodlands, Transported Grassland Soils, and Concept of Grassland Climate in the Great Plains Region." *Science* 148 (April): 246–249.

Wicklein, John. 1994. "Spirit Paths of the Anasazi." *Archaeology* 47 (January/February): 36–41.

2

THE UNRAVELING OF
THE WILD GRASSLANDS

I n 1870 Brevet Major and Assistant Surgeon George M. Sternberg stood at the end of a four-century ecological drama. The ecological unraveling of grasslands that he was observing had begun at a time when, to many humans, its occupation appeared most ideal. This army officer had taken careful note of the ecological effects of Indian burning practices, and at the same time he could think of those people as uncivilized and living in a wilderness, which implied a place lacking in conscious, purposeful human management. Obviously, Sternberg was blind to any inconsistency in his logic, and despite this, he, too, thought firing the grasslands in the early spring an effective technique for controlling insects and fungi. He had written to John Martin, the editor of *The Union*, a Junction City, Kansas, newspaper, and asked the publisher to print his insights about what it would take to transform the wild grasslands into a domesticated landscape. Prior to this, Sternberg had ridden with the Seventh Cavalry all through the mixed-grass and short-grass biomes and had developed a keen interest in the natural history of these ecosystems. Consumed with curiosity about the land, he studied its fossil remains scientifically and would, in time, become a nationally renowned paleontologist. In February 1870, however, his letter to the editor treated the subject of prairie fires.

In his article Sternberg expressed no concern over the removal of Indian peoples from the Sunflower State, yet he recognized how important they had been to maintaining the wild grasslands. He also understood the importance of soil types, climate, and animal wildlife to the potential success of creating "wheat countries." Sternberg astutely labeled Indian peoples' grassland management practices the "heroic farming of the American Indian," and his insights about this style of "farming" are worth noting.

The wild game of the country is his crop. Autumnal fires were his reapers, to aid in collecting and harvesting. Much evil was done; also some good. Let us examine the matter a moment.

> Indian countries are *clean* countries. No muddy roads . . . No underbrush or decayed logs and rubbish in their woods, for the annual fires clean up everything, leaving but the greenest trees with thick bark. . . . This style of farming is exhaustive and destructive, tending to sterility where sterility is possible.

This 1907 Frederic Remington painting catches the dazzling sight of Indian-set fires in the tallgrass prairies. (Library of Congress)

Yet, though he exhausts the surface and banishes the rains, the Indian does not exhaust the soil *below* the surface, for he does not stir it. And in destroying everything and seeding nothing, he invariably delivers his country into the hands of white men, free from those noxious insects, which prey upon the grains and fruits of civilized culture. (Sternberg 1870)

Sternberg was witnessing the final scenes in the unraveling of the wild grasslands. To understand this breakdown it is necessary to recognize the broader set of historic relationships that had intertwined soils, animals, grasses, climate, cultures, economies, and the struggle to control solar energy over the previous four centuries.

ANIMAL DISTRIBUTIONS, GRASSLAND PATTERNS, AND SOIL TYPES

A verdant mask in spring has always disguised a myriad of intricate ecological relationships in the grasslands. Undulating grass stems with their head seed spikes tossed by the winds, appearing like the surface waves of an ocean, hid from Sternberg's view a teeming diversity of life woven through and above their stalks and entwined within their supportive root systems. The grasslands were like a rich tapestry that seemed uncomplicated from a distance, but upon closer examination revealed an intricate binding of innumerable threads forming subtle patterns of color and design. Just as the human mind defines, and then creates, the patterns that the threads take in a tapestry, the patterns of plants and animals in the grasslands Sternberg observed were largely shaped by the distribution of soils throughout the region. Soils are rich in diversity, and scientists have categorized them into many types and subtypes.

Several factors and forces interact in the formation of soil types. The mineral composition of any one area is formative, and this "parent material" will have distinct properties as a result of its interaction with local climatic conditions. This relationship of climatic conditions and parent materials in the soil distinctly favors certain plants over others. Some life-forms thrive in acidic, dry conditions, and others in moist, warm, alkali situations. At the same time, whatever flora flourishes in any given clime and parent material will ultimately add to the composition of the soil itself through its own decomposition, nutritional uptake, water use and transpiration, and carbon and oxygen exchanges (USDA 2005).

The types of plants in any one locale have always had a distinct effect on which animals could thrive in a particular area, and these same animals became part of the relationships shaping the formation of soils. The way in which animals have harvested, digested, and excreted plant material contributed ingredients that have composed important elements of soil formations. Moreover, when animals adapted themselves to certain plant species, they in essence reflected the location of a particular soil type.

The relief of the land also has shaped soil characteristics. Steep slopes have shallow soils, and deep, wide river valleys have rich, deep, highly productive loams. In some places, such as the Flint Hills of Kansas, thin, shallow soil profiles sit atop thick limestone substrata. Humans found this a difficult to impossible region to farm. The sand hills of Nebraska also proved punishing to farm, as the deep parent material and slopes yielded themselves poorly to either Indian or Euro-American farming practices of Sternberg's times. Yet other animals would find these same places wonderful abodes. Altogether, the longer any set of relationships forming soils had to co-evolve, the more complex and nurturing the soils became for reproducing that set of relationships. Of course, external forces could intervene at any given time and completely disrupt the intricate web of relationships forming any given soil type. Changes in climate, animal populations (including humans themselves), or plants could, and did, change the nature of a soil entirely or even destroy it.

The soil that Sternberg described is a rich mixture of many types. The most common first-order soil types found throughout the grasslands in North America are mollisols and entisols. Mollisol is derived from Latin roots: *mollis*, meaning soft, and *sol*, a shortened form of *solum*, meaning soil. This order comprises nearly 7 percent of all soil on Earth and nearly 22 percent of the soil covering the United States. Entisols, formed by *ent*, either a meaningless prefix or perhaps referring to recent, and *sol*, again meaning soil, blanket around 16 percent of the planet and over 12 percent of the United States. These soils came into being with the retreat of the great glaciers and upon them grew rich patterns of grasslands where certain animal populations proliferated and horse-borne human cultures prospered for a fleeting time. Naturally, herbivores, or plant-eating animals, prospered nicely throughout the grasslands. Nonetheless, they did not thrive everywhere in equal numbers. Calculating the numbers and the distribution of herbivores such as bison (*Bison bison*), pronghorn deer (*Antilocapra americana*), elk (*Cervus elaphus*), or white-tail deer (*Odocoileus virginianus*) in the grasslands prior to 1870 is a matter of informed speculation. Assuredly, those animal ranges and numbers fluctuated dramatically throughout the grasslands from 1500 to 1870 when Sternberg wrote.

Many factors accounted for these demographic changes. First, rich grass-producing mollisols and entisols neither underlaid all of the grasslands nor existed under the same climatic or geological conditions. These simple facts made grazing conditions in some areas lush, and in others lean. As a result, animal populations varied across time and space depending upon local conditions. Examine, for example, Bent County in present-day Colorado (see map on p. 35). The majority of the soils in the county are entisols, marked by a single profile, and of these approximately 230,000 acres are a combination of sandy soils or prairie sandhill sage dunes bordering the south side of the Arkansas River. The sandy soils supported grasses but not in the same prolific manner as the rest of the soils throughout the county. Bison and other ungulates would have avoided these areas unless other factors compelled them to graze there. Taken as a whole, under superb growing conditions, grass production in the entire county could have sustained a population of around 64,200 bison. In an unfavorable year for grass production, no more than 22,370 bison could have sustained themselves. Now when considering the highly sandy soils and dunes separately, together they could have supported at most 11,460 bison in a favorable year of grass production, and no more than 3,800 animals in an unfavorable year (Lott 2002).

Unknown to Sternberg, over the years as grazing conditions changed, the populations of grazing animals shifted throughout the grasslands. Prior to 1820, bison and elk stood in about equal numbers in the tallgrass regions whereas

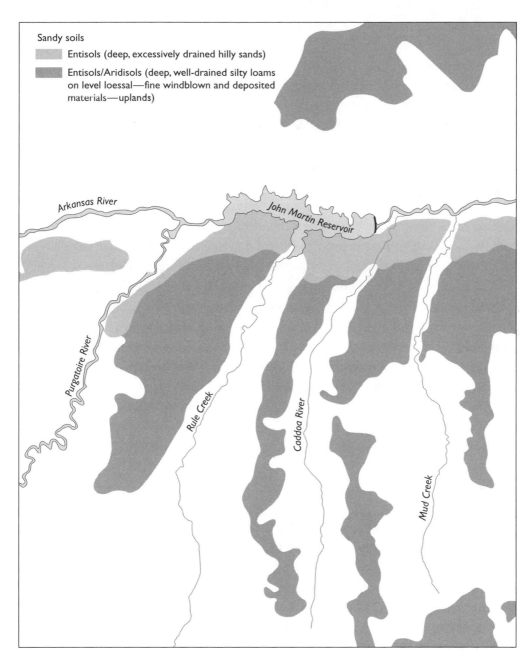

Sandy soils

Entisols (deep, excessively drained hilly sands)

Entisols/Aridisols (deep, well-drained silty loams on level loessal—fine windblown and deposited materials—uplands)

Soils in Bent County, Colorado.

pronghorn number around a third fewer. In the mixed-grass areas, bison out-numbered either elk or pronghorn by nearly a factor of three. This same pattern was more pronounced on the shortgrass High Plains where bison outnumbered elk and pronghorn by nearly a factor of six. Moreover, the bison population in the shortgrass regions was larger by a factor of ten than the numbers in the tallgrass prairies to the east. By 1860, the distribution and numbers of animals little resembled that prior to 1820. In the tallgrass prairies elk had disappeared altogether, and their numbers had fallen by around 85 percent. In the mixed-grass range, bison had retained their numbers, but elk were a mere 20 percent of their former population and pronghorn had decreased by around 20 percent. On the shortgrass High Plains, only 15 percent of its former population still remained, elk had disappeared, and pronghorn numbers had nearly doubled. By 1880, all three of these species were nearly gone completely from the grasslands (Shaw and Lee nd.).

Obviously, forces were working to alter the populations and distribution of these animals. Human hunting practices certainly stand out as a conspicuous reason for this precipitous population descent. Other factors, however, also contributed. Climate changes, shifting economic patterns, pandemics, habitat destruction, and overlapping species competition for the same sustaining resources also worked to destroy the wild ungulate populations in the grasslands. When examining climate, by 1500 weather patterns produced highly favorable conditions for the increase of ungulate populations throughout the grasslands. Only recently have climatologists begun to realize how significant these shifts were. The current term for this period of weather marked by cooler global temperatures in the northern hemisphere is the "Little Ice Age," which many climatologists date from 1400 to around 1870. Now scholars are beginning to understand its effects on the course of human history.

LITTLE ICE AGE

When Antonio Stradivari fashioned some of the world's finest tone-producing instruments, he owed a great debt to the southern Italian sunshine. In the early 1700s, he employed spruce harvested from the southern Italian Alps from an area called the "Forest of the Violins." The growth rings of the trees he used were unusually tightly spaced, indicative of cooler temperatures and short, warm, summer months. Over a hundred years later, while Sternberg hunted fossils, Edward Walter Maunder was reviewing sunspot records from the fifteenth, sixteenth, and seventeenth centuries. As an astronomer at the Royal Greenwich Observatory, he knew that typically somewhere between 40,000 to 50,000 sunspots might normally be recorded in any given thirty-year time span. Yet something odd had occurred during those three centuries, as in one twenty-seven-year period, 1672–1699, less than fifty sunspots were sighted. He verified these numbers, and after long analyses, he announced his findings in 1893, the same year Frederick Jackson Turner proclaimed the end of the American frontier.

Stradivari, Maunder, and Turner all shared something in common. Their work was shaped by the effects of a period of global cooling often referred to as the "Little Ice Age." Rare violins, the American frontier, and chilly terrestrial temperatures were all shaped by the waning of sunspots over a 450-year time span. Noteworthy in this story is the rise and flourishing of the horse-borne bison economy and the quintessential American image of an unbroken sea of grass covering broad regions of the present-day United States. These regions, shaped in the cool decades of the Little Ice Age, formed the last frontier, the conquest of which, according to Frederick Jackson Turner, gave rise to American exceptionalism. Perhaps the real "exceptionalism" had more to do with sunspots, or the lack thereof, than the cultural exceptionalism purported by Turner.

The effects of the Little Ice Age may have had a greater influence on the shape of economic and social relationships throughout the grasslands than is generally recognized. Was it mere coincidence that the Athapascan-speaking peoples, the Diné (now commonly known as the Apaches and Navajos) left their northern habitats and began arriving in the grasslands at the beginning of the Little Ice Age? What were the consequences of the seaports in Iceland and Greenland becoming frozen solid and inaccessible to the ships that had supplied Norse colonization efforts in Vineland? Is it too much of a stretch to think this change in global climate hindered the Norse and assisted the Spanish who lived in warmer climes to the south? Was there any relationship between the collapse of the Anasazi, Hohokam, and Cohokian civilizations and the onset of the Little Ice Age? Certainly, the emergence of a trade nexus connecting the Pueblos to the Dinés through exchanges of agricultural goods and bison products commenced with the dawning of the Little Ice Age. This trade nexus would establish economic barter patterns involving thousands of humans who arrived in the grasslands during this time span.

What actually caused the temperature decreases across the northern hemisphere during the Little Ice Age is still subject to inquiry and debate. Some scientists believe that as little as a few tenths of one percent of the sun's energy output can have significant consequences for global climates. Sunspots are certainly an indicator of solar energy yields, and the lack of them may have indicated lower solar energy production. Other factors may have contributed as well. For example, volcanic eruptions spewing thousands of tons of particulate matter into the atmosphere could have triggered global cooling for spans of several years. These dust clouds, not enough by themselves to cause an event such as the Little Ice Age, could have combined with lower solar energy yields and thereby prolonged cooling trends long after the particulate matter had dropped out of the atmosphere. For example, increased snow cover over large portions of any hemisphere will reflect a significant amount of solar energy back into space and tend to form a reinforcing feedback loop keeping global temperatures cool. Whatever the ultimate cause of the Little Ice Age, its effects were certainly an important factor in shaping the grasslands from 1400 to 1870 (Tkachuck 1983; Fagan 2000).

HUMAN POPULATION AND SUBSISTENCE PATTERNS

At the beginning of the Little Ice Age, a significant human population lived in the grasslands. Calculating population estimates for this region is a dicey undertaking, and archaeologists have been struggling to devise an effective way to do it. When they excavate village remains they carefully note the size of the dwellings and the types of food eaten. From these clues, population estimates are made and then projected over the surrounding area with a similar ecology. Taking a very conservative estimate of 0.8 people per square kilometer throughout the Central Grasslands, a population of over 2 million people is highly reasonable. It is very likely that by 1800 over 90 percent of that population had disappeared from the region as a result of diseases introduced from Europeans such as plague, smallpox, measles, influenza, diphtheria, scarlet fever, typhus, and cholera. In short, as one anthropologist has noted, the remnant populations represented a highly impoverished people compared to their number prior to European contact (Schlesier 1994).

As in the Central Grasslands, the Desert Grasslands had a fairly abundant human population prior to contact with Europeans and Euro-Americans. Of course, estimates of this population vary greatly. For example, an early estimate by Carl Sauer pegged the population of present-day southeastern Arizona at around 30,000 people. More recently, scholars believe this number is a pale reflection of a much larger population. The 30,000 figure appears more representative of a population severely lowered by introduced European diseases and debilitating attacks by Apaches. The intermountain grasslands, or the Great Basin area, held the least population per square mile of either the Desert or Central Grasslands. Whereas perhaps as many as .8 people per square mile may have lived in the Central Grasslands, the estimate of many scholars places a mere 2 people per 20 to 30 square miles throughout the Great Basin. With a population ratio of 1 person per square mile, the Utes may have prospered better than any of the other groups who lived in the region. A great variety in population densities existed throughout the area. Other researchers have calculated that somewhere between 22,000 to 45,000 people once lived in the region. All together, this population is a fraction of the number of people who once lived in the grasslands east of the Front Range.

The ways the peoples of the Great Basin struggled to meet bare subsistence requirements contributed largely to this low population density. Large-game hunting, so characteristic of peoples living east of the Rocky Mountains, was fairly rare among those living in the intermountain grasslands. They took advantage of small game and insects, besides utilizing over 100 plant species. Those people had to know a variety of edible plants, as reliance on any one food source made life precarious. Moreover, limited food sources made the maintenance of a large population impossible. The Eastern Shoshones, who migrated throughout the Great Basin to the north of the Great Salt Lake, would be the ones later called the Comanches. Between 500 BCE to approximately 1600 CE, they spread north of the Great Salt Lake and had abandoned much of the earlier agricultural practice they had once pursued in the Gila River Valley of present-day Arizona. Even to maintain their sparse existence in the Great Basin they practiced infanticide to keep their population within their means (Flores 1991).

For those living in the grasslands east of the Rocky Mountains, life was much easier than it was for those living in the Great Basin. The Central Grassland peoples practiced both agriculture and hunting, and this allowed them to flourish by 1500. Especially representative of this were those practices of the Wichitas and the Pawnees. These two peoples shared linguistic traits, and so it should come as no surprise that they practiced similar agricultural techniques. The Pawnees lived between the Kansas and the Platte Rivers, while the Wichitas lived along and south of the Arkansas River. Both groups engaged in an agriculture dominated by women while the men organized two annual bison hunts, one in the fall and the other in late spring. Men also hunted other game throughout the year. These people lived on a fairly rich diet of beans, melons, squash, and corn supplemented by gathered plums, grapes, and nuts. Besides hunting game animals, both large and small, they fished.

Athapascan-speaking peoples, as already noted, were occupying many regions of the Southern and Central Grasslands. They also pursued a mixed hunting and agricultural lifestyle. In fact, they relied more on agriculture than hunting. They occupied river valleys like the Pecos, Purgatory, and Upper Canadian and lived in relatively permanent villages of hide- or brush-covered lodges. Most of the peoples who lived to the north hunted game animals for their main means of subsistence. The most notable nation, called Mandans by Europeans, also engaged in farming. Their villages were semi-permanent earth lodges of considerable size. For protection they often surrounded their villages with a log palisade. Like so many others living throughout the grasslands, their village sites were near major rivers.

CADDO, JUMANO, APACHE, AND PUEBLO EXCHANGES AND TRADE PATTERNS

The peoples occupying the grasslands produced enough surplus to engage in active trade with each other. In several ways the economic relationships between the Caddos, Jumanos, Apaches, and Pueblos typify many, if not most, of the emergent trade patterns throughout the grasslands as they had developed by the time of European contact. The Jumanos, a group of people who lived in the Rio Grande River Valley south of the Pueblo villages and in the Pecos River Valley to the east, engaged in irrigated agriculture and bison hunting in grasslands to the east. They also controlled salt, an exceptionally important trade item. Their primary village was on the Rio Grande River, about fifty miles above the Big Bend where the Conchos River joins it. The village, called La Junta de los Rios by the Spanish, was the nexus for Jumano trade. From there, traders took routes leading north toward the Pueblos. They exchanged primarily bison ribs with the Pueblos, who reciprocated with corn, beans, blankets, and pottery (Anderson 1999).

The exchange of meat products for agricultural and manufactured products formed the key elements of trade throughout the grasslands. As a result, there were scores of major agricultural trading locations throughout the grasslands. Most thrived along the major river valleys in the region. To the north the Mandans had their towns along the Missouri River. In the middle portions, the Pawnees had major villages along the Republican and Platte rivers. The Caddos had sizable centers along the Canadian and Red rivers. And of course, the Pueblos based their major communities throughout the northern reaches of the Rio Grande River. All of these major river valleys were important centers of agricultural production and manufactured goods such as pottery, jewelry, leather goods, and cloth from the cotton grown in the southern range. In the uplands well beyond these large river valleys, hunters ranged seeking mainly bison but taking other animals as well, such as deer, pronghorn, and much smaller game such as rabbits. What emerged was a complementary exchange of meat protein for agricultural and manufactured goods.

Before long, well-worn trade routes crisscrossed the grasslands. These routes bound those people together in diplomatic, social, kinship, and economic relationships. The majority of traders traversed the more numerous routes south of the Arkansas River, but at least one major route led through the Great Basin desert grassland. Barterers also ranged south and north along another major route following the Missouri River Valley. As a result of this trade based upon sizable herds of game animals and fairly reliable agricultural production, these people were well nourished, populous, and largely free of diseases at the time of European contact (Tanner 1995).

ARRIVAL OF EUROPEAN DISEASES

The arrival of Europeans to the grasslands fundamentally altered the evolving relationships of the inhabitants. Europeans introduced unfamiliar diseases, animals, plants, religions, social structures, economic systems, and technologies, and these would fundamentally alter the ecologies of the grasslands. No other ecological force had so dramatic an effect on the resident peoples as European-borne diseases. The spread of these pathogens may have had an unexpectedly early and devastating effect on the peoples in the grasslands. The Spanish brought with them diseases never before known in North America. Smallpox (*Variola vera*) was undoubtedly a highly efficient killer of grassland peoples. Even among Europeans the disease often had a mortality rate as high as 40 percent. Among Indians, who completely lacked any natural resistance to this disease, the mortality rates far exceeded anything ever experienced by Europeans or Asians. Some scholars believe rates could have soared as high as 80 percent.

Smallpox, most authorities agree, arrived in Mexico in 1519. After that it spread with great rapidity throughout the ancient Aztec empire and in short time demolished that population. The disease could have traveled, so some authorities think, along the trade routes leading north to the desert grasslands.

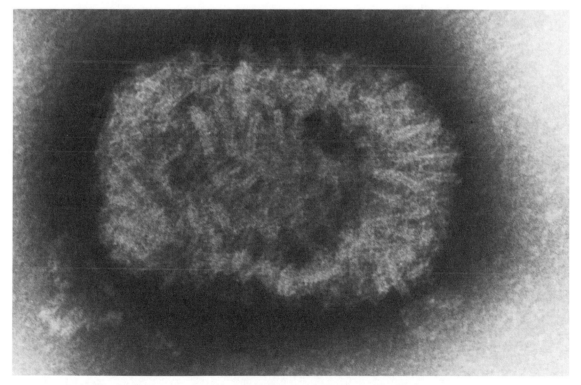

A smallpox (variola) virus particle, or single "virion." Due to Indian peoples' lack of immunity, smallpox plagued Indians throughout the grasslands. (Centers for Disease Control and Prevention)

From there it could have radiated out through the trade networks linking the bison-hunting cultures with the more agricultural Pueblo peoples. If this were the case, the disease may have made its first appearance in the grasslands as early as 1525, nearly fifteen years before any of those people would encounter Francisco Vasquez de Coronado and his troops. By 1530 European diseases had certainly made their appearances in the southern portions of the grasslands. Measles in 1531 followed by another disease in 1535 began taking their tolls. In 1545, just four years after Coronado arrived in the region, the bubonic plague struck. When Juan de Oñate explored the grasslands through present-day Kansas in the early 1600s, he noted a succession of abandoned villages, very likely the results of pandemics (Hall 1989).

Besides simply killing people in massive numbers, these diseases had other important ecological effects. Consider first that perhaps 90 percent to 95 percent of the pre-1500 grassland population had vanished by the late 1700s. First, the disappearance of these people opened niches throughout the grasslands. When any new arrivals ventured into the grasslands they would have found a largely depopulated land with vast areas readily accessible for immediate occupation. They could become strong and retain control over large parts of the grasslands if they could devise strategies enabling them to cope with the same diseases that had carried off the former residents. Also, such large numbers of vanishing hunting peoples would have had consequences for game animals. Obviously, with no one around to hunt them the herds would have increased. The only factors limiting their growth would have been the carrying capacity of the grasses and the concurrent rise in predators such as wolves. In the 1700s bison herds might have been as large as they ever were, given the effects of European-borne diseases.

CONTACT AND CONSEQUENCES

Other ecological forces, although not as deadly and destructive as European-borne diseases, also had dramatic consequences for grassland ecosystems. By the early 1600s, the Spanish had established a stronghold in the Rio Grande River Valley. There they worked to control the Pueblo peoples by converting them to Christianity and making them loyal subjects of the King of Spain. The Spanish also worked to bring Pueblo lifestyles more in tune with Spanish customs.

The Spanish introduced many novelties to the Pueblos and Apaches, but none more important than horses. Up to 1600, Pueblos and Apaches had engaged in a peaceful, kin-based system of trade. The Spanish demands for tribute and their slaving expeditions put a severe strain on these relationships, and soon the Apaches were raiding Pueblo and Spanish villages. As this commenced, the Apaches acquired horses and become mounted hunters and warriors. Soon they had extended their bison-hunting range and developed effective horse-borne raiding of agricultural villages. In 1680 an event occurred with far-reaching ramifications for the history of the grasslands. The Pueblos, tired of religious oppression, economic domination, and cultural hegemony by the Spanish, rose up and pushed the Spaniards out of the Rio Grande River Valley to present-day El Paso, Texas. The Pueblos' success, short-lived as it was, had one especially important consequence. They began freely trading horses to whomever they wanted.

The Utes lived to the north of the Pueblo villages. Now with the Spaniards gone, they found open, unfettered opportunities to trade with the Pueblos. One object of immediate desire was horses. The Utes, in turn, introduced these animals to their distant, related kin, the Comanches. The Comanches took to horses quickly and realized the great utility these animals held in pursuing bison. The Utes and Comanches had nearly twenty years to work out their adaptation to horses before the Spanish reconquest, and by the time the Spanish had control of the upper Rio Grande River Valley again, the Comanches had begun a mass migration onto the grasslands.

The acquisition of horses seemed a positive force for the Comanches, but they quickly learned the liabilities of horse tending, too. It is easy to imagine the sense of freedom a horse rider might have bounding across an open grassland. It is important to understand that horses came with their own set of needs, and if left unmet these had serious consequences for both horse and rider. Water was one of the two controlling resources establishing how effective a horse might be for its owner. A reasonable, average daily intake of water for a horse is between 10 to 12 gallons a day. Moreover, horses fare badly if they have no water for twenty-four hours. Consequently, tending horses or riding them necessitated being constantly near a water source. This limited the freedom of riding horses and using them in the hunt more than one might expect. For example, bison can go four days without drinking water, and this allowed them much greater freedom to roam the grasslands than horses had (Sherow 1992).

For bison, the introduction of horses onto the plains spelled bad news. While horses had a more limited range than bison, horses did give bison hunters greater reach and more mobility throughout the grassland than they ever possessed before. This certainly made hunting bison easier and more productive for people such as the Apaches and Comanches. As horse herds grew, their very numbers meant competition for the same resources that bison relied upon. Horses and bison shared a dietary overlap of around 80 percent. Perhaps as many as 2 million mustangs and around 500,000 domestic horses lived below

Many artists captured the exhilaration of horse-borne bison hunting, including Swiss-born Karl Bodmer in his "Indians Hunting the Bison" (1833). (Historical Picture Archive/Corbis)

the Arkansas River Valley before 1850. This fact alone would have reduced the possible number of bison from around 8.2 million to under 6 million (Flores 1991; Isenberg 2000).

More mobility brought the Comanches into contact with other Europeans to the north. By the early 1700s French Canadians had established trade relations with the Pawnees and introduced guns, another new item with far-ranging consequences. The Spanish became greatly alarmed by this when General Juan de Ulibarri sought to return to the Rio Grande Puelbos around sixty Picurís from their exile in a place called El Cuartelejo. This particular Apache ranchería was most likely located in present-day west-central Kansas. In a protected valley, with exceptionally good water sources, those villagers thrived on their agricultural production. But they also were exposed to Pawnee raids from the east, and the pillagers were armed with French-made guns. When Ulibarri arrived at El Cuartelejo in the summer of 1706, the Apaches showed him the guns they had captured. While guns were nearly useless in hunting bison and other game animals, they certainly were effective in killing humans. Those without access to guns quickly realized their vulnerability and sought access to guns themselves for their own protection (John 1975; Thomas 1932).

Intertribal warfare had definite ecological consequences sometimes overlooked. The only way Indian peoples could acquire guns was through trade. That meant they needed to be able to produce something desired by those who had guns. Furs certainly became one commodity eagerly sought by traders like the French Canadians who were trading with the Pawnees by 1700. All kinds of animal furs were bartered, but certainly bison robes and beaver pelts were the most common. Of course, this trade would have a significant bearing on the populations of those animals and on the ecological relationships the animals had with other species.

THE RISE OF THE COMANCHE EMPIRE

The combined effects of European-borne diseases, cultural conflicts over religious beliefs, climatic changes concurrent with the onset of the Little Ice Age, and the dispersal of introduced European animals, especially the horse, created an ideal situation for the success of aggressive, expansionist peoples such as the Comanches to the south and the Lakota Sioux to the north. In the case of the Comanches, they began appearing on the grasslands around 1700. More than a mere coincidence, they ventured out onto the grasslands at a most opportune time.

How marvelous the grasslands must have seemed to them. The place provided everything they needed. There seemed no limitation on the size of anyone's horse herd given the range of grasslands and the apparently superabundance of its production. Most important, the entire reach of the grasslands south of the Arkansas River appeared sparely populated, and this provided the Comanches with plenty of room to increase their own numbers. The lack of dense human population may have had another important consequence for the Comanches. With the population of the grasslands in rapid decline as a result of European-introduced diseases after 1500, a niche for the Comanches opened. They were able to exchange their old practice of hunting and gathering on foot for a horse-mounted bison-hunting culture. Also, hunting pressures on game animals, especially bison, had undoubtedly lessened prior to their arrival. Consequently, the numbers of bison and other wildlife would have increased. Upon reaching the grasslands, the Comanches gazed upon what must have appeared to them a protein-laden cornucopia. What they failed to realize, and there is no reason to expect that they should have, was that they had entered the grasslands under unique circumstances. Grazing conditions had improved with the onset of the Little Ice Age, and under these conditions, horse and bison herds flourished. Moreover, immigrating Comanches faced little human opposition to their occupation of the grasslands, as disease had severely reduced much of the population prior to 1500.

The Comanches acquired and mastered horse-tending practices quickly. Within a hundred years the northern Comanches especially would be noted for their large herds of horses. In addition, they were breeding for horses and mules as well as capturing wild horses and stealing horses from other tribes and the Spanish. They understood one thing about mules quite well: mules were a more durable animal for work on the plains than were horses. By the early 1800s, travelers often noted mules and Spanish donkeys beside horses in Comanche herds.

Mounted on their horses, Comanche men—and some women—became effective bison hunters. Riding into bison herds, while dangerous work, rendered rich rewards. The tribe feasted on bison meat and took the hides for their own needs and trade. So well did bison meet their subsistence needs that they dropped their former practices of infanticide and did whatever they could to increase their population. Moreover, they had enough surplus meat to trade for agricultural and manufactured products with their neighbors such as the Pueblos and New Mexicans to the west, or the Wichitas and Caddoes who lived in the river valleys cutting through the southern grasslands.

One group of people, the Apaches, stood in their way, and only for a short time. They had arrived too early to make the same adjustments to the grasslands as had the Comanches. The Apaches had come prior to the advent of the horse, and they had established agricultural villages in several river valleys. Even after they had acquired horses they elected not take up a fully mobile existence. As a result, these villages became death traps for the Apaches as the Comanches easily found and raided them. The Comanches had adopted the horse but also a mobile lifestyle. They had a good sense of the territory that they controlled, but they could ill afford to remain in fixed village sites. Their horse herds, the movements of bison, and coping with diseases necessitated a roving mode of existence.

In the span of a few decades and well into the mid-1800s, the Comanches controlled the largest horse herds in the North American grasslands. They may have collectively owned as many as 10 horses per capita. A census taken in 1855 pegged the number at 6.5 horses per capita. Such large herds meant the constant need for fresh pastures, as none of the grassland peoples harvested and stored grasses. The need for ample forage became more acute in winter months. They adapted by breaking into smaller groups with fewer numbers of horses. These villages were more easily moved once surrounding pasturelands had become depleted. Also, the Comanches moved farther south to where the rangelands remained green throughout most of the year. This strategy worked out so well in terms of maintaining large herds that the Comanches were known for regularly eating horses when bison were difficult to find.

Besides providing transportation, horses were their main form of wealth. This became nearly a universal standard of measure among all grassland peoples. The Comanches became the main suppliers of horses to nearly all of the other nations to the north of them. Huge trade fairs were conducted at a locale on the Arkansas River in present-day eastern Colorado called Big Timbers. The peoples living to the north had considerable difficulty keeping their horses alive through winters and were in constant need of refreshing their herds. The Comanches easily supplied this need in return for the manufactured goods arriving by way of French Canadian and English traders who dealt with the nations to the north of the Comanches (Hämäläinen 1998).

Sheep also became an important introduced animal. This ungulate, which the New Mexicans raised in the hundreds of thousands, required extensive grazing lands. The manner in which sheep grazed also altered the grasslands for scores of miles around the villages and small cities of New Mexico. By some estimates, from 200,000 to 500,000 sheep were being traded in markets around Chihuahua. As early as 1806, when Spanish authorities held Zebulon Pike and his troops under arrest, Pike noted in his journal that over 30,000 sheep that year had been traded in southern markets (Hall 1989).

Soon, well-established trade networks emerged throughout the grasslands. The Sioux, Pawnees, and Osages to the east controlled the flow of manufactured goods into the grasslands, whereas the Comanches, and their allies the Kiowas and Plains Apaches, controlled the horse trade flowing south to north across the grasslands. By the late 1700s, the Comanches had chased the Apaches out of the grasslands east of the Rio Grande River, were raiding throughout Tejas, and were attempting to displace Spanish control of the Rio Grande Valley.

Don Juan Bautista de Anza, one of the most effective governors of Spanish New Mexico, countered Comanche expansionism by defeating Cuerno Verde (Green Horn), so named by the Spanish. Cuerno Verde led some of the most effective Comanche warriors, but in 1779 Anza successfully attacked and killed him, routing his force. This brought about a rapprochement with the northern Comanches, called the Yamparicas, and others called the Jupes. As a result, Anza initiated a plan to settle a large group of Jupes in an agricultural village located at the junction of the St. Charles and Arkansas rivers. Anza was of the opinion that agriculture was "most proper for the instruction of man born to live in Society" and that only through farming would the Comanches truly become "civilized." Governor Fernando de la Concha continued the work of Anza by building irrigation works and pueblos on the site and by providing livestock and seed. This social experiment ultimately proved unattractive to the Comanches, who abandoned it in less than two years (Kavanagh 1996).

Nonetheless, Anza and his successor, Concha, clearly recognized some of the liabilities of living on the grasslands. They counseled Paruanarimuco, the Comanche chief, that the bison herds were being depleted. The governors also recognized the need for irrigated agriculture in order to sustain reliable crop production in a land of unpredictable rainfall. In short, to raise domestic grasses such as corn, wheat, and oats and tend domestic ungulates would require a transformation of the grasslands. Nearly another century would pass before Anza's vision for the grasslands would take hold.

NEW ARRIVALS AND WAYS OF ECOLOGICAL ADAPTATION

Social and political revolution in Mexico had profound consequences for reshaping the social and economic memes throughout the grasslands, and concurrently its ecological dynamics. Revolutions in Napoleonic Europe gave rise to an independent Mexico. As early as 1810, Miguel Hidalgo y Costilla, an influential radical priest, called for Mexican home rule. For the following decade there was political and military turmoil in Mexico that came to an end only with the rise to power of Agustin de Iturbide. His forces drove out the last remaining supporters of Spanish rule, and Iturbide proclaimed an independent Mexico in 1821.

Under the heavy-handed rule of Iturbide, who assumed the title of Emperor Agustin I, trade was opened with the United States. Before, the Spanish had been very careful to exclude American traders from New Mexico, Texas, and California. For over a decade, American traders had been unsuccessful in trying to open markets with New Mexico. Now trading routes opened, and Americans soon began pouring into New Mexico. This policy was continued even after the overthrow of Agustin I in 1824 when Mexico became a republic and Guadalupe Victoria became its first elected president.

Even though there were several well-worn roads leading into New Mexico, the most notable one was the Santa Fe Trail. William Becknell, a resident of Franklin in the recently formed state of Missouri, loaded some pack animals in September 1821 and headed west to Santa Fe. His route took him over well-traveled paths used by Indian peoples for hundreds of years and later by the Spanish. The road had some exceptionally good qualities for draft animals. Water sources abounded all along the trail, whether it was crossing the headwaters of rivers and creeks, skirting springs, or following rivers. In season, grass was also plentiful along the entire length of the trail. Often enough grass remained after the growing season to sustain draft animals in fairly good condition if the drivers were careful about the placement of their camps.

Soon the exploits of many of these traders were published, and interest in commerce with the Republic of Mexico grew considerably. Many Americans saw their main opportunities lying south in the states of Chihuahua, Durango, and Zacatecas. The Central Plateau of Mexico was one vast, rich grassland where owners of incredibly large estates raised hundreds of thousands of horses and cattle. Mountains ringed the edges of the plateau, and these were the source of some of the richest gold and silver mining lodes in North America. The mining towns of Durango and Zacatecas promised rich rewards for any enterprising American trader who could make his way to those lands. Many in the federal government were excited about the prospects of this commerce and wanted to facilitate its potential. By 1825, Congress, through the lobbying efforts of Missouri senator Thomas Benton, appropriated money for mapping and marking the trail. George Champlain Sibley led this expedition in 1825, and his journals recorded the resources of the entire route.

ENTER THE CHEYENNES, ARAPAHOS, AND SIOUX

The Bent brothers, Charles and William, pursued a golden opportunity along the Santa Fe Trail when they established their trading post, Bent's Fort, near the confluence of the Purgatory River with the Arkansas River by 1834. They had been working for quite some time with the Missouri Fur Company (MFC). Jean Pierre Chouteau managed this company out of its headquarters in St. Louis, Missouri. His enterprise exerted a strong influence in the bison robe and beaver pelt trade. The reach of the company extended up the Missouri River to the Mandan villages. Later, the Chouteau family would extend its reach west along the Kansas and Platte rivers where the Kaws and Pawnees lived, and to the southwest among the Osages. The Bents realized the dangerous nature of the work in the northern portions of the grasslands. The Blackfoot Confederacy was a powerful group who were supplied with guns and goods by rival operations out of Canada. The Hudson Bay Company traders were the great nemesis of the MFC's operations, and the Blackfeet fiercely contended with all of its rivals whether Euro-Americans or Indian peoples. Understanding this, the Bents were on the lookout for a safer port in the storm (Christian 2004; Binnema 2001).

Many among the Cheyennes and Arapahos were also looking for a better and safer place to live. They were well acquainted with the burgeoning trade to the south, and chiefs such as Black Kettle and Yellow Wolf encouraged the

From the early 1830s to around 1850, Bent's Fort served as an international emporium for people living in the Central Grasslands. (Library of Congress)

Bents to move with them to the south and establish a stake in the region. For over a century, the Cheyennes had been migrating westward, forced to leave their original homeland along the upper reaches of the Mississippi River Valley. They had been making a gradual transition from an agriculture-based economy to a horse-borne, bison-hunting culture. In the early 1800s, the bison herds still looked large, accessible, and indestructible to the Cheyennes, Arapahos, and Bents. Hunting these animals and trading the robes would form the backbone of their relationship. Also, horses would figure prominently in this arrangement. If successful, the Cheyennes and Arapahos could position themselves as brokers between the horse-rich Comanches to the south and the horse-poor nations to the north. Moreover, they could make guns more accessible to the Comanches and trade bison robes to the Bents for all the manufactured goods they could desire. Notably, too, the Cheyennes, and their allies the Arapahos would be positioned to raid the large horse herds of the Comanches, Kiowas, and rancherías on the Central Plateau of northern Mexico.

The Bents, on the other hand, could become the exclusive agents for all of the manufactured goods flowing into these markets. Moreover, with a good site for their company, they could position themselves profitably in the recently opened Santa Fe and northern Mexico trade. Both Charles and William enhanced their prospects through kin relationships. William married Owl Woman, and after her death, her sister Yellow Woman, in 1847, and thereby cemented his ties to the Southern Cheyennes. In either 1835 or 1836, Charles

Yellow Wolf as portrayed by Lieutenant Abert (1845). (Corbis)

married Maria Ignacia Jaramillo, a wealthy widow whose kin ties attached her to the most prominent political and business leaders in Taos, New Mexico. This was a common pattern among Euro-American traders throughout the grasslands. The men would often marry into prominent families of Indian peoples or other ethnic groups. The officials in the American Fur Company and several among the Chouteau family, such as Auguste, who married a woman of an important Osage family, all typify those sorts of trade and kin relationships. The forging of these relationships and the opening of new trading posts throughout the grasslands facilitated trade in horses and furs. While this intercourse made many people rich and enhanced the social prestige of a great many others, it also had disruptive ecological and attendant social consequences.

THE MEMES OF TRADE AND CULTURE

Before coming to grips with the ecological ramifications of the horse-borne hunting cultures and the fur trade, we need to know the nature of the trade itself. It is crucial to understand that the commerce on the prairies was global in nature. Whether conducted at the many posts of the American Fur Company, or through ventures such as Bent's Fort, or by the independent free traders who roamed throughout the grasslands, all trade shared some common traits. Most important, two forms of economic exchange, barter and money, made this trade work. Next, the goods exchanged mixed those of the grasslands with items drawn from throughout the globe (Wishart 1979).

When considering the economics of this trade, it might be useful to think about it in terms of memes. A good way to envision a meme is to compare it to a gene. Whereas genes are the biological building blocks of an organism, memes are the conceptual building blocks of culture. A meme might take the form of the root understanding of God such as Allah, Yahweh/Jehovah, Ra, or Maheo, the god of Muslims, Jews and Christians, Ancient Egyptians, or Southern Cheyennes. These building blocks share a similar nature with genes in that each must be reproduced in order to survive. They also adhere to the laws of evolution, especially natural selection.

Barter and monetary exchange on the grasslands might usefully be seen as two memes. They were the building blocks of two forms of economic exchange, and each was reproduced within cultural constructs. Indian peoples, in varying ways, used barter, which for this analysis includes gift giving, and Euro-Americans used monetary exchange to build their economic relationships. These concepts had to be reproduced through education and practice to remain viable components of these relative cultures. When either became displaced through competition, it faced the possibility of extinction, and such became the fate of barter.

Monetary exchange would in time became the ultimate form of economics in the grasslands. As humans in all likelihood were the keystone species of the grassland ecosystem, the way in which they thought about resources, and then used resources in their economic exchanges, would have a significant effect on the biotic and abiotic relationships within the ecosystem itself. For example, when one species, such as bison, became defined as an economic commodity, the hunting pressures on it had effects on the other relationships it shared with other species. The ill effects on wolves were certainly pronounced as the numbers of bison declined. When cottonwood trees suddenly became an important resource to nourish horses throughout winters, the stripping of this tree had ecological consequences for any relationships with other species developed prior to this time. For example, bison quickly found that riparian ecosystems where cottonwoods grew were a poor place to overwinter when populated by horses eating twigs and cottonwood bark, and by humans seeking shelter for themselves and their steeds. Also, humans were more than prone to hunt bison. All together, the links between bison and riparian ecosystems began to break apart.

Consequently, the way in which any culture viewed resources had important repercussions for the ecosystem in which these resources were found. In essence, the memes which defined any culture became important parts shaping the energy flows creating the grassland ecosystems. Always, generalizations about any culture overshadow important variations within it. However, generalizations are informative in how they give light to the thinking and values driving the actions of a group of people. Euro-Americans had particular ideas about the grasslands that in time would immensely alter it. Most important, they saw it lacking "civilization." Strikingly, Euro-Americans often viewed the grasslands as a desert, a place devoid of human occupation. Certainly they understood that Indian peoples lived in it, and they understood that they planted and harvested crops. Euro-Americans just considered Indian peoples as lacking civilization and, as such, were somewhat less than human. Paradoxically, Euro-Americans could also wax eloquent about the beauty and fecundity of the grasslands while in the same breath describing the place as a desert.

Certain things were absent from the place that would give it civilization as they understood it. They wanted to see Protestant Christianity take hold and displace all forms of Indian peoples' religious practices. They desired to replace all forms of Indian agricultural practices with their vision of people living in houses, building barns, and harvesting commercial crops grown in fenced fields marking off their privately owned lands. They pictured a "domesticated" landscape overturning a "wild" one where wheat and corn replaced tall bluestem and buffalo grass, where cattle grazed in place of bison, and the American beauty rose grew in manicured gardens where the prairie rose once had blossomed.

The Hispanic-Pueblo worldview in Nuevo México advocated a less rigorous human-centric notion of life. This world was a highly isolated one. These people occupied around 1 percent of the land in Nuevo México, and most of this in the Rio Grande River Valley. Theirs was a community-based, agro-pastoral system of life. They raised crops of corn and cotton, and had large herds of horses, cattle, sheep, and goats. They had a difficult time maintaining ecologically sustaining practices in such a confined setting. The population of the region increased dramatically over time, domestic herds became larger, and around the major cities rangeland denuding brought serious problems. Moreover, the need for wood for building and fuel resulted in deforestation along the foothills with increased erosion in arroyos and other watercourses. Clearly, their herding and agricultural practices, even though not rooted in a capitalistic economic system, had some real ecological liabilities (MacCameron 1994).

Some of their social practices helped to mitigate these problems. Following the reconquest, through intermarriage especially, Hispanic and Pueblo cultures began a mutual and reciprocal cross-acculturation. The Nuevo Mexicano view of life became distinct from those prevalent in Mexican states to the south. Perhaps these inhabitants began to adopt some elements of the Pueblo fertilization practices used to maintain both their human communities and the plant and animal communities surrounding them. The state recognized landholding patterns conforming to resources and geographical features that held the promise of sustaining communities. After the United States conquered Nuevo México, the cultural differences between Americans and Nuevo Mexicanos became sharp. In the poetry called *décimas*, Nuevo Mexicanos made their dislike of Americans quite clear. Those poets disliked American capitalists and the *ricos* among them, who championed capitalism as economic reform. The poets saw American women as weak and thin, and the Mexican-American War as unjust. The American occupation brought little more than suffering and the starvation of *faithful* Christians— that is, Catholics as opposed to Protestants (Campra 1971).

Indian peoples throughout the grasslands had their own ways of visualizing the world around them, and these various views shaped the way they managed and used the resources in their environments. It is tempting to generalize far too much about what appear to be the similarities in the spiritual practices of these peoples. For example, all of the horse-borne hunting peoples but the Comanches in the grasslands to the east of the Rocky Mountains practiced a Sun Dance. But the different Sun Dances represent about as much similarity in beliefs as do the Protestant rituals and doctrine of Congregationalists, Southern

Baptists, and Unitarians. Still, perhaps a close look at one nation's practices will give some insights into how the views of Indian peoples were apart from those of the Nuevo Méxicanos and Americans.

The Sun, *K'o* (Grandfather), bathed Kiowas with their main and most potent source of power. Kiowas sought their *dwdw*, personal power, more from this source than any other. They regularly offered sacrifices of their most important sustenance, bison meat, and often their own flesh, when supplicating *K'o*. Warriors painted *K'o* images on their shields as an effort to take the *da*, power, of the sun into battle. During the Sun Dance participants often had depictions of *K'o* painted on their bodies. Bison were the only ones with the ability to acquire their own *dwdw* directly from the sun. As such, bison had more power than any other living being. One Kiowa song went something like this: "I am a buffalo, / I'm standing here without water, / Right in the middle of the day, / That's the reason I'm getting power." Unlike any human being, a bison could stand out alone on the grassland and while bathed in sunlight, collect the power of *K'o* (Kracht 1989).

Great warriors among the Kiowas commanded the respect of their entire community. Those men, by virtue of their great accomplishments in slaying enemies or counting coupe (touching an enemy with a special stick thereby indicating the enemy not even worth killing), exuded mighty *dwdw*. In a similar manner, Kiowas normally regarded killing bison, the most powerful of living things because it took its *dwdw* directly from *K'o* himself, akin to requiring the *dwdw* necessary to kill any enemy. In hunting and warfare both, hunters and warriors demonstrated the strength of their *dwdw*. While success in warfare commanded the highest respect among Kiowas, a hunter's feat in killing bison received nearly equivalent recognition among the people.

All cultures had different ways of viewing their environments, but they all engaged in trade. This trade, while propelled by different motivations, had some drastic ecological effects. Two trade items had particular ramifications for the grasslands and the people who lived in them. These were whiskey and guns. Of the two, whiskey flooded the grasslands. Wholesalers in St. Louis might buy whiskey out of the Ohio River Valley for as little as $.31 a gallon, and traders on the grasslands would sell it for $28 or more per gallon. Obviously traders such as the Chouteaus, the Bents, distillers in Taos, and freelancers could reap tremendous profits on whiskey if not caught by the army, which enforced prohibition in the grasslands, especially if they could sell their entire stock without loss, which seldom happened.

Certainly the outfitters at the AFC posts and the Bents were engaged in this trafficking. In 1838, the Bents alone purchased 1,145.5 gallons of trade alcohol from the Chouteaus in St. Louis, and they acquired another 1,120.5 gallons the following year. What was the implication for fur-bearing animals such as bison? To illustrate, say all of this whiskey was purchased with bison robes. The market for this amount of whiskey would have resulted in the deaths of at least 27,000 bison. Undoubtedly, fur-bearing animals throughout the grasslands suffered given the far-reaching nature of the whiskey trade.

The other item, guns, often made the difference in whether a nation could successfully protect itself or expand its reach. The right guns also enhanced the chances of a warrior proving himself and thereby rising in social esteem. The introduction of guns to the Blackfoot Confederacy through the Hudson Bay Company enabled them to become the dominant force along the northern reach of the grasslands (McGinnis 1990). The gun trade in the central portions allowed the Southern Cheyennes and Arapahos to reign supreme there for around a half century. So important was acquiring arms that the Comanches and Kiowas entered into a peace accord with the Southern Cheyennes and Arapahos in 1840. For their part, the Comanches and Kiowas allowed the Southern Cheyennes and Arapahos access to southern bison hunting grounds and to horses, either through trade with themselves or with deep raids into northern Mexico. In return, the Southern Cheyennes and Arapahos opened the gun market to the Comanches, a market that Cheyennes and Arapahos had ready access to through their own trade connections with Euro-American merchants such as the Bents (Jablow 1951).

Another important object of trade, one with accompanying ecological ramifications, was the horse. Horses provided great mobility, but they came with ecological constraints. The daily nutritional needs of a mustang ranged from 10 to 12 gallons of water and 10 to 25 pounds of hay. He also needed a pound of salt a week. In addition, mustangs were susceptible to sunstroke, parasites, and freezing when improperly tended. So keeping horses healthy in the grasslands was difficult for Indian peoples. They lacked any kind of enclosed shelters for their horses in winter aside from their own dwellings, if large enough. About the only structures that accommodated horses in the winter were the large earth lodges of the Mandans or Pawnees. And even here, only a few horses could be placed inside along with the human occupants.

Indian peoples also lacked the means to harvest, cure, and store fodder for overwintering horses. This meant relying on rangeland pastures at all times of the year. In winter villages, women and youngsters had the onerous tasks of collecting cottonwood bark and twigs for supplemental feed for the herds. This, in conjunction with cutting trees for heating fuel, began to denude the riparian woodlands, the same resources that gave wintertime protection for people and horses alike. In summertime other ecological difficulties emerged. Of course, grazing conditions improved with the onset of spring—provided that adequate rains fell. When dry conditions occurred, that is, less than ten inches of rain in the year, 1,000 horses could easily require over forty-two acres of rangeland a day for adequate foraging. Coupled with a daily requirement for water, this meant horse-borne hunting cultures found themselves frequently on the move in search of grazing and water. In dry years it was far more difficult for groupings within the nation to remain together, and the more separate and smaller a group, the more susceptible it was to raiding.

Despite these obstacles, horses were in great demand throughout the grasslands. Unsurprisingly, given the more difficult horse-tending climate of the northern reaches, nations there, such as the Mandans, Blackfeet, Crows, Arikaras, Gros Ventres, Sioux, and Pawnees, were in constant need each spring of replenishing their herds due to wintertime losses. In July 1843, John C. Frémont met an Oglala man who was scouting ahead for his village. He was looking for Arapahos who were reportedly hunting in the region. He represented some Sioux who had lost their entire horse herd due to the severity of the previous winter and who hoped to purchase replacements from Arapahos (Frémont 1845).

Simple transportation needs also required an abundance of horses. When the time arrived for the Pawnees to begin their spring bison hunts, normally a wife would need six horses just to haul the family belongings required for the quest. Not all horses were used for transportation, hunting, or as a means of wealth. They were also used for food. Comanches and Kiowas, as already noted, would often turn to their horses whenever other meat sources failed. The Northern Paiutes and the Yokuts, who lived in the desert grasslands of the Great Basin, herded horses nearly exclusively for food, and then for transportation. But they hardly ever used horses for hunting. Of course, the complete lack of bison herds in the Great Basin contributed to this. Still, different people valued horses for different reasons depending upon location and ecological conditions (Fountain 2004).

In 1855, the number of horses per nation became more bountiful the farther south into the grasslands one traveled. Of the Northern Comanches, Kiowas, Apaches, Southern Cheyennes, and Southern Arapahos, all possessed an average of 6.25 horses per capita except for the Cheyennes, who owned 5.55 horses per capita. The Comanches had the easiest time of tending horses, as the southern grasslands seldom, if ever, experienced the harsh winter conditions more common to the north. Moreover, when exceptionally cold weather blew in, it seldom remained for long.

Horses certainly enabled their riders to gain access to bison herds throughout the grasslands in the first half of the nineteenth century. The importance of controlling these areas was enough to spark fierce intertribal warfare. The works of Dan Flores and Andrew Isenberg provide clear evidence that the overhunting of bison was occurring several decades before Euro-American hide hunters nearly obliterated the herds in the 1870s. In 1834, when Colonel Dodge visited with the Comanches and Kiowas, he warned them that they were ill advised to rely on hunting as their main resource. In 1846, at Bent's Fort, Yellow Wolf asked Lieutenant Abert for help in making the transition from a horse-borne, bison hunting culture to agriculture. In 1851, Thomas Fitzpatrick wrote to the commissioner of Indian affairs that the peoples living in the Central Grasslands were in a "starving state."

Indian agent W. D. Whitfield, while at Bent's New Fort, just west of present-day Lamar, Colorado, completed a census of his agency in August 1855 (see Table 1). He recorded how many people lived within his jurisdiction, how many animals they hunted, and how many horses they kept. Assuming his numbers are an accurate reflection of reality, the Comanches, Kiowas, Plains Apaches, Southern Cheyennes, and Southern Arapahos were either not taking in enough bison to sustain themselves or failing to reach their trade potential.

The people of those nations hunted bison for two main reasons: sustenance and trade. Consider sustenance first. One estimate suggests that a person living among these nations required at least six bison per year for sustenance (i.e., for food, shelter, and clothing). Given a total population of 11,470 people in the five nations, meeting normal sustenance needs would have translated into 68,820 bison.

TABLE 1
1855 Census by Agent Whitfield, Upper Arkansas Agency

Tribe	Lodges	Maximum robe potential for trade	Population	Subsistence requirement	Bison killed	Remaining potential for trade if subsistence needs were fully met	Per capita kill	Per capita trade potential	Per lodge trade potential
Comanches	400	8,000	3,200	19,200	30,000	3,600	9.4	1.1	9.0
Kiowas	300	6,000	2,400	14,400	20,000	1,867	8.3	0.8	6.2
Plains Apaches	40	800	320	1,920	2,000	27	6.3	0.1	0.7
Cheyennes	350	7,000	3,150	18,900	40,000	7,033	12.7	2.2	20.0
Arapahos	300	6,000	2,400	14,400	20,000	1,867	8.3	0.8	6.2
Totals	1390	27,800	11,470	68,820	112,000	14,394			

Source: W. D. Whitfield. Census of the Cheyenne, Comanche, Arapaho, Plains Apache, and Kiowa of the Upper Arkansas Agency, 15 August 1855. U.S. Department of Interior, Bureau of Indian Affairs, Letters Received by Office of Indian Affairs, 1824–1881, Record Group 75, M234, Roll 878.

Next consider trade. Seasoned traders pegged the average robe production per lodge at around twenty per year, and they estimated that three bison were killed for each one selected for tanning. In other words, normally sixty bison were killed in order to produce twenty marketable robes. The 1,390 lodges among the six nations would have had a production capacity of 27,800 robes, which under usual hunting practices would have meant killing around 83,400 bison.

Given these estimates, hunters needed to kill 152,220 bison to satisfy both the normal sustenance requirements of their people and to fulfill their robe trade potential. Agent Whitfield, however, estimated that the hunters had killed only 112,000 bison. Something had to give. They could have either met their production capacity and gone hungry, or met their normal sustenance needs but suffered economically. Only one nation, the Southern Cheyennes, could have met its sustenance needs while simultaneously reaching its maximum potential in robe production in 1855. For all other nations, attaining their potential robe production took food from their mouths, and meeting their sustenance needs curtailed their trade. Other factors need to be kept in mind. Hunters took trade robes from November through March and then from cows, which had the softest and best furs for trade. During the other months the hunt was primarily for food and leather. The summertime hunts were becoming more and more difficult each passing year as ecological conditions in the grasslands dramatically changed from 1840 onward. As noted by many scholars, the combination of several factors, including horse dietary overlap, bovine diseases, wolf predation, natural mortality rate, climatic change, habitat disruption, subsistence hunting, and market hunting, all led to a plummeting bison population and the end of the horse-borne hunting cultures in the grasslands. Still, there were a few ways these grassland residents could supplement their economic and subsistence needs other than through bison. Other fur-bearing animals were also harvested in great numbers in the region. For example, 5,650 elk, 44,000 deer, and 7,825 bear were killed in the same year. These nations derived $38,000 in revenues from the chase, or an annual per capita income per lodge, which might be thought of as a kin group, of just over $15 per year.

This was not nearly enough to provide for the material needs of most grasslands Indian peoples. An 1867 inventory of a Southern Cheyenne Dog Soldier and Oglala Sioux village by the U.S. Army reveals a rich material culture made up of items of indigenous and industrial manufacture (see Table 2 on p. 60). These items were the ones left behind by the occupants of 251 lodges who fled the advance of the 7th Cavalry. Certainly, their possessions were more numerous and elaborate than the ones enumerated by General Hancock's troops, but a

TABLE 2
Inventory of the Dog Soldier and Oglala Village, April 14, 1867

Dog Soldiers		Oglala Sioux	
Indigenous	Industrial	Indigenous	Industrial
144 par fleches	35 water kegs	159 par fleches	63 water kegs
522 bison robes	13 coffee mills	420 bison robes	239 saddles
48 rawhide ropes	142 sacks of paint	140 door mats	216 tin cups
142 head mats	197 saddles	94 horn spoons	142 axes
55 horn spoons	22 hoes	145 head mats	54 brass kettles
14 wooden spoons	49 axes	19 wooden spoons	59 coffee pots
111 door mats	19 brass kettles	61 stone mallets	15 coffee mills
13 stone mallets	8 coffee pots		3 tea kettles
22 meat stones	152 tin pans		141 kettles
111 lodges	1 oven		4 curry combs
	6 hammers		3 pitch forks
	4 scythes		51 chairs
	49 kettles		8 bridles
	12 tea kettles		70 sacks of paint
	34 frying pans		9 drawing knives
	152 tin pans		1 lance
	12 crow bars		1 U.S. mailbag

Source: *Junction City (Kansas) Union*, 4 May 1867.

sampling from this inventory provides an accurate reflection of the material cultural of those people.

Clearly, these people were engaging in significant trade. Also, they were becoming more and more reliant on industrial manufactured goods. They lacked the means to fashion a coffee mill much less the ability to grow and harvest coffee beans, but undoubtedly, grassland peoples enjoyed their coffee. The Kiowas even fancied flavoring their brew with New Mexican chilies. In short, the plains economy worked because it was part of a global system. The coffee on the shelves of Bent's Fort or Fort Union came from Sumatra or Africa, and guns from England. The bison robes tanned by women in Assiniboin or Lakota villages kept women in their carriages warm in the streets of London.

COLLAPSE OF THE GRASSLAND BIOME

By the mid-1840s, if not before, people throughout the grasslands understood that the ecosystem could no longer support their cultural and economic aspirations. Rather than an epiphany, the realization that a faltering ecosystem was eating away at cultural and economic supports emerged like the unfolding of a pleasant dream slowly becoming nightmarish without any way to awaken. While travel across the grasslands increased yearly from 1820 to 1840, with the advent of the Mexican-American War traffic increased exponentially. Josiah Gregg recorded what certainly must be considered a minimal numerical increase of trade wagons plying the Santa Fe Trail after the beginning of the war. Perhaps over 500,000 people made the trek across the Oregon and California trails in the years 1840 up to the beginning of the Civil War (Unruh 1979). Add to these numbers the additional traffic of military expeditions and regular troop movements and the gold rushes to the Rocky Mountains in the late 1850s (Ball 2001; West 1998).

In conjunction with the waning of the Little Ice Age, all of this traffic had severe ecological consequences for the grasslands. The climate throughout the grasslands was becoming a bit more erratic during the 1850s and 1860s. There were years of little rainfall and high temperatures, and other years of high rainfall and very cool weather. For example, in 1855 and 1860, the grasslands received little rainfall and high temperatures, and this resulted in certain ecological effects. The Kiowas had great difficulty in tending their horse herds during the spring and summer of 1855. In Set'tan's calendar, he referred to that season as the "Sitting Summer" because the Kiowas had to stop often to rest their quickly fatigued, dehydrated horses (Mooney 1898). Bison herd movement in 1860 definitely tended eastward to the tallgrass prairies in search of better grazing conditions. This change of location could hardly have proven beneficial, as it placed the herds closer to a more dense human population. Moreover, the effects of overhunting cows was becoming more evident. Eugene Bandel, who was stationed at Fort Leavenworth, noted that the herds nearby had a cow to bull ratio of 1 to 5, hardly an auspicious range for maintaining the reproduction of the herds (Bandel 1931–1943).

In 1867 the spring and summer were cool and rainy. Grasses do poorly when they receive too much water. The mycorrhizal fungi die when too much water saturates the soil. When this happens, grasses lose the protein nourishment that the fungi would normally transfer from the soil into their root systems. Any grazing animal would have received less sustenance from such plants and would have been unable to maintain its weight. Cattlemen called grasses in this condition "washy" because of their high water content. This was the range condition when the first Texas cattle herds were driven to Abilene, Kansas, in 1867 (Sherow 2001). The year before, snow had fallen at Fort Dodge in May, a very late snowstorm by anyone's account.

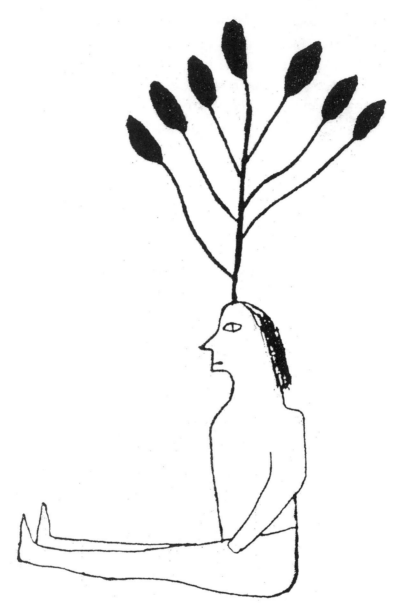

The "Sitting Summer," as depicted in the Kiowa calendar. (U.S. Government Printing Office)

When the increasing traffic through the grasslands is added to the effects of climate change, ecosystems undergo rapid alterations. Colonel Doniphan's Army of the West occupied New Mexico in the fall of 1846, and his troops noted the large herds of goats, sheep, cattle, and horses of the villages and how denuded the land was for miles around any of these sites. When the army occupied Santa Fe, Doniphan had his sizable horse herds grazed nearly thirty miles away. The addition of supply trains and the vast standing herds of army horses required soldiers to tend the herds farther and farther away each passing day. By the end of October, the situation had worsened to the point that George Gibson wrote in his diary about the impossibility of acquiring feed for army mules, beef cattle, or oxen within a fifty-mile radius of Santa Fe, as the entire countryside had been "literally eaten up" (Gibson 1935).

Compounding ecological change was the disappearance of important riparian woodlands. These were the sites for winter camping grounds of Indian peoples where they could provide some shelter for themselves and their horse herds, and gather heating fuel. Where these places were unoccupied by humans, bison overwintered. Both Indian peoples and bison migrated out of these systems onto the grasslands proper as soon as green blades broke the ground. In their wake came the overlanders, soldiers, and gold rushers. They had gathered at the staging cities on the eastern borders of the grasslands. Places such as St. Joseph, Independence, and Westport, Missouri, were all important. And in 1854, when Congress formed the Nebraska and Kansas Territories, other cities such as Leavenworth and Atchison, Kansas, and Omaha, Nebraska, became prominent disembarkments as well. Thousands of people and hundreds of thousands of domestic animals entered the grasslands each spring with the fresh appearance of green grass.

They arrived in the riparian woodlands after Indian peoples and bison had headed out onto the uplands. When the Euro-Americans entered these places, they also chopped the trees for fuel, and soon many of these woods disappeared altogether. This was a frequent occurrence all throughout the heavily traveled Platte River Valley, especially when the army began establishing outposts along the entire route. Other spots, such as Big Timbers along the Arkansas River Valley in the eastern portion of present-day Colorado, became severely reduced in size and substantially depleted of trees.

Certain areas of the grasslands were less affected by this great movement of people. The northern reaches of the grasslands, those well north of the Platte River, would remain intact for another couple of decades. Not until the mid-1870s would those grasslands begin undergoing some of the same transformations that had so dramatically altered the grasslands to the south. But in time, disease, overhunting of the bison, the rush of Euro-Americans into the area, whether to search for gold in the Black Hills of present-day South Dakota, or to look for better cattle-grazing lands in present-day eastern Wyoming and Montana, or to seek farming opportunities in present-day Nebraska and the Dakotas—all would have the same devastating results for Indian peoples and bison as had occurred earlier to the south. For a while, the Lakota Sioux and their Northern Cheyenne and Arapaho allies could dominate the lands north of the Platte River. In fact, through aggressive warfare and shrewd negotiations, they forced a removal of Euro-American forts along the Bozeman Trail. But the rush of gold seekers in Dakota Territory and the overwhelming power of the American Army with its ability to cut off access to trade, grazing grounds, and other resources left the Sioux in an impoverished state regardless of their temporary military victories in such engagements as the Battle of Greasy Grass in June 1876, or, as otherwise known by Euro-Americans, Custer's Last Stand.

As early as the 1840s, Euro-Americans and Indian peoples were seeking alternative means for living in the grasslands other than reliance on a horse-borne hunting economy. The conversation between Yellow Wolf and Lieutenant Abert at Bent's Fort in 1846 indicated a growing awareness that hunting would not sustain Indian peoples indefinitely. Yellow Wolf requested assistance from the United States government so that the Southern Cheyennes could make a transition to an agricultural economy. He realized this would be a difficult undertaking, but he saw few alternatives. The federal government, regardless of the prodding of agents and commissioners, moved slowly in addressing the situation. In 1860, however, the negotiations of the Fort Wise Treaty with the Southern Cheyennes and Arapahos promised them the wherewithal to make the transition to farming. The Bureau of Indian Affairs undertook building an agricultural outpost along the Arkansas River. The contractor was charged to build flood irrigation works, a schoolhouse, blacksmith shop, and other structures necessary for the Indians to thrive on a much smaller range of land. Regardless of any good intentions of the bureau, the entire project completely failed. The Civil War diverted funding from the effort and delayed its start until late 1863 and early 1864. By then, the war and its spillover effects had disrupted hunting, further destroyed bison and horse habitats, altered trade and diplomatic relations, and spread disease and social decay throughout the grasslands. Anger with Euro-Americans among Indian peoples mounted, and deadly conflicts increased.

Over two decades of armed conflict between Euro-Americans and Indian peoples ensued beginning with the Sand Creek Massacre in November 1864 when John Chivington and his Colorado volunteers destroyed Black Kettle's peaceful village. For all practical purposes, this warfare came to an end when Sitting Bull headed south out of Canada to Fort Buford where he and his followers surrendered in July 1881. In the meantime, the army began forcing Indian peoples throughout the grasslands onto reservations where they were to be "civilized," which meant making them farmers, practitioners of Protestant Christianity, and speakers of English. Euro-American policymakers saw this destruction of the wild as the saving grace for the grasslands, but in actuality, it ultimately destroyed one form of grassland ecosystem and replaced it with another.

THE BEGINNINGS OF THE DOMESTICATED GRASSLAND

The end of the Civil War only speeded up the rate of American "domestication" of the grasslands. In 1867, Alexander Gardner, a renowned portrait photographer during the Civil War, had been hired by the directors of the Union Pacific to photograph the route of some of its lines. Gardner's photographs capture in great detail the results of unleashed transformative forces. But he also captured the remnants of the older landscape once managed by Indian peoples. Seldom had anyone depicted the path of "American Civilization" as well as Gardner did in his photographs. Indian peoples with their horse-borne hunting cultures and American farmers, city dwellers, and industrialists could not occupy the grasslands simultaneously because of the different ways the respective cultures em-

Alexander Gardner's photograph of Lawrence, Kansas, shows the advance of Euro-American domestication of the grasslands in 1867. (Library of Congress)

ployed the region's energy and water sources. Euro-Americans always held the balance of power with their ability to draw on remote, abundant sources of fuel, fiber, and food. This capacity overwhelmed Indian peoples who could not disentangle themselves from an increasingly weakened grassland ecosystem and trade networks.

Gardner's photographs capture this transformation of the grassland in its first stages. He also caught on his glass negatives the last remnants of the grassland that Indian peoples had managed. His prints of Lawrence, Kansas, show a small city with substantial commercial buildings. Shown also is the Kansas River bottom where tree stumps dot a landscape dominated by railroad tracks, fields, and service buildings. When he took his shots of Manhattan and Abilene, these small outposts of Euro-American culture and economics were surrounded by an open prairie. But this landscape would undergo a rapid transformation as prairie fires were suppressed and exotic plants and domesticated animals were broadcast over the surrounding area. Traveling with Gardner was a fellow named Josiah Copley. Copley made an astute observation about the transformation of the grasslands. The monetary, market-driven nature of his culture found undisputed expression in his assertion that "the locomotive must precede the plough, and the town the farm." Copley's thinking about the settlement of the frontier was a complete 180 degree reversal of Turner's frontier thesis. Moreover, Copley rightly understood this monied commerce was fully underwritten by railroad connections to the commercial powers in the East. The new meme of trade, he realized, had much greater strength and staying power in the grasslands than the meme of commerce bound up in bartering for furs, agricultural produce, and livestock.

Copley's vision of the grasslands, like Sternberg's and so many others', was thoroughly enmeshed in ambivalence. On one hand, he could look at the short-grass plains and see "a beautiful and cheerful-looking country—gently undulating, and here and there presenting hills or buttes that partake of the character of ruggedness" (Copley 1867, 20). On the other hand he witnessed the "wrecks" or "the stark remains of the [grassland's] monarchs, the buffaloes, bleaching in the sun and wind" (Copley 1867, 20). This was a place of waste and desolation in its "natural" state, awaiting only "the hand of enlightened industry and taste to make it beautiful and home-like" (Copley 1867, 20). Copley understood that for the memes of Euro-American culture to gain a hold in the grasslands required institutions capable of reproducing them. In Manhattan, Kansas, he saw one of the first institutions ever created for that purpose: Kansas State Agricultural College. The charter of this college, one of the first endowed by the Morrill Land Act of 1862, promised to educate an agrarian society in the principles of democracy and in the science of industry and agriculture.

The Englishman William A. Bell, who also accompanied Gardner, saw in this budding institution "The Paradise of Petticoats." He saw women receiving the same education in the same setting as men, and new for the times, the faculty supposedly made no distinction in their teaching based upon gender. Bell exaggerated the case when he said "bold, indeed, would be the man . . . who dared to oppose openly this phalanx of political Amazons" (Bell 1870, 16). In October of the same year, less intimidated Kansas male voters defeated the first proposal in the nation to make women's suffrage part of the state's constitution. Even though Bell engaged in hyperbole, he likened the importation of ideas like civic equality for women, a meme if you will, to a living entity that quickly took root in "virgin soil to an extent hitherto unprecedented." In much the same way that horses had quickly occupied open niches in the grasslands, Euro-American memes easily occupied the open niches left by the retreat of Indian peoples and traders. Memes such as the creation of money markets for trade and commodities, along with their distinctive technological infrastructure, would blossom in the grasslands. This form of trade, so its champions proclaimed, would not only transform Indian peoples, but it would also open an avenue for the flow "of true Christian civilization." Americans such as Bell, Copley, and Sternberg saw this as the "right plan of human progress."

THE BRIDGE BETWEEN THE WILD AND DOMESTIC GRASSLANDS

The development of the Texas cattle trade is an excellent example of how the Euro-American memes of culture and economics combined to begin the transformation of a wild grassland into a thoroughly domesticated one. The Chisholm Trail can be more readily understood as an ecosystem itself rather than simply as a pathway for Texas cattle herds. As an ecosystem, or a dynamic community of life-forms and physical forces, the Chisholm Trail was a short-lived system, an ephemeral ecosystem bridging a previous one largely shaped by the presence of Indian peoples, and a later one formed when farmers dominated the landscape. The botanist Daniel Botkin once astutely wrote: "Life and the environment are one thing, not two, and people, as all life, are immersed in the one system" (Botkin 1990, 188). From 1860 to 1885, Texas drovers endeavored to control the water and solar energy sources of the Chisholm Trail in culturally shaped pursuits. Doing so they altered the dynamic properties of the Chisholm Trail environment, or its surroundings. These changes affected more than just the plants, animals, and water resources of the trail; they shaped the culture and lives of the people who occupied the region.

The trail environment encompassed a great extent of territory. All of the Texas rangelands should rightfully be considered part of its environment. Miles of rangeland on either side of the trail leading north were parts of it, too. And the rangeland amounting to nearly the western half of Kansas was at one time part of it. Drovers realized their cattle economy rested upon a few ecological pillars. They needed to control the solar energy reserves wrapped up in grasses; they required access to water sources throughout the grasslands; and they required markets and a transportation system that would make cattle trading economically profitable. The Kansas Pacific Railway Company provided the transportation link necessary to take Texas cattle to markets in Chicago, from where the meat and associated products were shipped worldwide. The growing cities of the Midwest and East, along with those in Europe, provided the nearly un-

Alexander Gardner's 1867 photograph of Abilene stockyards displays the initial railroad shipping point for Texas cattle herds. (Library of Congress)

quenchable market demands for meat. Drovers such as W. W. Sugg joined with cattle brokers like Joseph McCoy to link Texas cattle to market outlets such as Abilene, Kansas, where McCoy built stockyards for routing cattle to Chicago. This operation helped to resolve a difficult problem associated with Texas cattle. Texas tick fever, caused by a microscopic protozoal (*Pyrosoma bigeminum*) living inside ticks that infested Texas longhorns, proved exceptionally deadly to domestic cattle herds in the Midwest. Midwest farmers and cattlemen violently resisted having their herds come in contact with Texas longhorns, and the Abilene depot and stockyards were far enough west of these domestic herds to resolve any potential conflicts.

So long as drovers remained beyond the farms and large cities, their system worked fairly well. But climatic and ecological forces often threatened to destroy the system. Drovers often practiced "overwintering" their herds when Chicago cattle markets were glutted and depressed. Putting a herd out on the open range over a winter near a shipping point placed it close enough to be one of the first shipped out the following spring when prices rebounded. But the winter of 1871–1872 showed the dangers of open-range cattle ranching more than a decade before the devastating blizzards of 1888. In the fall, prairie fires, which drovers had been trying to suppress, had burned off the bluestem prairies; this meant drovers had to push their herds farther west onto the shortgrass plains for adequate forage. In November, an unseasonable storm with snow and sleet blanketed the grass in ice. The cattle, scattered by strong gales and unable to reach any free-flowing water, dehydrated quickly, weakened, and froze to death in a matter of days. Joseph McCoy estimated that of 350,000 cattle, over 250,000 died in the storm—along with fourteen cowboys.

As farmers entered the Chisholm Trail environment, they shaped an entirely new biome, one that was neither conducive to open-range cattle operations or to Indian hunting practices. The long-standing biomes of Indian peoples, prairie grasses, and animals disappeared. The grassland biome, necessary to Texas drovers, gave way to agricultural biomes, which displaced the wild grasslands all together. In effect, the sort of short-lived ecosystems shaped by Texas drovers and their longhorns intersected with two incompatible biomes, the prior one of Indian peoples and the later one of farmers. For both Indian people and Texas drovers the results were the same: they could not maintain their material cultures (Sherow 2001).

The emergence of the farming ecosystem out of the previous ones is clearly depicted by a writer for the *Wichita Eagle* in 1872. The correspondent had counted over seventy thousand Texas cattle in a twenty-five-mile radius around Wichita. Herders were fattening these animals while waiting for the first opportunity to secure a good price at the stockyards. The columnist also simultaneously noted maturing cornfields in the region intermixed with luxuriant wild-grass pastures. In fifteen years' time, only one of these elements dominated the region: domestic grain fields. After 1900, J. J. Roberts of Del Rio traveled north and visited Wichita, Newton, and other former cattle shipping points on the Chisholm Trail. "Where the old Trail passed through in those early days . . . the change that meets your eyes is but little short of marvelous. Where saloons and dance halls stood are now substantial school buildings and magnificent churches and the merry prattle of happy children. And it was a deep feeling of pride that came to me, to know that I had had an humble part in bringing about this wonderful change" (Roberts 1925, 785–786).

By 1900, the Little Ice Age was a climate of the past. And how to define the grasslands then? It had become a domesticated grassland in the main. There were remnants of the grassland ecosystems formerly managed and shaped by Indian peoples, but fewer acres persisted each passing year. Indian peoples were confined to reservations where mostly well-intentioned people taught them "civilization" and others not so well intended took every advantage of their situation to enrich themselves. Only a few hundred bison remained, and wolves were a thing of the past. In Kansas, no deer ran across the land. The soils of the grasslands still retained their patterns, but people treated them far differently than they had a century before. Euro-Americans expected more from them and when those expectations were not met, they tried to change the very nature of the soils to meet their preconceived expectations. Sternberg's notions of progress had arrived and taken hold in the grasslands.

References

Anderson, Gary. 1999. *The Indian Southwest, 1580–1830: Ethnogenesis and Reinvention.* Norman: University of Oklahoma Press.

Ball, Durwood. 2001. *Army Regulars on the Western Frontier, 1848–1861.* Norman: University of Oklahoma Press.

Bandel, E. 1931–1943. *Frontier Life in the Army, 1854–1861.* In *The Southwest Historical Series,* Vol. 2, edited by Ralph P. Beiber and LeRoy R. Hafen. Glendale, CA: A. H. Clark.

Bell, William A. 1870. *New Tracks in North America: A Journal of Land and Adventure whilst Engaged in the Survey for a Southern Railroad to the Pacific Ocean during 1867–8.* New York: Scribner.

Binnema, Theodore. 2001. *Common and Contested Ground: A Human and Environmental History of the Northwestern Plains.* Norman: University of Oklahoma Press.

Botkin, Daniel B. 1990. *Discordant Harmonies: A New Ecology for the Twenty-First Century.* New York: Oxford University Press.

Campa, Arthur L. 1971–1972. "Protest Folk Poetry in the Spanish Southwest." *Colorado Quarterly* 20, 355–363.

Christian, Shirley. 2004. *Before Lewis and Clark: The Story of the Chouteaus, the French Dynasty That Ruled America's Frontier.* New York: Farrar, Straus and Giroux.

Copley, Josiah. 1867. *Kansas and the Country Beyond, on the Line of the Union Pacific Railway.* Philadelphia: J. B. Lippincott.

Fagan, Brian. 2000. *The Little Ice Age: How Climate Made History, 1300–1850.* New York: Basic Books.

Flores, Dan. 1991. "Bison Ecology and Bison Diplomacy." *Journal of American History* 78 (September): 465–485.

Fountain, Steve. 2004. "A Horse Is a Horse of Course: Multiple Horse Cultures and Ethnocultural Change in the North American West." Paper presented at the Forty-Fourth Annual Conference of the Western History Association.

Frémont, John Charles. 1845. *Report of the Exploring Expedition to the Rocky Mountains in the Year 1842, and to Oregon and North California in the Years 1843–44.* Washington, DC: Gales and Seaton.

Gibson, George Rutledge. 1935. *Journal of a Soldier under Kearny and Doniphan, 1846–1847,* edited by Ralph P. Bieber. Glendale, CA: A. H. Clark.

Hall, Thomas E. 1989. *Social Change in the Southwest, 1350–1880.* Lawrence: University Press of Kansas.

Hämäläinen, Pekka. 1998. "The Western Comanche Trade Center: Rethinking the Plains Indian Trade System." *Western Historical Quarterly* 29 (November): 485–513.

Isenberg, Andrew C. 2000. *The Destruction of the Bison: An Environmental History, 1750–1920.* New York: Cambridge University Press.

Jablow, Jacob. 1951. *The Cheyenne in Plains Indian Trade Relations, 1795–1840.* New York: Monographs of the American Ethnological Society.

John, Elizabeth A. H. 1975. *Storms Brewed in Other Men's Worlds: The Confrontation of Indians, Spanish, and French in the Southwest, 1540–1795.* Lincoln: University of Nebraska Press.

Kavanagh, Thomas W. 1996. *The Comanches: A History, 1706–1875.* Lincoln: University of Nebraska Press.

Kracht, Benjamin R. 1989. "Kiowa Religion: An Ethnohistorical Analysis of Ritual Symbolism, 1832–1987." Ph.D. Dissertation, Southern Methodist University.

Lott, Dale E. 2002. *American Bison: A Natural History.* Berkeley: University of California Press.

MacCameron, Robert. 1994. "Environmental Change in Colonial New Mexico." *Environmental History Review* 18 (Summer): 17–39.

McGinnis, Anthony. 1990. *Counting Coup and Cutting Horses: Intertribal Warfare on the Northern Plains, 1738–1889.* Evergreen, CO: Cordillera Press.

Mooney, James. 1898. "Calendar History of the Kiowa Indians." In *Annual Report of the Bureau of American Ethnology, 1895–96.* Washington, DC: Government Printing Office.

Roberts, J. J. 1925. "Fifty Years Ago." In *The Trail Drivers of Texas,* 2nd ed., edited and compiled by J. Marvin Hunter. Nashville, TN: Cokesbury Press.

Schlesier, Karl L., ed. 1994. *Plains Indian, A.D. 500–1500: The Archeological Past of Historic Groups.* Norman: University of Oklahoma Press.

Shaw, James H., and Martin Lee. Nd. "Ecological Interpretation of Historical Accounts of Bison and Fire on the Southern Plains with Emphasis on Tallgrass Prairie: A Final Report to the Nature Conservancy of Oklahoma."

Sherow, James E. 1992. "Workings of the Geodialectic: High Plains Indians and Their Horses in the Region of the Arkansas River Valley, 1800–1870." *Environmental History Review* 16 (Summer): 61–84.

Sherow, James E. 2001. "Water, Sun, and Cattle: The Chisholm Trail as an Ephemeral Ecosystem." In *Fluid Arguments: Five Centuries of Western Water Conflict,* edited by Char Miller. Tucson: University of Arizona Press.

Sternberg, George M. 1870. "The Causes of the Present Sterility of Western Kansas and the Influences by Which It Is Gradually Being Overcome." *Junction City (Kansas) Union,* February 5, 1.

Tanner, Helen Hornbeck. 1995. *The Settling of North America: The Atlas of the Great Migrations into North America from the Ice Age to the Present.* New York: Macmillan.

Thomas, Alfred Barnaby. 1932, 1969. *Forgotten Frontiers: A Study of the Spanish Indian Policy of Don Juan Bautista de Anza, Governor of New Mexico, 1777–1787.* Norman: University of Oklahoma Press.

Tkachuck, Richard D. 1983. "The Little Ice Age." *Origins* 10 (No. 2): 51–65.

United States Department of Agriculture. Natural Resources Conservation Service, "NRCS Soils Website" http://soils.usda.gov/ (accessed June 26, 2005).

Unruh, John D. 1979. *The Plains Across: The Overland Emigrants and the Trans-Mississippi West, 1840–1860.* Urbana: University of Illinois Press.

West, Elliott. 1998. *The Contested Plains: Indians, Goldseekers, and the Rush to Colorado.* Lawrence: University Press of Kansas.

Wishart, David J. 1979. *The Fur Trade of the American West, 1807–40: A Geographical Synthesis.* Lincoln: University of Nebraska Press.

3

THE URBANIZED AND DOMESTICATED GRASSLANDS

Strange as it might sound, the grasslands after 1870 became a thoroughly urbanized and domesticated ecosystem. Clive Ponting once noted an important global phenomenon. Unlike Frederick Turner's explanation for the development of Euro-American settlement patterns, Ponting illustrates how the rapid growth of cities across the face of the planet began determining or at least shaping the economic, social, and ecological contours of agriculture (Ponting 1991).

The creation of a domesticated grassland east of the Rocky Mountains was no simple matter. Educational institutions such as land-grant colleges were instrumental in providing the expertise that would power the aspirations of Euro-American market culture values. Passage of laws, such as the 1862 Homestead Act, encouraged colonization of the grasslands, and new plants were introduced that would become marketable commodities. Innovative laws and court suits encouraged and regulated water development in a land of limited rain. Attempts at corporate agriculture in the form of bonanza farms, even if unsuccessful in and of themselves, pointed the way toward twenty-first century corporate agribusiness techniques. All of this resulted in a vast ecological transformation of the grasslands. Bacteria and fungi were displaced; wild forbs, grasses, and animals were destroyed to make room for crops and domesticated animals; and invasive plant species such as salt cedar, juniper, Russian thistle, just to name a few, found openings in a simplified ecosystem and soon flourished. Much of this initial domestication effort would prove unsustainable as illustrated by the collapse of open-range ranching and the Dust Bowl. These failures would set the stage for massive federal intervention dedicated to saving and enhancing the domesticated grassland.

In contradistinction to the focus on agriculture in the eastern grasslands, when Euro-Americans first took serious note of the Southwest and intermontane grasslands, many recognized the grazing potential of it but few thought it worth farming. Prior to the American Civil War, Hispanos had made significant advances in spreading their sheepherding practices in some parts of the Southwest grasslands. They found the grasses around them nicely suited for grazing. Blue Grama, Sprucetop Grama, and Slender Grama provided excellent and durable forage when sensibly stocked. Some grasses, such as Hairy Grama and Black Grama, while nutritional were susceptible to overgrazing, and were soon replaced by Six Weeks Needlegrass and Six Weeks Grass, key indicators of overgrazed pastures. Army railroad surveyors certainly noted the grazing potential of the area, and some recorded the existence of agriculture along irrigated river valleys. Captain John Pope's expeditionary accounts published in 1854, for example, gave high praise to the wine production near El Paso, Texas (Meinig 1971).

Unlike their "optimistic" views about the Southwest grasslands, Euro-Americans' judgments were not far off the mark in predicting that the Great Basin grasslands would prove a harsh place to eke out a living. In time, only the irrigated agriculture practiced by Mormons along the eastern fringes would show any bounty. In July 1847, the first immigrants began preparing land for irrigated farming. Possibly one group of Mormons acquired some familiarity with New Mexican irrigation techniques while wintering at the confluence of Fountain Creek and the Arkansas River near tiny Pueblo in 1846–1847, and then employed these methods along the western fringes of the Wasatch Mountains. Quickly those enterprising people began damming and diverting as many streams as possible to create an agricultural base that would allow them to become self-sufficient and thereby immune from the brutal persecution they suffered when they lived in Illinois, Indiana, and Missouri. By 1850 they had over 16,000 acres in irrigated farmland that adequately supported the swelling numbers of Mormons flooding into the area. The spread of these communally based colonies thrived as highly localized developments dependent upon a scant number of small streams flowing out of the mountains. The greater extent of the Great Basin was dominated by Bluebunch Wheatgrass and Big Sagebrush. At the southern and hottest end of the basin the grasslands hugged the lower slopes of the mountain ranges, and as one traveled farther north the bunchgrasses began to occupy more fully the various smaller basins within the Great Basin itself (Vale 1975).

While Euro-Americans may have seen limited economic potential in both the Southwest and Great Basin grasslands, they would treat these regions as they had grasslands to the east of the Rockies. These desert and intermountain grasslands were in need of domestication or "civilization," which meant removing Indian peoples and confining them on small reservations, then placing their former domain into economic production and integrating its production into national and international markets. Policymakers and speculators had complete faith in the power of applied science and technology to transform the desert into a garden. By 1900, however, it appeared to many that those grasslands were exceptionally fragile despite their harsh, forbidding exteriors. A few farsighted scientists would advocate conservation techniques for ranges and farmland both, a plea that went largely unanswered until the 1930s (Robbins 1993).

THE URBAN ECLIPSE OF THE AGRICULTURAL ECONOMY

Changing demographics across the globe had a direct influence on the grasslands. Before 1800, a mere 2.5 percent of the world's population lived in cities, but by 1900 over 1 in 10 did, and over two-thirds of this population was centered in Europe and North America. New York, for example, was only 1 of 10 cities in the United States to have a population over 20,000 in 1830. Soon thereafter, the urban population of the United States doubled every decade until 1860. As this trend continued, more than 50 cities, such as New York, had populations over 100,000. Industrialization explains the large growth of urban areas, and by 1900, for the first time ever, the urban industrial economy outpaced the agricultural economy.

This urban growth, along with its hungry mouths, placed exponential demands on agricultural production. Cities relied on those living in rural areas to grow or raise their food, and their immense pull shaped the direction, technological developments, and pace of farming. The agricultural forces that had once given rise to cities were now shaped and guided by the sheer growth of its urban offspring. The grasslands fell fully within an urban gravitational sway, and Chicago, more than any other city, stimulated the growth of domesticated grasslands (Cronon 1991).

Before the Civil War, entrepreneurs in Chicago realized that they sat poised to capitalize on two aspects of the new corporate-industrial economy. First, they could channel raw agricultural products to their city, then convert these into finished products and control the flow of these into the burgeoning eastern urban corridor of the United States and the urban economy of Europe. Second, they realized the economic potential of marketing manufactured products to the rural populations and shipping them through the same transportation system that moved the agricultural production of the countryside.

These businessmen, and by and large they were men, saw controlling railroad lines as the key for their likely success. The construction of the Illinois Central Railroad showed the potential of this system for channeling trade and for domesticating the ecology and geography of the tallgrass prairie. The directors of the company created towns as collector points for agricultural production and spaced them about ten miles apart along the tracks. This placement was intentional based on how much time it took a farmer to load a wagon, harness a team to it, and drive it to town and return home in one day. Moreover, through the vigorous efforts of

Transcontinental railroads drew the grasslands into an urban economy. This photograph shows the completion of the first transcontinental link at Promontory Summit, May 10, 1869. (National Archives and Records Administration)

Senator Stephen Douglas of Illinois, the Illinois Central Railroad became the first land-grant system in the nation. In short, the federal government provided the railway company a set amount of public land for every mile of track constructed. The company was then free to sell the land to retire the cost of building the system (Stover 1975).

The passage of the Kansas-Nebraska Act in May 1854 was intended to make way for railroads, most importantly those radiating out of Chicago. This would have two effects: Indian peoples, along with the "wilderness" that they inhabited, would be eliminated and the rails would transport the farmers who would domesticate the grasslands. Indian grass and tall bluestem would give way to wheat and corn, bison and deer to cattle and hogs, earth lodge villages to cities with streets lined by buildings of stone, brick, and lumber. The goal of this act took a little longer to implement than its originators such as Senator Douglas had hoped. The issue of whether Kansas would be a free or slave state complicated the transformation of the grasslands, and it would take a civil war to settle that question. But once the Civil War had put the issue of slavery to rest, the transformation of the grasslands east of the Rockies began in earnest (Sengupta 1996; Miner and Unrau 1978).

Similarly, throughout the Southwest grasslands and the Great Basin, railroad construction transformed the economies of those regions by connecting them to national and international markets. After the end of the Civil War, the rapid spread of cattle ranching and irrigated and dryland farming all followed in the wake of rail construction. This, in time, indelibly transformed the land. Energetic, if not at times reckless, men initiated these iron lines to create an economic empire. On May 10, 1869, at a lonely place in the Great Basin called Promontory Summit, a crowd of reporters and workers watched as Thomas Durant, the vice president of the Union Pacific, and Leland Stanford, president of the Central Pacific, pounded in the last two spikes linking the rails that bound the country east to west. Now, travelers could cross the North American continent from New York to San Francisco in a mere ten days, and the produce of the Great Basin could be shipped to markets on either coast. To the south, construction crews had begun laying the rails of the Atchison, Topeka, and Santa Fe Railway Company with Cyrus K. Holliday in command of the corporation. He set into motion the creation of a web of iron that connected the Southwest grasslands to national and international markets.

EURO-AMERICAN MARKET CULTURE VALUES AND LAND-GRANT COLLEGES

The prevailing American worldview sought to domesticate and civilize the grasslands. Several assumptions sustained this value system. The vast majority of Euro-Americans believed the world had come into existence exclusively to serve them. Furthermore, they exhorted "improving" and "dominating" the landscape surrounding them. In short, as many newspaper editors and promoters phrased it: their mission was to "create a garden in the grasslands." Clearly, this phrase indicated a belief that the grasslands were lacking in enough bounty to support the cultural expectations of Americans. In transforming their environments, these immigrants expected to create profitable farms, industries, and cities (Emmons 1971).

The memes shaping this new industrialized agriculture found their most secure procreative environment in a unique American institution—the land-grant college. The United States Congress enacted the legislation creating these schools in May 1862. The ostensible mandate was to create in each state at least one place of education for the "industrial classes" of Americans where they would acquire a "liberal and professional education . . . in secular pursuits and professions of life" (An Act 1862). In such a setting, students were to engage in scientific and classical studies besides learning military tactics and "subjects related to agricultural and mechanical [read engineering] arts." Congress funded this initiative by providing 30,000 acres of federally owned land for each representative and senator in a state. In those states where no public lands existed, Congress gave them scrip representing federally owned lands in other states or territories.

Improving the economic wealth of the nation through a more knowledgeable exploitation of agricultural resources became the prime underlying purpose behind the bill. As Senator Justin Morrill of Vermont put it: "Agriculture undoubtedly demands our first care; because its products, in the aggregate, are not only of greater value than those of any other branch of industry, but greater than all others together; and because it is not merely conducive to the health of society, the health of trade and of commerce, but essential to their very existence" (Key 1996, 211). For the "proprietors of the soil," land-grant colleges would provide the "knowledge which will prove useful in building up a great nation—great in its resources of wealth and power, but greatest of all in the aggregate of its intelligence and virtue" (Key 1996, 212). In essence, the Congress enacted the legislation because it possessed economic potential in promoting the rapid colonization of the public lands, which would, in turn, contribute to the national treasury.

Two other congressional acts further strengthened the mission of land-grant colleges. The first was the Hatch Act, which Congress enacted in 1887; it established a federally funded system of agricultural experiment stations under the auspices of the land-grant colleges. Station scientists experimented with crops, insect control, technological innovations, and animal husbandry. Each school published its results in a series of agricultural experiment bulletins, and these were distributed or made available to farmers throughout the nation. The Smith-Lever Act of 1914 was the second act of Congress to increase the public outreach of land-grant colleges. In this, Congress created the extension service with the mission to provide people not in attendance at a college or university with information on how to improve their agricultural practices or their home economics. At the same time, with the staff support of the extension service, 4-H became an integral component of extension work (Gates 1979).

Land-grant schools quickly became the incubating force for the Euro-American farm practices that would transform the grasslands. In these centers scholars, scientists, and researchers worked in acquiring the knowledge that would further strengthen the Euro-American hold on the grasslands. Judging from the mottos of several schools, to those who taught and researched at them, the idea of dominating and subduing Nature for human progress remained preeminent. The motto of Iowa State University is quite specific in this context. It reads:

"Science with Practice." The motto of Kansas State University clearly reflects the same sentiment: "Rule by Obeying Nature's Law." The domination of Nature through the scientific study of the physical world became a guiding light for these colleges.

Many important scientists, whose names in the annals of U.S. history remain fairly obscure, were instrumental in the propagation of Euro-American practices. Ainsworth Blount of Colorado State Agricultural College experimented with hundreds of varieties of grain crops and the development of experiment stations throughout the eastern portions of the state. Elwood Mead pioneered irrigation engineering at the school while his successor, Louis Carpenter, occupied the first chair of irrigation engineering in the nation. At Kansas State Agricultural College, A. S. Hitchcock conducted his fieldwork on wild grasses. He later took his findings to the Smithsonian Institution, which published his *Manual of Grasses of the United States* (1935). Mark Carleton, a student of Hitchcock, became enthralled with the study of grasses, and especially of domestic ones such as wheat. His work led him to the Russian steppes where he studied wheat varieties, and back to the United States and Kansas where he cultivated and experimented with scores of winter wheat species as he strove to find varieties most adaptable to the varied climate of the grasslands. His research had important results in the development of hard winter wheat production throughout the region (Isern 2000).

In the Southwest and Great Basin grasslands, creating sustainable, domesticated forms of ranching and agriculture proved exceptionally rough going, and in this endeavor state agricultural colleges demonstrated their worth. The thinking within these institutions followed the same established lines that gave form and direction to the colleges in the grasslands to the east of the Rocky Mountains. The curriculum revolved around the practical arts and sciences, those studies that yielded techniques for improving irrigation works, for increasing the technological efficiency of farm equipment, and for increasing knowledge about soils, grasses, and water resources so that they could be employed more effectively in economic production.

Hiram Hadley had ambition and vision. Inspired by his Quaker faith, he believed in the power of education and established teacher colleges in the Midwest before heading to New Mexico for the health of one of his sons. After arriving in 1887, Hadley settled in the small town of Las Cruces, and he quickly took up the work of establishing a college. He opened the college doors in the fall of 1888, and by the end of the year he had over sixty students enrolled. His chance to enhance Las Cruces College came when the New Mexico Legislative Assembly accepted the provisions of the Morrill and Hatch Acts, transforming the small preparatory school into a land-grant agricultural college. By the end of 1889, the Agricultural College of New Mexico was a reality with a functioning experiment station (Kropp 1967). By 1890, federal funding had reached the campus, and A. E. Blount, the agriculturalist at the station, began his study of irrigation and farm crops in earnest. Hadley understood well that "the experiments performed in New York or Georgia are of little value to New Mexico," and research was needed to understand this "region somewhat peculiar to itself" (Hadley 1890, 4). The goal was to make commercial agriculture a sound proposition through the application of science and technology, or in Blount's own words: "Agriculture is broadly founded in the laws of Nature." Understand and master these and the results would "become of incalculable value" (Blount 1890, 2). Soon research was under way on crop-destroying insects, domestic grasses and forage plants, wild grasses, introduced weed species such as Russian thistle, and the suitability of the Rio Grande water for irrigation. Arthur Goss noted how "the water supply is probably the question of most vital importance in New Mexico agriculture," and he set out to analyze the chemical properties of the flow in the Rio Grande and published his findings in 1893 (Goss 1893, 33). His main concern centered around alkali concentrations, which if too high would effectively destroy plant growth. By 1900 the college had a well-established mission with a student enrollment of over 200, with women representing 30 percent of the students.

The origins of Utah State University, the "people's college," while appearing similar to those of Kansas, Colorado, and New Mexico, came into existence with a theocratic underpinning. Brigham Young understood the value of an educated agricultural class not only for learning the applied science of soil conservation and crop production, but also to achieve what he called "spiritual wealth." As he phrased it: "Go and fence a field, sow and water it for the kingdom of God on Earth . . . the aim of the Latter-day Saints is to beautify the earth for the reception of the Son of Man" (Ricks 1938, 9). Years followed before Anthon H. Lund would introduce a bill into the Territorial Legislature of Utah for the creation of a land-grant college. The "Folk Schools" of Denmark, in essence rural agricultural high schools, were his initial inspiration, and in 1888, the legislature passed a bill to create and fund an agricultural college and experiment station under the provisions of the Morrill and Hatch Acts. In Logan, Utah, construction of the college began in 1889, and under the direction of college president Jeremiah W. Sanborn, classes opened for the "education of a rural western democracy—a rural college for the 'masses'" in September 1890. While certainly the focus of the college remained firmly fixed, as Lund eloquently expressed it, "in furnishing our farmers with such methods as may facilitate their labors and bring them the greatest result"

(Ricks 1938, 30), President Sanborn made it a point to cultivate the spiritual side of his students. He required daily attendance at chapel, conducted, as Sanborn put it, free of "sectarianism or from any attempt directly or indirectly to change the church alliance of any student" (Ricks 1938, 31). In appearance, Sanborn's directive appeared democratic and was implemented, perhaps, more for the faculty who were drawn from non-Mormon institutions than for the students, who nearly all were adherents of the same religion. By World War II, the college had an enrollment of over 3,800 students who studied an expanded progressive curriculum covering four main components—agricultural and civil engineering, teacher training, economics, and liberal arts.

MARKET CULTURE VALUES AND COMMODIFICATION

While land-grant colleges provided the applied science for agriculture, the Board of Trade in Chicago serves as the best example of how the economics of agriculture evolved. Just as the New York Stock Market Exchange represented the economic strength and vigor of American corporate capitalism,

The Chicago Board of Trade, as depicted in this 1900 photograph, transformed farm production into abstract commodities. (Library of Congress)

the Chicago Board of Trade became the center of agricultural marketing in the United States. The Euro-American meme of agricultural marketing evolved through the trade mechanisms developed by the members of the Board of Trade, and this, in time, would have consequences for the type of farming actually practiced in the grasslands.

It might be well to reflect on the major concepts underlying a Euro-American economic view of agriculture. These might all be summarized under the rubric of the "market culture," which has at its core three principles. First, Euro-Americans saw the streams, rivers, grasses, and soils as potential commodities rather than as parts of an intricate ecosystem. Next, they had the divine obligation, they believed, to use the resources in the pursuit of progress and economic gain. Moreover, the government had the duty to assist this exploitation of resources (Worster 1979).

More than any other institution, the Chicago Board of Trade, created in 1848 by eighty-two Chicago merchants, turned discrete components of ecosystems into marketable commodities. Wheat, oats, corn, pigs, and cattle became the most important resources collected, traded, and sold into national and international markets centered in Chicago. After the Civil War, railroads brought an ever-increasing number of animals and harvested grains to Chicago. These yields were gathered, stored, weighed, graded for quality, and refined in the mills and stockyards of the city. In 1865, traders devised new marketing practices that made it easier for them to abstract the economic value of the goods and to sell "futures" rather than the goods themselves. This contrivance allowed traders to fill demands for flour and meat while at the same time supposedly creating a more stable economic environment in which farmers and traders alike would prosper. Some questioned just how well this system has benefited either farmers or the consuming public, but certainly a great many traders who skillfully bought and sold futures became fabulously wealthy (Williams 1982; Cronon 1991).

In its simplest form, a futures contract is a legally binding agreement made between a buyer and seller for a set quantity and quality of a good for a certain price at an agreed upon future date. This procedure allowed brokers to trade farm products from grains to hogs without ever having to deal physically with flour or hams. The contracts for the products became more important than the products themselves, and in this manner the value of farm animals and harvests became abstract, brought into being apart from the goods themselves. In such a system, farmers often found themselves with little control over the market forces establishing prices for their products. The economics of farming had become an urban enterprise.

FARMING SYSTEMS

In urbanizing and domesticating the grasslands, homesteading (purportedly a system providing enterprising people without means the opportunity to own land and become self-sufficient) played an important role. George Henry Evans, an antebellum reformer, saw free land as a cure-all for American social problems, and through his leadership, the National Land Reform Association sought to stop land speculation and to create a nation of small, independent farmers. By the late 1850s, Horace Greeley, the influential editor of the *New York Tribune*, became the main voice of the organization. With the departure of Southern Democrats from Congress in 1861, the Republican Party passed the Homestead Act in May 1862. The act, which went into effect in January 1863, allowed any "person" who was over twenty-one or who was head of a family, and who was a citizen, to file claim to 160 acres of public land already surveyed. After five years the homesteader could gain clear title to the land if he or she had improved it by building a residence, putting up a fence, or planting crops. "Free land" looked good on paper (Opie 1987).

The reality of farming "free" land in the grasslands, however, seldom matched the promise. For one, the law failed to consider that a farmer might need more than 160 acres to make a living. Also, a poor person could never afford the capital costs of "improving" 160 acres, as the average farm had a capital value of around $2,700 in 1870. For many it was cheaper to buy 160 acres for $200 under the provisions of the Preemption Act than it was to bear the cost of improving homestead land. The amount of land available for homesteading was fairly limited, too. While Congress made available 84 million acres for homesteading, in comparison, three times that amount went to railroads; 140 million acres were donated to the states; and 175 million acres were bound in Indian reservations.

The success rate for homesteaders varied considerably. In the Dakotas fewer than half received final title to their land. In Kansas, 7 percent to 10 percent decided to buy their claims outright rather than bear the cost of improvement. Another 41 percent stayed on to gain title to the their land, and 49 percent relinquished their claims altogether. A beneficial aspect of the act arose as it opened the door to land ownership beyond just men. Because the act referred to homesteaders as "persons," many women acquired land in their own right. Before 1900, in Trego County, Kansas, four women were among the thirty-nine homesteaders who had acquired clear title to their land. In eastern Colorado, in a couple of counties, women accounted for 18 percent of the claimants, and of these, 55 percent ultimately "proved up" their claims. Obviously, the success rate of women homesteaders compared very favorably with that of men.

This notion of civilizing the grasslands and its occupants, while potentially beneficial for Euro-Americans, had harsh consequences for a great many Indian peoples. Senator Henry Dawes saw it as his mission to convert Indian peoples to a Euro-American farming way of life in order to "solve" the Indian problem. The reservation system proved an expensive burden on the federal treasury, and Euro-Americans were encroaching upon these large tracts of land where they illegally farmed, grazed, and/or stole livestock. Dawes sponsored the General Allotment Act of 1887, under which Indian peoples forfeited their reservations and individuals supposedly acquired legal title to 160 acres of land upon which they were to become self-sufficient. The remainder of the land, the "surplus," was to be sold, with the proceeds going into a trust fund for the future use of the tribes.

This legislation proved exceptionally unpopular with most Indian peoples throughout the grasslands. Take the situation of the Kiowas, for example. They had worked out profitable leasing arrangements with some Texas cattlemen and had begun developing their own cattle and horse ranching economy. The leadership of the Kiowas made it perfectly clear to the Jerome (Dawes) Commission that they did not want to have their reservation allotted. They knew that making the transition to farming would prove difficult and that they lacked the resources to make it happen, too. Komalty, speaking during the negotiations, warned: "Remember that we have horses and cattle. These will in a few years die of starvation. We have no machines to put up hay. . . . Eight years from today every one will be afraid to own ten horses and fifty cattle because there will be no grass to eat." Some tribal members took the initiative to resist legally the breakup of their lands. In one of the most famous cases to come before the United States Supreme Court, *Lone Wolf v. Hitchcock*, the court asserted the plenary power of Congress over Indian affairs. This court decision rendered treaties worthless and gave Congress the power to deal with Indian peoples as it saw fit (Lynn-Sherow 2004).

Now forced onto small allotments of land, many Kiowas made a good faith effort to adopt Euro-American farming ways. However, as they feared, they had few means to make the move a success. Unable to use their land as collateral, they lacked capital for buying farming equipment, and the Kiowa, Comanche, and Apache Agency at Anadarko, Oklahoma, provided little in the way of adequate means to help these people with the transition. When they did plant and raise successful wheat crops, they had to rely upon custom crews to make the harvest. This often proved disastrous, as the crews worked Euro-American fields first, and more often than not, the crews left the Kiowa wheat to stand and go to seed in the field. Given all of the factors working against the Kiowas, it is a wonder that any of them retained title to even an acre of their allotted lands (Lynn-Sherow 2004).

American culture, like a mighty tsunami, rolled over and submerged the Indian peoples of the Southwest and intermountain grasslands. However, their experiences differed considerably from those peoples living in the grasslands to the east of the Front Range. The wave lost some of its momentum by the time it reached peoples such as the Navajos and Utes, and while they still encountered hardship and suffering, they fared better than those who had faced the initial sweep. Contemporary Navajos and Utes have retained their reservations, and while enduring unrelenting poverty and forced assimilation, they have kept and nourished their cultural identities. The Navajos, for one, had a particularly devastating experience after the Mexican-American War. Euro-American men bound for the goldfields of California often abused the Navajos, and this led to violence between the two nations. With the outbreak of the Civil War and abandonment of the forts established to keep the peace, the Navajos took advantage of the situation and raided with some degree of impunity. In 1863, Colonel Kit Carson, under the direction of General James Carleton, launched a merciless assault against the Navajos that resulted in the infamous "Long Walk," a forced march that killed untold numbers of men, women, and children together with the destruction of their vast sheep herds and other livestock. After a disastrous four-year experiment designed to convert the Navajos to irrigated agriculture, the federal government entered into a treaty that returned the Diné to their original homeland, and from there they renewed their pastoral practices with thousands of sheep and goats supplied by the federal government as part of the agreement. By 1880, they were well known as sheep herders, master rug weavers, and jewelry artists (Dutton 1975, 1983; White 1983).

The Utes, who lived north of the Navajos, had developed a highly diversified subsistence strategy prior to 1850. Women's labor accounted for a significant amount of food production gathered from plants, fish, and insects. The Utes traded with the other agriculturalists for corn and supplemented their diets through bison hunting. The robe and meat trade became more profitable as the acquisition of horses significantly expanded their hunting range. By 1860, Mormon colonies had largely destroyed the Utes' rounds of hunting, gathering, and trade, and the federal government attempted to provide the Utes with an alternative means of subsistence in the form of irrigated farming. This effort fell into disrepair, and in 1865 the government negotiated another treaty that placed the Utes of the Great Basin on a reservation where they would be taught Euro-American–style farming. However, Ute men were unenthusiastic about farming, as digging in the earth was the proper role of women, and soon they regarded

Euro-Americans, and their allotments, as "useful for providers." The men among the Colorado Utes, who signed their treaty in 1868, cleverly used the corn harvest to condition their race horses and ultimately to retain their migrating hunting practices, something Ralph Meeker, their agent, fervently opposed. All along, the Utes passively resisted farming while the men strove to maintain their horse herds. By 1885, the Great Basin group possessed over 7,000 horses, and the Colorado group over 5,300, far more than either needed to do their farming. As late as 1895, the Utes were securing over 60 percent of their subsistence from government rations. Clearly, the Utes' success in maintaining their cultural identity rested upon an ever-greater reliance upon Indian Bureau aid (Heaton 1995; Lewis 1994; Smaby 1975).

Racism also functioned as a Euro-American meme that had ecological and social consequences for African-American farmers. More than any other group, African-Americans dreamed of securing their own freeholds. Unlike the Kiowas, they wanted their own land to farm, yet this dream was harder for them to realize than it was for any other group. Prevented from acquiring good farmland in close proximity to Euro-American farmers, they often took up ill-suited sites for monocropping. On these lands they developed specialized truck crops such as peaches in Oklahoma Territory and placed the pickings into urban markets as far away as England. Others eked out a bare subsistence through small-time diversified farming (Lynn-Sherow 2004).

Unimaginable success did come to a few lucky souls. Junius G. Groves arrived in Kansas penniless in 1879. He took his first job at $.40 per day, and out of that came his board. By 1910, he was a millionaire farmer and recognized as the "potato king" of Kansas. He had specialized in truck crops and found his niche in potatoes. One of his sons, Charles, graduated from Kansas State Agricultural College and continued the success of his father. In a speech delivered in 1906, he captured the ideals and aspirations of African-Americans throughout the grasslands. "My prayer to God is, that while we are yet young and full of vim and vigor, we will not lose an opportunity until we have established a solid founding in the soil of our God, thus making the burden for the future generation lighter, for it is a familiar principle in political economy that a people owning the land they live upon can with difficulty be cajoled or forced to leave" (Hawkins 1999, 211). Yet success in Kansas came to few others over the decades. In 1900 African-Americans worked 1,782 farms, but by 1930 the number had fallen to 941. In Oklahoma they were effectively prevented from attending the agricultural college in Stillwater and were relegated to Langston Colored Agricultural and Normal University in 1897 (Hawkins 1999).

Undoubtedly, one of the most successful groups to bring European farming methods to the grasslands were the Mennonites. They migrated as complete

Junius Groves and his wife Matilda numbered among the few exceptionally successful African-American farm families in the grasslands. (Kansas State Historical Society)

communities, leaving the steppes of Russia where they had resided since the reign of Catherine the Great. As pacifists, they sought a sanctuary where they could avoid military service. In 1874, the governor of Kansas made them such a commitment. These farm families had wealth, and they readily purchased some of the best potential farmland in the state from the Atchison, Topeka, and Santa Fe Railroad. Their farming methods, especially the growing of hard winter wheat, were well adapted to the grassland climate of North America. Living in tightly knit communities, they flourished and built new lives revolving around their churches and family farmsteads (McQuillan 1990).

DOMESTIC GRASSES

The Mennonites who arrived in Kansas in the 1870s are often credited with launching the practice of growing hard winter wheat when in actuality, many people were involved in its development and proliferation. When Theodore C. Henry arrived in Abilene, Kansas, he immediately recognized the potential of wheat farming in the area as opposed to its serving as the shipping point for Texas longhorns on their way to the Chicago stockyards. Henry had engaged, rather unsuccessfully, in growing cotton in Alabama after the Civil War; being a carpetbagger from New Yorker had only made matters worse. So in 1867 he left for Abilene, Kansas, and in 1873 he began experimenting with soft winter wheat on an extensive range of land just east of town after experiencing poor results in raising corn and spring wheat. Moreover, the 1874 grasshopper plague had left his fields largely untouched, and his 500 acres bordering the tracks of the Kansas Pacific railroad line advertised itself as an unparalleled success. Besides, given the cost of planting and harvesting of $.50 a bushel, and a return of twenty bushels to the acre, Henry boasted of a nice return on the dollar, or as he phrased it: "I farmed in kid gloves, without horse or hoe." Soon people dubbed him the "wheat king" of Kansas.

Prior to his success with winter wheat, Henry and researchers at Kansas State Agricultural College had experimented with growing several varieties of soft spring winter wheat, but the success rate varied. Spring wheat matured in the late months of summer and harvesters reaped the crop in the fall, or what was left of it after the normally hot, dry months of summer had taken their toll on the plants. Winter wheat, on the other hand, was planted in the late fall, sprouted in winter and matured in the mild, wet months of spring. It ripened just at the beginning of the hot, dry time of summer, at which time farmers harvested their crops. Still, winter wheat, less so than spring wheat or corn, failed when the weather of the central grasslands deviated beyond the range in which

Henry Worrall's 1875 illustration of T. C. Henry's winter-wheat fields near Abilene, Kansas. (Courtesy of James Sherow)

the crop thrived, and this happened too frequently for reliable harvests. Henry and researchers experimented with hundreds of varieties of this crop until a few promising varieties were identified. Professor E. M. Shelton, at Kansas State Agricultural College, conducted much of this early research on different wheats.

Henry knew alternatives to soft winter wheat existed, and beginning in 1877 he began experimenting with a hard winter wheat called Red Russian or Turkey. As he readily recognized, it possessed excellent winter hardiness and produced higher protein levels than either soft spring or winter wheats. The trouble was in milling the grain, as the hard berries rapidly wore away stone grinding wheels. Technological refinements in sifting by a French miller, Edmund N. La Croix, who settled in Minneapolis, led to the development of hard, steel rollers for grinding and gave rise to the first mills adapted to processing hard winter wheat into flour. The first successful steel roller mill began operating in Minneapolis, Minnesota, in 1878. Production quickly spread to other states, and Kansas had its first working mill in Enterprise, a small town ten miles east of Abilene, in 1882. As a result, farmers throughout the grasslands began growing hard winter wheat (Henry 1906).

Still, as was the case for many Indian peoples before 1870, Euro-American farmers relied primarily on corn. So long as farmers grew corn in areas that averaged twenty-five inches of precipitation or more annually they could count on a successful crop. West of the 100th meridian rainfall averages fall below twenty-five inches a year, and north into the Dakotas the growing season was often too short for corn. Growing corn meant success for farmers when they took up lands east of the 100th meridian and south of the Dakotas. But as they approached the 100th meridian, or north into the Dakotas, growing corn became an ever more risky proposition. As farmers entered these more unpredictable climes, they embraced winter wheat. This explains why it was not until 1914 that wheat surpassed corn as the dominant grain crop in Kansas. In Oklahoma, it was 1918 before the wheat harvest toppled corn. In Nebraska, in the early decades of the state, wheat harvests exceeded corn, but by 1880 corn had become the reigning grain crop giving its residents the familiar moniker of "the cornhuskers." Settled well after the introduction of hard winter wheat to the south, Dakotan farmers, in stark contrast, relied on winter-wheat farming almost exclusively. By 1900 Euro-American farms, whether with corn, wheat, or another domestic grass, had replaced much of the Indian-managed grasslands. Kansas was a good example with over 20 percent of all its acreage in corn and wheat harvests by 1900.

Euro-Americans introduced other domestic grasses to the region beyond winter wheat. Many of these were imported from Africa, and like the Turkish varieties of wheat, were well suited to grassland climates. From the sorghum genus came broomcorn, grain sorghum, great millet Kaffir corn, and sweet sorghum. These crops proved exceptionally good for domestic animal feed and as nonglutenous grains for milling. Of course, more traditional grain crops such as oats, rye, and barley were planted and harvested, too. Euro-American farmers also quickly discovered the usefulness of importing certain legumes such as alfalfa, clover, and later soybeans. Alfalfa and clover became exceptionally valuable pasture crops, and while "native" grassland pastures were still commonplace before 1940, domestic pastures became more important to most farmers. According to the United States Department of Agriculture, Kansas farmers stacked around 2.5 million tons of alfalfa in comparison to 1.6 million tons of "other" hays in 1919. While farmers eventually decreased the amount of land in hays, alfalfa harvests continued to exceed those of other hays.

Euro-Americans brought other exotic species to the grasslands that replaced the wild ones as habitats supporting wild species disappeared. Cattle replaced bison, pigs outnumbered bears, and Leghorn chickens outpaced prairie chickens. Even native species of fish were extirpated. Early on, the first fish commissioners in Kansas explored ways of stocking rivers with exotics as a potential alternative food source. Carp did very well but proved an exceptionally unpopular dish. The experiment to stock the Kansas River with California salmon flopped. By 1900, deer were completely absent from the state as hunters and the destruction of habitat took their toll. Antelopes, elk, bears, wolves, bison, and cougars had all virtually disappeared from the state. This same pattern prevailed all across the grasslands as Euro-American land uses took root and grew (Sharp and Sullivan 1990).

As Euro-American farmers streamed into the Southwest grasslands and the Great Basin, they found even less success in dryland farming than had homesteaders to the east of the Rockies. Only two areas of the Southwest grasslands showed any initial promise, and both of these were in New Mexico. One was in the northeast corner near the cities of Roy and Mosquero on the El Paso and

Northwestern Railroad, and the other was located along the east-central portion of the state near Clovis on a spur of the Atchison, Topeka, and Santa Fe Railroad. In general, all the attempts to do dryland farming in these areas failed to meet expectations, and as quickly as people had arrived they left. By 1930, only those who had shifted to either ranching or some form of irrigated agriculture remained.

Similarly throughout the Great Basin the same pattern of failure repeated itself. Nearly all reliable farming by anyone in the Great Basin relied on irrigation; however, a few attempted dryland farming. Such was the case in Sahara, Utah, and Afton-Metropolis, Nevada. Having settled there toward the end of World War I, the farmers in these two locations met disaster as desert rabbits, feral horses, and cattle routinely destroyed their crops. One of the more unusual attempts at creating an agricultural community occurred when a group of Hawaiians, who had converted to Mormonism, migrated to Utah and founded the town of Iosepa. Their efforts were only slightly more successful than those at Sahara, and after some extreme hardships they, too, abandoned their efforts. By 1930, Euro-Americans generally considered it pure folly to attempt dryland farming anywhere in the Great Basin.

While dryland farming proved fruitless, irrigated farming had greater success. The most notable people to transform the ecology of the Great Basin have been the Mormons. Beginning with their arrival near the Great Salt Lake in 1847, their irrigation works established the agricultural and economic base of their society. Their ditches and fields dramatically transformed the ecology of the eastern edge of the basin and many other locales near flowing watercourses descending within various mountain ranges. Their enterprising and tightly bound communal ways, emboldened by their belief that they were fulfilling God's will, were further enhanced with the discovery of gold in California and later the Comstock mining operation. The thousands of people who made their way into or through the Basin relied on the Mormons for food supplies and livestock, creating prosperous markets for the Mormon enterprise. By the 1870s, the arrival of the Union Pacific Railroad had tied the Mormon economy into the greater national one.

By 1900 what an astonishing feat the Mormons had achieved. They had brought under irrigation over 250,000 acres of cropland on which over 10,000 farms flourished. Up until the 1890s, the church hierarchy guided nearly every aspect of "tending a rose in the desert." Yet as farmers engaged in more market-oriented enterprises, such as sugar beet production, they encountered an inability to raise sufficient capital to build ever more expensive irrigation systems. In order to achieve more lucrative production they became reliant on outside market forces and capital, and consequently the control of their systems became less dominated by the church and operated more akin to the market-driven mutual stockholding companies so commonplace east of the Rockies. Engineering expertise, such as that provided by John Widtsoe, who served as president of Utah State Agricultural College from 1907 to 1916, became indispensable as Mormon systems became more technologically sophisticated in order to compete in national and international markets. Theology, science, Mexican irrigation traditions, and world markets would all interact in creating an agricultural landscape little resembling the one inhabited formerly by the Shoshones and Utes (Worster 1985; Alexander 2003).

THE IRRIGATION CRUSADE

As in the Great Basin, but to a lesser degree, an important key to Euro-American occupation and domestication of the grasslands east of the Rockies was overcoming the unpredictability of rainfall. Rainfall on the grasslands is one of the best examples of a chaotic system. Farmers, obviously, fared poorly during extended times of little rain, as their forms of crop production required a reliable application of water. William Smythe, a newspaper man from Omaha, Nebraska, thought he had the solution for solving the water problem faced by farmers. Irrigation, he believed, was the cure-all, as it promised social, economic, and ecological benefits for all of its practitioners. He came to this conclusion when as a newspaper editor in Kearney, Nebraska, he witnessed the effects of a severe drought in 1888. While visiting pueblos and villages in the Southwest at the same time, he observed how irrigation kept crops flourishing regardless of the lack of rainfall. Smythe thought this technology had great potential for the grasslands as a whole, and he would dedicate the rest of his life to spreading the word (Smythe 1969).

In 1891, Smythe organized and led the National Irrigation Congress. Its first congress was held in Salt Lake City, Utah, and was attended by representatives from every western state. He also began publishing and editing the *Irrigation Age*, a clarion call to garner support for his ideas. He put his money where his print was, and he heavily invested in irrigation colonies, such as New Plymouth, Idaho. Many of these floundered, but he kept the faith and wrote *The Conquest of Arid America* in 1899. This work fully and clearly expressed the idea of domesticating and transforming the grasslands into profitable homesteads that would support a flourishing American way of agriculture (Pisani 1992).

The capital expense of flood irrigation, however, made these enterprises dubious investments. T. C. Henry found this out as he developed several irrigation companies in Colorado. Every single one of them fell into an economic morass, resulting in huge monetary losses for investors and farmers alike. In fact, around 90 percent of all irrigation companies had either gone bankrupt or were in severe debt by 1900. Only a few seemed to be profitable as the operators of those systems had come to specialize in truck crop farming that served nearby urban markets. George Washington Swink, who lived in Rocky Ford, Colorado, was masterful in developing this type of irrigated agriculture. Onions and melons, especially the regionally famed Rocky Ford Cantaloupe, made him a very wealthy individual.

Swink also bore a major responsibility for launching the sugar beet industry in the region. This type of irrigation farming, which spread to Nebraska and Kansas, required exceptionally careful management. The plants had to be thinned as soon as three or four leaves emerged; this required significant stoop labor, which was provided initially by German immigrants and later by Mexicans. The irrigation systems had to be efficiently managed and well maintained, as the crop required precise watering in the spring. Raising beets was a labor- and capital-intensive operation, but it also yielded great profits and anchored the economic base of several cities throughout the High Plains.

Irrigation also wrought ecological changes throughout the watersheds of the grasslands. Salinity levels began to rise in the flows of streams. As ditches channeled flows across farmlands, the water washed out minerals, especially salts, in the soil and returned these to creeks and rivers. As the water was reused downstream, the return flows concentrated the mineral buildup in the streams, and this salt-laden water would be deposited throughout the next system downstream. Salinity buildup in the soils would over time make farming more and more difficult. It also facilitated the spread of salt cedar (*Tamarix aphylla*), originally an ornamental tree imported from the Middle East to Texas in the early 1870s. These trees rapidly gained a roothold throughout the river valleys of the grasslands where saline conditions were favorable, and that was just about anywhere irrigation was practiced south of the Platte River. The copious water consumption of these trees, as much as 200 gallons a day, further complicated the operation of irrigation systems throughout the southern reaches of the grasslands (Sherow 1990).

The irrigation crusade throughout the rest of the intermountain grasslands little resembled the community values of the Mormons and placed more faith in commercial aspirations with little, if any pretense, of fulfilling God's will. The science and economics of irrigation were infinitely more important than establishing a yeoman's paradise. The labors of folks who built and managed the irrigation works in the Boise River Valley near Middleton, Idaho, certainly bore witness to this trend. The town itself arose in response to nearby mining, and the people flooding into the area held an entrepreneurial bent of mind right from the

Salt cedar, Tamarix L., *one of the most invasive, water-consuming plants plaguing irrigation operations throughout the grasslands. The trees choked the Arkansas River near Garden City, Kansas, in 1986. (Courtesy of James Sherow)*

start. Thrifty, industrious, and calculating, these farmers continued to increase their wealth and farm operations in an economically successful manner through the trying decade of the 1890s. This was a fast-paced, self-seeking society tethered by few community bonds (May 1994).

The ecological transformations associated with irrigation were as important and longer lasting than the social, political, and economic ones. Mark Fiege calls these ecosystems "hybrid landscapes," ones farmers found impossible to dominate and shape fully according to their will. The dams along the Snake River could not completely regulate its flows, and while indeed these structures destroyed habitat for some species, they created new haunts for white pelicans, great blue herons, and numerous ducks. The ditches became home to pocket gophers that destroyed the retaining walls. Also, the ditches provided a home for insects such as weevils and grasshoppers, and these attracted greater numbers of birds. Even the russet potato grown in Luther Burbank's field "reflected the reciprocal interplay of culture and nature that created the irrigated landscape as a whole." This fungus-resistant tuber was at the center of a "dynamic relationship . . . of economics, climate, soil, water, plants, and animals" (Fiege 1999, 206).

Russet potatoes and sugar beets are two introduced crops that solidified the economics of irrigated agriculture in the grasslands. (Corel)

The practice of irrigation also led to the development and refinement of prior appropriation water law, which rapidly abstracted the value of water and turned that into an economic commodity. This doctrine, which arose in irrigation practices in New Mexico and was incorporated into the state constitution of Colorado in 1876, became known as "first in time, first in right." Ostensibly, the water in rivers and streams belonged to the state as public property, and the state issued rights to people for the use of the water. A person acquired a "water right" by establishing a "beneficial use" of the flow. States in the grasslands have defined beneficial uses nearly exclusively in terms of economic pursuits—domestic, agricultural, or industrial. The person filing for a right must show that he or she is putting all of the water applied for to beneficial use. For example, Farmer A might file for a right to 1,000 cubic feet per second of water flow, and if his system can accommodate that amount of water and apply it to his fields, then he has established a beneficial use. The date when the farmer first started using the water sets the date of his right. Again, if Farmer A began operating his irrigation system on May 7, 1880, then that became the date of his water right. Under this system, any farmer with a *prior appropriation* predating May 7, 1880, would receive his water before Farmer A, and any user establishing a water right after May 7, 1880, could divert flows only when Farmer A had his right fully served.

This sort of water right system initiated a rapid economic development of water resources wherever it was applied. If one wanted to farm successfully in eastern Colorado or the High Plains of Wyoming, it behooved him or her to establish an appropriation right to water as quickly as possible. Latecomers faced the prospects of farming without reliable flows into their irrigation systems, or cities without enough water to support a growing urban population. By 1900, in several states the river courses had more water rights than the normal flow of the stream could support, or as the locals referred to this situation, their rivers were "over appropriated." The right to water had in many respects become more valuable than the water itself, and some rights were far more valuable than others as a result of their dates. As futures had abstracted the value of farm production, water rights abstracted the value of stream flows throughout the grasslands.

IRRIGATION COURT CASES

The economic value of water became an issue in one of the most important water cases ever decided by the United States Supreme Court. This suit, *Kansas v. Colorado* (1907), arose over the division of interstate river flows in the grassland. Well before 1900, Kansans around Garden City had watched the diminution of the Arkansas River flows, and they saw Colorado irrigators upstream as the culprits. Marshall Murdock, a powerful newspaper editor in Wichita, Kansas, also feared the effects of stream flow depletion on his vision of an inland port. He held both irrigators in western Kansas and eastern Colorado as culpable. Believing Coloradans more to blame, he used his enormous political influence to have Kansas bring before the U.S. Supreme Court a suit of original jurisdiction in May 1901. Murdock's hope was to roll back water development in Colorado so as to create a greater flow of water through Wichita.

The justices agreed to hear the case, and in 1904 they allowed the United States government to intervene in an effort to protect the interests of the newly created Reclamation Service. This case involved complicated questions about hydrology, economic development, state water laws, and state-federal relationships over the development of interstate stream flows. To sort out these issues, the justices appointed a special commissioner to take and organize testimony. By the end of 1905, he had collected and recorded more than 120 exhibits and 8,500 pages of testimony from over 300 witnesses. With this mass of information, attorneys made their respective positions clear. Kansans believed the riparian doctrine gave them the right to an undiminished river flow; Coloradans claimed the prior-appropriation doctrine gave them the right to use all the water within the boundaries of the state regardless of the effect to downstream users in another state; and the United States Attorney General made a case for congressional control of non-navigable rivers that crossed state lines.

The justices' decision affirmed a commodity view of river flows and paid little heed to the ecology of rivers. The opinion, written by Justice David Brewer, a former Kansan, favored neither state's water doctrines nor the position of the Reclamation Service. Instead, it elaborated a doctrine of equity, essentially an accounting procedure weighing the relative economic gains each state had made in using the flows of the Arkansas River. As a result, each state could maintain its own water doctrines and the federal government was

What's The Matter With Colorado? SHE'S ALL RIGHT!

In this cartoon, Coloradans celebrate the U.S. Supreme Court decision of Kansas v. Colorado *(1907). (Courtesy of James Sherow)*

prevented from asserting congressional control over the river. The evidence, Brewer noted, had shown that both states had made economic gains all along the river valley. As Kansas could not show economic losses resulting from the upstream developments in Colorado, then it lacked any reason for restraining water uses in Colorado. More than just settling this case, the doctrine of equity became the guiding precedent for nearly all future interstate water suits (Sherow 1990).

One other important U.S. Supreme Court case, *Winters v. The United States* (1908), emerged over irrigation practices in the northern portions of the grasslands. Henry Winters was experiencing some hard times farming on his irrigated land along the Milk River in northern Montana. He and others thought the prior appropriation doctrine of water rights had perfected their uses of the river flow. Downstream from their farms and ranches was the 600,000-acre Fort Belknap Reservation for the Assiniboines and Gros Ventres. A few years earlier, both nations had been on the brink of starvation, so in the Fort Belknap Treaty of 1888, their leadership relinquished their right of occupancy to over 17 million acres of land for the reassurances of a place where their children would be educated, and they would be taught and provided all means for stock raising and farming. By the turn of the century, their agent, William R. Logan, thought their progress in irrigated agriculture looked promising. This rosy picture turn grim when in 1905, little rain fell and the ditches of Fort Belknap carried no water. Over 200 upstream users, like Winters, took the little flow that found its way into the Milk River bed, and nothing reached the headgates at Fort Belknap. Winters, as did the others around him, thought he could do this because most of their irrigation rights predated the construction of the Fort Belknap Reservation irrigation system—first in time, first in right.

Agent Logan rightfully feared for the well-being of the 1,300 Gros Ventres and Assiniboines under his jurisdiction if they failed to produced adequate crops to see themselves through the approaching winter. He asked his contacts in the Office of Indian Affairs to assist him, and soon Carl Rasch, the U.S. attorney for the district of Montana, asked the federal courts for an injunction to stop the non-Indian irrigating upstream from the reservation. Each federal court complied, and Winters and supporters, feeling aggrieved, appealed their case every step of the way to the U.S. Supreme Court. In 1908, a nearly unanimous court decided in favor of the Indians and created a water right, "reserved water right," that was neither based in riparian common law nor followed prior appropriation doctrine. Without deciding how much water was required, the court concluded that the tribes must have "a sufficient amount of water from the Milk River for irrigation purposes." In brief, it mattered not whether upstream users put the flows of the river to work before those on the Belknap Reservation did, nor did the Indians' "reserved right" depend upon their actually diverting the water at all, as the right was "reserved" with the creation of the reservation. With this decision in hand, the work of civilizing Indians and domesticating the land could continue (Hundley 1978).

ENTER FEDERAL RECLAMATION

In the minds of many, the desert and intermountain grasslands seemed suitable only for grazing, and the immense difficulties in dryland and irrigated farming only reinforced this point of view. If ever the rose were to bloom in the desert, it would require far greater resources put to irrigation than private capital could provide. Consequently, people turned to, and picked, the deeper pockets of the federal government to underwrite expanded projects throughout the region. The Snake, Pecos, Rio Grande, and Salt River valleys were just a few of the places where massive amounts of federal subsidies were employed. The objective, as tangible as any desert mirage, was to use scientific expertise to guide centrally controlled projects to create profitable cornucopias tended by democratically principled yeomen. Such a vision was fraught with social, economic, political, and ecological contradictions. Despite erratic achievements, the will to dominate "Nature" overrode any hesitation to use the federal government to achieve what individual enterprise could not.

Today, given the actual accomplishments of reclamation, it is hard to fathom the depth of its supporters' faith in their power to transform the grasslands. At best, over-optimism in the potential of federally sponsored irrigation fueled the thinking of promoters and engineers in New Mexico. In the early 1890s, men such as Thomas B. Catron, a powerful entrepreneur, confidently envisioned somewhere between one-quarter to one-half of New Mexico placed into profitable, irrigated farms. In the 1890s, Senator Stewart, who chaired a Senate committee investigating the possibilities of federally supported irrigation, imaged the entire breath and length of the Rio Grande River Valley covered in productive farms watered, in his own words, by "mountain and storm supplies which now run to waste." Who needed free-flowing streams when the same should be put to economic production? At the same time, John Wesley Powell, explorer and at the time the director of the United States Geological Survey, took a far more cautious view of the situation, as he believed the potential of irrigation in New Mexico was highly inflated by its ardent supporters. He feared economic and social catastrophe on projects without adequate water sources. He also warned that inadequate planning could lead to speculative and

Aerial view of the Roosevelt Dam on the Salt River, Arizona; the first operational project constructed by the Reclamation Service. (Library of Congress)

monopolized projects. After the passage of the Reclamation Act in 1902, the events unfolding on the Rio Hondo Project near Roswell in the Pecos River Valley certainly bore witness to Powell's worst fears. Originally, the project was to have irrigated over 15,000 acres, but it never provided for more than 1,200 before frittering away over $375,000. This economic loss says nothing about the human hardships and suffering also endured. Other projects were more successful such as the Elephant Butte project on the Rio Grande River. However, it involved complicated international and interstate issues with Mexico and Texas that were only resolved when agreements were reached with the federal government and New Mexico (Clark 1987).

Some projects worked quite well, but not the way they were originally planned. In the Salt River Valley of Arizona, where the Hohokams once prospered, Euro-Americans had rebuilt an irrigated farm system sometimes using the very same ditches dug in ancient times. Federal reclamation there could not bring in new farmers to reclaim public lands, as most of the valley was well settled and cultivated by 1900. Roosevelt Dam, built both to provide supplemental water to in-place irrigators and to produce hydroelectric power, became one of the key building components in securing the prosperity and economic growth of Phoenix, which arose again not out of ashes but from the depths of an artificial reservoir. Moreover, the tensions of scientifically engineered social and technological planning with agrarian aspirations were partially resolved as local users congealed in cooperative arrangements, and through this mechanism they shaped important policy decisions regarding the operations of the system. While in no conceivable manner did this project reclaim additional public lands, nor did it produce an agrarian democracy, nor was it funded by agricultural production, it demonstrated at least the economic and technological feasibility of federally sponsored reclamation (Smith 1986).

BONANZA FARMS

In domesticating the land east of the Rockies, some turned to an industrialized factory model of farming, but before 1940 the technology, economics, and ecological adaptations of grassland farming were all too immature to support such an undertaking. Still, the factory model of agriculture had gained a foothold in the eastern portions of the grasslands from Kansas to North Dakota. The most prominent of all of these enterprises, and there were over 8,000 farms in this corridor that contained over 1,000 acres, was the one managed by Oliver Dalrymple in the Red River Valley of present-day North Dakota. The managers of the land company of the Northern Pacific Railroad hired Dalrymple to apply his farming expertise in wheat growing to eighteen sections of land in order to display its productivity with the intention of using this demonstration to encourage settlement on, and the sale of, the land grant of the railroad company. This and similar operations became known as "bonanza farms."

Dalrymple, like T. C. Henry, enjoyed initial success. In 1875 his harvest on two sections of land proved the potential of wheat farming. The railroad advertised his accomplishments widely, and this stimulated a much-hoped-for land rush to the area. By 1880, Dalrymple managed around 100,000 acres of bonanza farms for a few companies. He organized his lands into 6,000-acre "farms" managed by a superintendent, who oversaw three divisions of 2,000 acres each. Each of these smaller units had its own outbuildings for farm machinery and draft animals, and a boardinghouse for the hired hands. Migratory work gangs handled over a hundred binders, scores of threshing machines, and hundreds of animals. So long as wheat prices remained steady throughout the 1880s, this factory model of monocropping produced fantastic profits for its absentee owners

Oliver Dalrymple managed huge wheat acreages known as bonanza farms. These were massive operations, as depicted in these two drawings. The top drawing shows the farm buildings; the bottom drawing shows the method of sowing wheat. (Courtesy of James Sherow)

and on-site managers like Dalrymple. But given any significant shifts in rainfall, prices, or property taxes and the system tilted out of control. By 1890, this was the case for the Dalrymple operations, and they came unglued at the seams while most smaller, diversified farmers weathered those climatic and economic changes. Nonetheless, the bonanza farms foretold the future of "get big or get out" (Fite 1974; Shannon 1945; Hammer 1979).

OPEN-RANGE RANCHING

The Texas cattle drives of 1867 proved the profitability of placing longhorns into Chicago markets. An idea quickly evolved out of this practice that was to keep cattle on the open range until markets were ripe for shipping the animals east. The grasslands of eastern Colorado became one of the first areas where this was practiced with some degree of success. John Iliff showed how the open-range cattle business could yield great profits. He had arrived in Colorado in 1859 as part of the gold rush that year. Like so many others, he failed miserably as a miner, but he fared better mining the miners. He did a little truck farming, and his vegetables and fruits commanded a miner's fortune. He amassed enough capital to establish a small trading post on the California Trail near present-day Cheyenne, Wyoming, and in bartering with the overlanders, he acquired the makings of a sizable cattle herd. Soon his herds grazed the still federally owned Platte River Valley, and those animals fattened well on free grass and found their way to eastern urban markets on the Union Pacific Railroad. By the time of his death in 1878, his herd size was estimated at over 35,000 animals.

Profits of 20 percent or more attracted a great many investors, both foreign and domestic. While farmers were quickly transforming the land of Kansas into domesticated grasslands, open-range farmers scooted their herds onto the public domain grasslands in eastern Wyoming, the Dakota Territories, and Montana Territory. The winters seemed to be mild enough for cattle to survive with some actual weight gains. By 1880, large amounts of Scottish and English capital flowed into these grasslands and transformed millions of acres into a grazing region for domestic cattle. For example, the Prairie Cattle Company, incorporated under British laws, controlled over 2.5 million acres of rangeland in Colorado, with smaller holdings in New Mexico and Oklahoma Territory. Cattle barons had to cooperate in order for such exploitation of the public domain to work, and soon, they regulated themselves in organizations such as the Wyoming Stock Growers' Association in 1879 (Atherton 1961).

Not all ranching practices resembled this "beef bonanza." While men such as J. P. Wiser of Prescott, Ontario, built huge baronial ranching estates in the Flint Hills of Kansas, others, like Welsh immigrants, took an entirely different approach. Unlike Wiser, who as an absentee owner hired "pasture men" to tend his holdings, the Welsh took jobs in mines and cities to amass enough capital to buy a few cattle and fence enough acreage to graze them. Many, like Walter Jones and his family, slowly increased their holdings while either paying in cash for whatever they bought or borrowing only what they could afford to lose. They also maintained a far more conservative stocking ratio, which meant having at least seven acres of grassland for every steer they grazed. In comparison, Wiser and his type often stocked on five acres or fewer per steer. Over the long run, the Welsh ranchers remained stable while the Wiser types of operations eventually fell into financial ruin (Hickey 1989).

Plenty of warning signs had occurred signaling to ranchers that open-range operations were a hazardous undertaking. One of the first object lessons came in 1871 when Texas drovers attempted overwintering their herds of cattle on the shortgrass regions west of Abilene, Kansas. Chicago markets were dull in the fall, so the drovers figured that by placing their herds on grasslands to the west over the winter the herds could be driven to the rail connections at the

This drawing by Charles Graham portrays the staggering consequences suffered by open-range cattle operations in the winter of 1887 to 1888. (Library of Congress)

first opportunity in the spring when prices rebounded and before the drives from the south reached the northern railheads. An early ice storm and blizzard quickly killed that idea, along with hundreds of thousands of cattle. Joseph McCoy estimated that over 250,000 animals, representing over 70 percent of the herds, perished in that storm.

Apparently, this warning was not enough. The boom times of the late 1870s and early 1880s made it appear to many open-range ranchers that the good times were there to stay. By the mid-1880s, a glut of cattle in the Chicago markets depressed prices nationwide. Public domain grassland was becoming harder to control as more of it was transformed into farmlands and other people fenced off portions as rangelands. In some areas, overgrazing had taken its toll, and ranchers had difficulty finding grass adequate to keep their herds fat. By the autumn of 1886, the economics and ecology of open-range ranching had weakened. Then came the blizzards of November and the ice storms of January 1887. A ranch hand in the Judith River Basin of Montana, named Charles Russell, drew a postcard symbolizing the devastation. Ranchers, such as Theodore Roosevelt, encountered huge, staggering loses. In January, or as the Sioux called it, the "moon of cold-exploding trees," Roosevelt lost around 65 percent of his herd and most of the $80,000 that he had invested in his Dakota venture. A thoroughly depressed Roosevelt wrote, "I am bluer than indigo about the cattle," and with that he left, fully realizing that open-range ranching had come to an end (Morris 1979, 373).

In the Southwest and Great Basin grasslands, strong market currents flowed through the communication and transportation networks radiating round the region, and Euro-American ranchers were pulled into their wake. Before the railroads, Euro-Americans were reluctant, at most, to try their hand at ranching. While early expedition reports indicated sufficient grasses to maintain domestic animals, the widely dispersed locales of water and the absence of large herbivores aside from pronghorns suggested a place of little potential. Spanish-style ranching, however, illustrated how Euro-Americans might successfully exploit the grazing potential of the basin. Essentially, Spanish-style practices included simply letting a large number of branded or marked cattle loose to roam freely throughout a common rangeland without any great concern for herd mortality. At a later time, a roundup would occur, and the animals would be slaughtered and butchered.

Still, this method was fraught with ecological difficulties. Despite the fine nutritional value of the perennial grasses in the region, climatic conditions relegated their growing season to a few weeks in the late spring. Cattle, not surprisingly, relished these plants especially during the brief time of their flowering and seed production. Consequently, it took little time to deplete the rangeland of its grasses, and big sagebrush rapidly filled the void. Unfortunately for cattle, they often died whenever they resorted to grazing the ever-present big sagebrush. While high in protein, sagebrush contains toxic oils that often lead to rumen stasis, paralysis of the rument walls. Compounding the problem, squashing wild range fires also encouraged the proliferation of sagebrush to the exclusion of grasses. By 1900, botanists working the Nevada experiment station recommended re-seeding the rangeland and promoted controlled burns to reduce the spread of big sagebrush.

What ultimately encouraged the growth of ranching in this harsh cattle environment was the availability of rail transportation, which began with the completion of the transcontinental railroad in 1869 and the development of water sources in the rangelands. Generally, ranchers understood that cattle should have available water within a four-mile radius of their grazing, and no more than ten miles if playa lake resources existed. Ranchers dug out springs, sought artesian wells, and built windmill pumps. The exceptionally cold winter of 1889–1890 demonstrated to ranchers the need to supplement winter grazing with stored hay reserves, and growing this crop was possible only in mountain meadows or irrigated pastures. The cold winter of 1889–1890 also opened rangeland to the influx of sheep, a herbivore that fared better on big sagebrush, had a better tolerance for water deprivation than cattle, and could overwinter on the range without resort to hay reserves. Besides, raising sheep required less capital investment than raising cattle. Of course, what resulted was the "tragedy of the commons"; with little effective range conservation practiced in the Great Basin, sheep and cattle quickly overgrazed the region and destroyed a great portion of all indigenous grass species (Young 1994).

In the Southwest grasslands a very similar pattern of rangeland destruction occurred during the transition from pastoralism to large-scale cattle ranching and was particularly evident in Arizona. Pastoralists emphasized herd increases on common rangelands whereas ranchers sought accumulations of monetary returns from herd production on privately controlled or owned lands. Stockmen had around 40,000 beeves in Arizona in 1870, and by 1890 they had increased the numbers in their herds to over 400,000 animals. While it first appeared profitable to graze what initially seemed to be an abundant perennial grassland, the extended drought of 1891–1893 brought crushing financial losses along with the destruction of the desert ecosystems. Botanists from the Arizona Experiment Station recorded denuded rangelands, extreme soil erosion, and starved cattle. Four critical factors converged to cause this catastrophe. Free and easily accessible common grasslands encouraged ranchers to put as many cattle on them as possible, and for a time both cattle and great profits multiplied. Second, this form of cattle raising held a competitive advantage over production on settled areas of the West or in the blizzard-devastated rangelands to the east. Third, developments in railroad transportation, communication, and international banking supported the free flow of massive amounts of capital into Arizona ranching, and together these combined to fuel cattle raising. By 1900, all of this had collapsed as soon as the grasslands could no longer support the economics of this arrangement (Sayre 1999).

Euro-American agricultural practices also suffered in the late 1880s and through the 1890s. Several factors combined to produce failures in many quarters. Lack of rainfall throughout the High Plains destroyed countless wheat crops. As William Allen White observed from his post in Emporia, Kansas, many demoralized farmers were driving wagons headed back east with the words painted on the canvas sides: "In God We Trusted, In Kansas We Busted." Moreover, wheat prices plummeted as American farmers began competing on a global scale with additional harvests coming from the Canadian prairies, the Russian steppes, and the Argentina pampas. During the 1880s, prices per bushel of wheat fluctuated between $1.05 in 1881 and $.45 in 1884 and averaged nearly $.70 over the decade. In the first half of the 1890s, prices per bushel averaged $.55. Hard-pressed farmers and the middle-class businesspeople who served them rose in political revolt and formed the people's party, the Populists, who campaigned for social and economic reforms. While the Populist movement itself was short-lived, many of its calls for social and political reform were eventually enacted into law. But the underlying ecological and economic causation of the movement went largely unaddressed. Monocropping still prevailed with its attendant depletion of soils. Market forces lay outside the ability of one or any group of farmers to influence, much less control. Moreover, while the myth of the yeoman farmer prevailed throughout the grasslands, the reality of farmers producing on ever-larger tracts of land in order to capitalize on efficiencies of scale dominated throughout the region.

Haymaking, one development more than any other, tended to save and perpetuate ranching throughout the Great Basin. As already noted, after the disastrous winter of 1889–1890, ranchers were well aware of how foolish it would be to continue wintertime open-range livestock operations on their clearly overgrazed desert grasslands. Prior to 1890, a few ranchers had piled hay for supplemental cattle feed; however, this was not a generalized practice. For example, in Elk County, Nevada, ranchers had around 16,000 acres in hay production in 1889, but by 1900 they had increased their hay acreage to around 239,000 acres, which supported fewer cattle than were on the range ten years earlier. This haymaking revolution in the Great Basin favored smaller operations that combined irrigated farming with ranching as opposed to dryland farming and open-range grazing, two systems that had proven indisputably maladapted to the environment of the Great Basin. However, a bottom line did exist on how small a ranch could be and still operate within this system, as it required a substantial capital investment in machinery and draft animals; a dependable workforce; and a stable, reliable source of irrigation water. In a 1922 report, social activist, suffragist, and first head of the department of history at the University of Nevada Anne Martin harshly criticized the social landscape of her state, one arising out of the ecological and economic conditions of ranching, a system that relied on single men as its workforce. According to Martin, this group of men, fully one-quarter of the population in the state, gave rise to a social environment that sustained bars, brothels, and bullies. Not until World War II engulfed the young, single men of Nevada and fossil fuel–powered machinery replaced draft animals would this way of life wane (Young 1983).

Interestingly, the shift in ranching also brought about a rapprochement between ranchers and the federal government. Between 1900 and 1920, representatives of the American National Live Stock Association (ANLSA) worked hand in hand with federal conservationists, especially those in Theodore Roosevelt's administration. The Public Land Commission, led by Gifford Pinchot, head of the Forest Service; Francis Newell, head of the Reclamation Service; and W. A. Richards, who administered the General Land Office, issued a report advocating federal regulation and leasing of all public rangelands. Ranchers began seeing themselves as "home-builders" or "cattle *farmers* [italics added]" to indicate that they regarded themselves as legitimate occupants of the lands just the same as any homesteader. They feared complete economic, social, and ecological ruin if federal policy in the Great Basin and Southwest grasslands allowed dryland farmers to homestead these lands. The future, so the ANLSA argued at the time, lay in the federal government's leasing of those same lands to ranchers, the ones who could build permanent homes in the region. On occasion, President Roosevelt advocated regulating the public rangelands in the "interests of the small ranchman, the man who ploughs and pitches hay for himself." While some in ANLSA may have championed the small "hay pitching" rancher, they were interested in establishing a place of permanence for themselves within the ideology of agrarianism. Eventually, this goal of federal leasing and regulation of the public rangelands would be realized in the Taylor Grazing Act of 1934; however, by that time the ANLSA had become a vocal opponent of federal management of the public domain (Merrill 1996).

Euro-Americans were not the only ones practicing ranching. The White Mountain Apaches, formally known for their alliance with the American Army, and the San Carlos Apaches, whom American forces had subjugated, found themselves confined to reservations by 1871. Although doing so was not easy, these people developed a flourishing ranching economy by 1900, and by the

1920s, they controlled almost all of their rangeland within their respective reservations in Arizona. Their land contained prime grasslands, and the Apaches were quick to realize the potential of cattle raising. At first, their aspirations were hard to achieve given how Euro-American ranchers had permits to graze large sections of both reservations. Moreover, as Alchcsay, a White Mountain Apache, understood all too well, often the breeding stock sold to them lacked enough teeth to even graze and soon died after arriving on the reservation. Still, with the hard efforts of sympathetic agents, C. W. Crouse on the White Mountain Reservation and James Kitch on the San Carlos, by the 1920s the Apaches had gained control of nearly all their rangeland and were raising profitable Hereford herds. Some, such as Wallace Althaha, became exceptionally wealthy. His Spear R cattle herd numbered as high as 10,000 head at one time, and in 1918 Althaha reportedly sold $45,000 worth of beef. Yet he strove to retain and practice a "traditional" Apache worldview and demonstrated how one could be both a successful rancher and Diné at the same time. The brand mark for the White Mountain Reservation herd was the "Broken Arrow," a symbol to show a breaking with war for an embrace of ranching. By 1930, their successes in ranching had won them the ire and envy of their Euro-American neighbors (Iverson 1994).

Regardless of the varied accomplishments in cattle ranching, Indian peoples and Euro-Americans both saw more success in raising flocks of sheep and goats. As already established, sheep proved more adapted to grazing conditions throughout the Southwest and intermountain grasslands despite some mounting ecological difficulties in maintaining range conditions. What is not so commonly known is the turn to goat herding by a great many people. More resistant to disease than either sheep or cattle, possessing the ability to eat a great diversity of food including small trees or shrubs, providing a rich source of milk, and useful in even threshing grains and beans with their sharp hooves, goats proved exceptionally valuable to New Mexicans and later to Euro-American and Indian peoples. These animals were even used to curb woody growth on rangelands. Angoras were introduced before 1900 and were popular for their long hair that was used in mohair products. Prior to the New Deal, the Navajos alone herded nearly 350,000 goats. The rapid increase in both goat and sheep herds gave rise to serious concerns among the rangers managing grazing on federal forestlands. In New Mexico, rangers observed soil erosion problems and the influx and proliferation of noxious plants such as sagebrush, creosote bush, prickly pear, broomweed, and tamarisk. The rangers laid the blame for mounting problems more often at the hooves of sheep and goats than cattle, and these officials effectively reduced herd sizes by two-thirds between 1910 and 1930 (Scurlock 1998).

WAR AND THE GRASSLANDS

As the economy of the United States improved beginning in the late 1890s, the farm economy followed suit. The agricultural economy especially prospered from 1909 until the beginning of World War I, a time that became known as the "golden age" of American farming. Soon, the business of mechanizing farms and growing crops for a burgeoning population replaced the business of farm rebellion. Moreover, the federal government began paying more attention to the problems of farming and addressed them with federal action. This included passage of the Newlands Act in 1902 and the formation of the Country Life Commission during President Theodore Roosevelt's administration and chaired by Liberty Hyde Bailey, a renowned professor of horticulture at Cornell University. The Newlands Act held out the prospects of more reliable farming through irrigation projects subsidized by the federal government, while Bailey's work ultimately led to improved education for farmers especially through the expanded work of land-grant universities and colleges.

All of the economics and ecology of farming the grasslands changed dramatically with the onset of World War I in August 1914. European crop production plummeted and American farmers gladly took up the slack. Throughout the southern portions of the Great Plains, wheat acreage increased by over 200 percent. The federal government asked Kansas farmers to increase wheat production by over one million acres. In the far western portion of the state, in one county alone operators drove over 100 tractors in an effort to increase wheat production by one-third. With the price of wheat soaring to over $2 a bushel, farmers required little further encouragement to increase their production levels. At the same time, few, if any, asked the tough questions about the ecological results of expanding monocropping into what would prove marginal lands along with the volatile economics of wartime production.

By 1920, European farm production had resumed, and international market prices for agricultural production reflected this trend. By 1921, wheat prices in the United States had fallen by one-half their wartime values, and this sent the agricultural economy throughout the grasslands into a sharp tailspin. At the same time, mechanization of farming proceeded apace, and those farmers who could afford tractors and harvesters could capitalize on the efficiencies of scale despite falling crop prices. But small farmers could not, and they went bankrupt by the thousands during the early part of the decade. Especially notable were the "dry farming" techniques designed to place into crop production what had been the shortgrass plains. New machinery made it fairly easy and quick for farmers with large amounts of capital to turn wild grasslands into domestic fields. With little effort, a Wheatland disk plow, often referred to as a oneway, could prepare more than sixty acres in one day. Three sections of a "duck-foot" cultivator could turn 125 acres of grassland into pulverized ground. This movement became known as the "great plow-up." From 1914 to the onset of the Great Depression, farmers brought an additional five million acres of wheatland into production in Kansas alone. Much of this newly cultivated land was in the western portion of the state. In roughly the same time period, farmers sowed around thirty-two million acres of new wheat fields. Throughout the southern plains, plowmen turned an area of shortgrass prairie the size of Rhode Island into wheat fields (Hurt 1994; Ham and Higham 1987; Worster 1979). In short, these farmers had set the stage for the creation of the Dust Bowl.

THE DUST BOWL

At first people throughout the grasslands little felt the shock of the stock market collapse in October 1929. The farm economy was already weak, and crop prices, despite the Wall Street disaster, remained stable through 1929. Trouble began mounting in 1930 as prices slid by a third, and by 1932 they were failing altogether, losing on average two-thirds or more of their 1929 values. Banks failed throughout the country, and farmers became hard-pressed for the credit

This Dust Bowl photograph reveals the consequences of prolonged dry weather combined with poor farming techniques and economics. (National Oceanic and Atmospheric Administration)

needed to maintain their operations. Worse, throughout large portions of the grasslands, the weather began turning hotter and rain became more infrequent. The exposed, plowed soils became parched, loose, and ready to blow away as soon as the winds could lift them into the air, and that did not take long to happen.

By 1933, soils throughout the domesticated grasslands from Montana to Texas began to blow. One of the first truly huge storms arose in Montana and Wyoming in May 1934. This storm carried an estimated 350 million tons of soil into the air and covered far-off cities where, as in Chicago, more than four pounds of dust for each resident in the city fell to the ground. Ship crews more than 300 miles offshore in the Atlantic found themselves cleaning Montana dust from the decks. The worst storms ever to strike the grasslands arrived in April 1935, and the worse of the worst came on the 14th, "Black Sunday." Car engines stalled as the static electricity in the storms short-circuited the ignition wires. People, as well as domestic and wild animals, suffered from "dust pneumonia" caused by inhaling fine silica, like tiny shards of sharp glass, that sliced their lungs. The nation became keenly aware of the tragedy unfolding in the region as the number and violence of these storms increased, and soon the entire area was known as the Dust Bowl.

THE NEW DEAL AND WORLD WAR II

Many individuals within President Franklin D. Roosevelt's administration took keen note of the ecological troubles mounting throughout the Great Plains. Among a plethora of problems, the ones most pressing revolved around soil depletion, lack of water conservation, and rangeland overgrazing. New Dealers began addressing these issues as they came to realize a connection between ecological adaptation and economic prosperity. The foremost proponent of soil conservation was Hugh Hammond Bennett. As a noted authority on soils, he became the first director of the Soil Erosion Service, an agency created within the Department of Interior. He wanted an agency much stronger in its ability to address soil conservation, and dust storms added urgency to his lobbying of Congress. In April 1935, Bennett's effort bore fruit in the passage of the Soil Conservation Act, which created the Soil Conservation Service (SCS) within the Department of Agriculture. Bennett became its director, and soon his agency began launching programs throughout the grasslands to conserve soil. Farmers formed districts in their states, and once these were done, they were able to draw upon the expertise and resources of the SCS to improve their farming techniques. Programs illustrated and underwrote new plowing techniques such as listing, or planting fields in al-

Hugh H. Bennett, the first director of the Soil Conservation Service. (Library of Congress)

ternating strips of grass and crops; others funded the planting of windbreaks, which were usually a quarter-mile line of trees placed to soften the strong winds over the plains (Brink 1951).

Edward T. Taylor, a congressman from Colorado, was another notable figure championing reforms for tending the grasslands; his concern resulted in passage of the Taylor Grazing Act, 1934. This act placed under the control of the Department of Interior all the remaining public domain suitable for grazing or crops. The department was authorized to create management districts and to charge ranchmen fees for grazing cattle on them. Also, 25 percent of these fees were to fund conservation of these same public rangelands. In the act, Congress withdrew all the remaining public domain lands from homesteading and authorized the Interior Department to purchase severely degraded or eroded lands for restoration. By 1938, the restoration program was transferred to the Soil Conservation Service, which began a program of creating and managing national grasslands.

The New Dealers also initiated dam-building projects under the direction of both the Bureau of Reclamation and the Army Corps of Engineers. While Congress funded the work of the corps ostensibly as flood control projects, many were in fact intended to be subsidies for irrigation farmers. For example, the John Martin Dam and Reservoir project in eastern Colorado was built as a flood control project; actually it served to store and regularize river flows into the far eastern portion of Colorado and the western part of Kansas as an attempt to provide the means to settle the continuing litigation over the Arkansas River. In this manner, the sugar beet industries in both states received a substantial boost.

In *The Future of the Great Plains: Report of the Great Plains Committee* (1936), New Dealers laid the most complete blueprint for addressing the problems of agriculture in the region. Morris Cooke, administrator of the Rural Electrification Administration, chaired the committee, which included people such as Hugh Hammond Bennett and Lewis Gray, assistant administrator of the Rural Settlement Administration who authored most of the report. The committee members held numerous public hearings throughout the grasslands and considered scores of letters from citizens making their own recommendations. An earlier report in August laid out many of the problems arising in the grasslands and outlined remedial action, and this was followed with the larger, more comprehensive final report published in December. While the committee fully recognized many of the causes of the problems, it failed to offer any solutions that included a critical analysis of the market culture. Moreover, the committee fundamentally misunderstood the grassland management practices and economic

Three panels from The Future of the Great Plains *(1936) illustrate the New Deal understanding of the grasslands. There is the "Great Plains of the Past," "Great Plains of the Present," and "Great Plains of the Future." (U.S. Government Printing Office)*

resource uses of Indian peoples, and assumed that Indian peoples lived in complete harmony with their environment without altering the "balance of nature" in the grasslands. The committee's solutions revolved around greater scientific study of the region and its resources, more federal acquisition of land for better conservation, increasing the size of farms for greater economic efficiencics, underwriting further development and exploitation of water resources, resettlement of failing farm familics, and the "control and possible eradication of insect pests which ravage periodically sections of the Great Plains." While the committee on one hand condemned the domination and control of nature, their recommendations furthered this approach on the other. In the main, these became the guiding principles for federal policy in the grasslands well into the twenty-first century (White 1986).

New Dealers also tried to apply the guiding principles of the report to conservation practices in the intermountain and desert grasslands. John Collier, head of the Bureau of Indian Affairs (BIA), had just as many concerns about soil erosion as did Hugh Hammond Bennett. Even before the New Dealers came to power, those in the BIA recognized overgrazed ranges and soil erosion problems plaguing the Navajo Reservation. According to the BIA, the Navajos' 1.3 mil-

This photograph shows a Navajo rug weaver. Najavos were well known for their weaving skills. (National Archives and Records Administration)

lion sheep and goats, 37,000 cattle, and 80,000 horses were twice the number that the land, of which over 15 percent was already denuded, could support. Conservation policies, which were hammered out between Collier and Bennett, focused on herd reduction and techniques to halt soil erosion. Immediately, problems with this policy arose. Women owned the sheep herds, yet the negotiations were done with men. A flat percentage herd reduction was ordered, and this benefited the large sheepherders to the detriment of the owners of small flocks. To Collier and others, it became evident that policy would be better directed if coordinated and administered on the reservation itself, and this led to the creation of the Navajo Service in 1936. One of the most important innovations was the creation of the Southwestern Range and Sheep Breeding Laboratory, which was to find the best sheep for the environment and that would produce the finest spinning wool. But the herd reduction plan ran into considerable opposition, especially from a Hampton-educated, fundamentalist Christian, a Navajo leader named J. C. Morgan. Consequently, Collier was able to achieve partial successes at best in improving grazing practices throughout the reservation (Parman 1976; White 1983).

In the Great Basin, ranching improved little on the public range until after the passage of the Taylor Grazing Act in 1936. Even with this legislation, little was initially accomplished in gaining control over livestock stocking rates, the elimination of soil erosion, and restoring grassland range conditions to pre-ranching times. With so little conservation practiced and with the diminution of indigenous grasses, alien species, such as Russian thistle and cheatgrass, rapidly occupied these open niches. By World War II, cheatgrass had become the dominant grass grazed by cattle, and the Great Basin rangelands little resembled their former selves (Young 1994).

At the same time that rainfall increased throughout the grasslands and federal policies began tackling some of the more egregious ecological maladaptations in the Great Plains, the economy of agriculture soared dramatically with the onset of World War II. Federal economic policy toward farmers also gave them assurances that the economic collapse that followed the end of World War I would not materialize with the end of fighting in Europe and the South Pacific. With soaring prices for their crops, farmers increased their holdings, mechanized more of their operations, and took advantage of federal programs such as those sponsored by the Soil Conservation Service. Also, the scientific research at land-grant colleges and the Department of Agriculture led to advances in insecticides and hybridized crops, all of which farmers readily applied.

An urbanized, domesticated grassland was thoroughly in place by the end of World War II. Subsidized by the federal government and by research and extension at land-grant institutions, farmers and ranchers were meeting both national and international urban demands through an elaborate market system centered in the Chicago Board of Trade and a transportation network crisscrossing the entire region. Farmers, for their part, expanded the size of their operations, mechanized them, employed greater amounts of petrochemical fertilizers and pesticides, and practiced better conservation techniques. Still, in the years to come they would face mounting social and economic problems rooted in their ecological relationships.

Similarly, in the Southwest and Great Basin grasslands, some economic gains had been made in places. Phoenix, Arizona, was becoming a prosperous urban center in large part because of federal reclamation in the Salt River Valley. Mormons had created an inland communal empire. Railroad lines crisscrossed the grasslands giving its residents access to national and international markets. On the surface, these landscapes appeared harsh, forbidding, and difficult to master. Indeed, these lands proved difficult to farm and ranch, but they were also, despite appearances to the contrary, exceptionally fragile places, and their ecological systems quickly broke apart given the Euro-American assault. And not only did the landscapes themselves prove vulnerable, but the cultures of the Indian peoples showed that they were just as unprotected as were the landscapes on which they lived. Their relationships to the land changed in order to adjust to rapid ecological change. As the desert and intermountain wild grasslands vanished, people began to wonder what had been lost, and how some little portion of this vanquished landscape might be preserved for posterity. No telling what lesson such places might hold for future generations in learning how to form mutually reinforcing relationships with the plants and animals around them.

References

An Act Donating Public Lands to the Several States and Territories Which May Provide Colleges for the Benefit of Agriculture and the Mechanic Arts. July 2, 1862. *U.S. Statues at Large*, 37th Cong., 2nd sess., pp. 503–505.

Alexander, Thomas G. 2003. "Interdependence in the Mormon Heartland: Mutual Irrigation Companies and Modernization in Utah's Wasatch Oasis, 1870–1930." *Mining History Journal* 10:87–102.

Atherton, Lewis. 1961. *The Cattle Kings.* Bloomington: University of Indiana Press.

Blount, A. E. 1890. "Announcements." *New Mexico College of Agriculture and Mechanical Arts, Bulletin No. 2.* Las Cruces, NM: np, 1–4.

Brink, Wellington. 1951. *Big Hugh: The Father of Soil Conservation.* New York: Macmillan.

Clark, Ira G. 1987. *Water in New Mexico: A History of Its Management and Use.* Albuquerque: University of New Mexico Press.

Cronon, William. 1991. *Nature's Metropolis: Chicago and the Great West.* New York: W. W. Norton.

Dutton, Bertha P. 1975, rev. ed. 1983. *American Indians of the Southwest.* Albuquerque: University of New Mexico Press.

Emmons, David M. 1971. *Garden in the Grasslands: Boomer Literature of the Central Great Plains.* Lincoln: University of Nebraska Press.

Fiege, Mark. 1999. *Irrigated Eden: The Making of an Agricultural Landscape in the American West.* Seattle: University of Washington Press.

Fite, Gilbert A. 1974. *The Farmers' Frontier, 1865–1900.* Albuquerque: University of New Mexico Press.

Gates, Paul Wallace. 1979. *History of Public Land Law Development.* New York: Arno Press.

Goss, Arthur. 1893. "The Value of Rio Grande Water for Purpose of Irrigation." *New Mexico College of Agriculture and Mechanical Arts, Bulletin No. 12.* Las Cruces, NM: The Independent Democrat, 31–58.

Hadley, Hiram. 1890. "Announcements." *New Mexico College of Agriculture and Mechanical Arts, Bulletin No. 1.* Las Cruces, NM: np, 1–4.

Ham, George E., and Robin Higham, eds. 1987. *The Rise of the Wheat State: A History of Kansas Agriculture, 1861–1986.* Manhattan, KS: Sunflower University Press.

Hammer, Kenneth M. 1979. "Bonanza Farming: Forerunner of Modern Large-Scale Agriculture." *Journal of the West* 18 (No. 4): 52–61.

Hawkins, Anne P. W. 1999. "Hoeing Their Own Row: Black Agriculture and the Agrarian Ideal in Kansas, 1880–1920." *Kansas History* 22:200–213.

Heaton, John W. 1995. "'No Place to Pitch Their Tepees': Shoshone Adaptation to Mormon Settlers in Cache Valley, 1855–70." *Utah Historical Quarterly* 63 (Spring): 158–171.

Henry, T. C. 1906. "The Story of a Fenceless Winter-Wheat Field." *Kansas Historical Collections* 9:485–497.

Hickey, Joseph V. (1989). "Welsh Cattlemen of the Kansas Flint Hills: Social and Ideological Dimensions of Cattle Entrepreneurship." *Agricultural History* 63:56–71.

Hitchcock, Albert S. 1935. *Manual of the Grasses of the United States.* Washington, DC: Government Printing Office.

Hundley, Norris, Jr. 1978. "The Dark and Bloody Ground of Indian Water Rights: Confusion Elevated to Principle." *Western Historical Quarterly* 9:454–482.

Hurt, R. Douglas. 1994. *American Agriculture: A Brief History.* Ames: Iowa State University Press.

Isern, Thomas D. 2000. "Wheat Explorer the World Over: Mark Carleton of Kansas." *Kansas History* 23:12–25.

Iverson, Peter. 1994. *When Indians Became Cowboys: Native Peoples and Cattle Ranching in the American West.* Norman: University of Oklahoma Press.

Key, Scott. 1996. "Economics or Education: The Establishment of American Land-Grant Universities." *Journal of Higher Education* 67:196–220.

Kropp, Simon F. 1967. "Hiram Hadley and the Founding of New Mexico State University." *Arizona and the West* 9:21–40.

Lewis, David Rich. 1994. *Neither Wolf nor Dog: American Indians, Environment, & Agrarian Change.* New York: Oxford University Press.

Lynn-Sherow, Bonnie. 2004. *Red Earth: Race and Agriculture in Oklahoma Territory.* Lawrence: University Press of Kansas.

May, Dean L. 1994. *Three Frontiers: Family, Land, and Society in the American West, 1850–1900.* New York: Cambridge University Press.

McQuillan, A. Aidan. 1990. *Prevailing over Time: Ethnic Adjustments on the Kansas Prairies, 1875–1925.* Lincoln: University of Nebraska Press.

Meinig, D. W. 1971. *Southwest: Three Peoples in Geographical Change, 1600–1970.* New York: Oxford University Press.

Merrill, Karen R. 1996. "Whose Home on the Range?" *Western Historical Quarterly* 27 (Winter): 433–452.

Miner, Craig, and William Unrau. 1978. *The End of Indian Kansas: A Study of Cultural Revolution, 1854–1871.* Lawrence: Regents Press of Kansas, 1978.

Morris, Edmund. 1979. *The Rise of Theodore Roosevelt.* New York: Ballentine Books.

Opie, John. 1987. *The Law of the Land: Two Hundred Years of American Farmland Policy.* Lincoln: University of Nebraska Press.

Parman, Don. 1976. *The Navajos and the New Deal.* New Haven: Yale University Press.

Pisani, Donald J. 1992. *To Reclaim a Divided West: Water, Law, and Public Policy, 1848–1902.* Albuquerque: University of New Mexico Press.

Ponting, Clive. 1991. *A Green History of the World: The Environment and the Collapse of Great Civilizations.* New York: St. Martin's Press.

Ricks, Joel Edward. 1938. *The Utah State Agricultural College: A History of Fifty Years.* Logan, UT: The Desert News Press.

Robbins, William G. 1993. "Landscape and Environment: Ecological Change in the Inter-montane Northwest." *Pacific Northwest Quarterly* 84 (October): 140–149.

Sayre, Nathan. 1999. "The Cattle Boom in Southern Arizona: Towards a Critical Politi-cal Ecology." *Journal of the Southwest* 41 (Summer): 239–271.

Scurlock, Dan. 1998. "A Poor Man's Cow: The Goat in New Mexico and the South-west." *New Mexico Historical Review* 73 (January): 7–24.

Sengupta, Gunja. 1996. *For God & Mammon: Evangelicals and Entrepreneurs, Masters and Slaves in Territorial Kansas, 1854–1860.* Athens: University of Georgia Press.

Shannon, Fred A. 1945. *The Farmer's Last Frontier, Agriculture, 1860–1897.* New York: Farrar and Rinehart.

Sharp, William, and Peggy Sullivan. 1990. *The Dashing Kansan: The Amazing Adven-tures of a Nineteenth-Century Naturalist and Explorer.* Lawrence, KS: Harrow Books.

Sherow, James E. 1990. *Watering the Valley: Development along the High Plains Arkansas River, 1870–1950.* Lawrence, KS: University Press of Kansas.

Smaby, Beverly P. 1975. "The Mormons and the Indians: Conflicting Ecological Systems in the Great Basin." *American Studies* 16 (1): 35–48.

Smith, Karen L. 1986. *The Magnificent Experiment: Building the Salt River Reclamation Project, 1890–1917.* Tucson: University of Arizona Press.

Smythe, William E. 1969. *The Conquest of Arid America.* Introduction by Lawrence B. Lee. Seattle: University of Washington Press.

Stover, John. 1975. *History of the Illinois Central Railroad.* New York: Macmillan.

United States Great Plains Committee. 1936. *The Future of the Great Plains: Report of the Great Plains Committee.* Washington, DC: Government Printing Office.

Vale, T. R. 1975. "Presettlement Vegetation in Sagebrush/Grass Areas of the Intermoun-tain West." *Journal of Range Management* 28:32–36.

White, Gilbert F. 1986. "The Future of the Great Plains Re-Visited." *Great Plains Quar-terly* 6 (Spring): 84–93.

White, Richard. 1983. *The Roots of Dependency: Subsistence, Environment, and Social Change among the Choctaws, Pawnees, and Navajos.* Lincoln: University of Nebraska Press.

Williams, Jeffrey C. 1982. "The Origin of Futures Markets." *Agricultural History* 56:306–316.

Worster, Donald. 1979. *Dust Bowl: The Southern Plains in the 1930s.* New York: Oxford University Press.

Worster, Donald. 1985. *Rivers of Empire: Water, Aridity, & the Growth of the American West.* New York: Pantheon Books.

Young, James A. 1983. "Hay Making: The Mechanical Revolution on the Western Range." *Western Historical Quarterly* 14 (July): 311–326.

Young, James A. 1994. "Changes in Plant Communities in the Great Basin Induced by Domestic Livestock Grazing." In *Natural History of the Colorado Plateau and Great Basin*, edited by Kimball T. Harper, Larry L. St. Clair, Daye H. Thorne, and Wilford M. Hess. Niwot: University Press of Colorado.

4

THE MOST ENDANGERED
ECOSYSTEM ON EARTH

Are grasslands the most endangered ecosystem on the planet? Many reputable ecologists such as Fred B. Samson and Fritz L. Knopf certainly think so. "Worldwide grasslands," they boldly contend, "are the most imperiled ecosystem." In their discussion of the United States, they cite figures indicating that in many states less than one-tenth of 1 percent is all that remains of wild grasslands. Given the highly fractured, island geography of contemporary grasslands, Samson and Knopf argue that at best, the future existence of wild grasslands is highly problematic. The urbanized, domesticated grasslands cover nearly all of the land once dominated by wild grasslands, and the great question may be to decide whether reversing this trend and saving what is left of its wild stands is worthwhile, or whether these dwindling ecosystems should be allowed to pass into extinction (Samson and Knopf 1996).

The fate of wild grasslands has not been sealed yet. Undoubtedly, several climatic, economic, and social forces have combined to destroy the wild grasslands, but in the years following the end of World War II, many forces have combined to preserve what is left of these biomes, too. A growing environmental movement has called for greater respect of the remaining wild grasslands. At the same time, Americans have exhibited an ambivalent attitude toward the grasslands, a divided view shaped by the economics of farming. Some people have seen the need for greater conservation, if not enlargement of wild grasslands, while others have viewed their future in the context of ever-larger farming and ranching operations making every use possible of the resources in the area for economic development. During the same years, the area has become highly urbanized, and the demographic trends in the region have brought in more people with less and less familiarity with wild grassland dynamics and communities, and little knowledge of the actual farming practices conducted in the region.

Since World War II, Americans have been anything but united in their visions for the future of the grasslands. Various worldviews have driven these contending notions, which highlight the fractious nature of American culture as it has related to living a proper life. Each of these various positions has played a part in shaping the ecosystems of the grasslands, too. In effect, the ecosystems of the grasslands are inherently cultural landscapes; that is, the shape of any given ecosystem in the grasslands not only reflects the values of the people who live in it, but is actually shaped in part by those values. This situation underscores the observation made by the ecologist Robert O'Neil, that humans are a "keystone" species in nearly all ecosystems on the planet.

ARTISTIC PORTRAYALS OF THE GRASSLANDS

Given the power of cultural values as expressed through work and recreation to shape ecosystems, it is important to understand some of the varying views people have had about grasslands. Cultural values might be likened to the software that guides the actions of a machine: they guide the way humans exploit, manage, and shape their environments. Since World War II, writers and filmmakers have certainly conveyed cultural values through their works, and their efforts have demonstrated the ambivalent attitudes held by the people who domesticated the wild grasslands. A few examples among the myriad of artists who have dealt with grassland stories will illustrate the point.

Roy Bedichek, a Texas naturalist, carefully observed and portrayed the prairies and Hill Country around him. In his book *Adventures with a Texas Naturalist,* written in 1947, he related a discomfort with the emergent transformations of his surroundings into urbanized landscapes. Fencing broke up the interconnectedness of the micro-ecosystems of the prairies and hills. The initiation of the "suburban chicken farm," highly "mechanized" in its techniques, led to "denatured" fowl. Bedichek caught an early glimpse of where post–World War II agriculture was headed, and he was exceptionally uneasy about its direction (Bedichek, 1947).

William Least Heat-Moon drew what he called a "deep map" of the Flint Hills, a somewhat intact tallgrass prairie spread north to south through the middle of the eastern half of Kansas. In *PrairyErth*, published in 1991, Heat-Moon captured the intertwined ecological and social history of Chase County, and he vividly depicted the wide-open feel of the Flint Hills and its seasonal moods. He reminded the reader that the place has not always looked the way it does today. He recalled how remnant mountains once higher in elevation than the Rockies remain buried under layers of rock deposited during a time when an inland sea covered the region. He also explained how the mark of the land has shaped the thinking of its human residents, a great many of whom have never seen a need for a national tallgrass prairie park.

Moviemakers, too, have pictured in striking ways how people have related to the grasslands—for example, director Peter Bogdanovich's movie the *Last Picture Show*. Released in 1971, the film illuminates on screen Larry Mc-Murty's 1966 book of the same title. The story centers on a cast of young people living on the Great Plains of Texas in the early 1950s. It is as much a lament for the passing of an open land once imbued with beauty and freedom as it is a denunciation of the morally crass, materially driven culture that destroyed the shortgrass range. Kevin Costner's *Dances with Wolves* makes the same point by embracing stereotypes. Certainly his film captures how an unbroken grassland and grand bison herds must have appeared prior to 1870. Costner, however, falls into stereotyping Indian peoples, the Lakota Sioux in this case, as living in complete harmony with their environments only to have their lives and the land they lived on threatened by the encroachment of a militaristic, expansionist Euro-American culture. In both of these movies, the eclipse of the open grasslands resulted in a corresponding loss of American freedom.

The question of whether Euro-Americans were a destructive force in the grasslands inspired some of the earliest environmental histories. In the 1940s and 1950s, James Malin's work integrated the findings of ecologists into his interpretation of environmental changes in the grasslands, and especially so when

James Malin (left) and Walter Prescott Webb (right) were historians of the grasslands. Their works were precursors to environmental history. (Kansas State Historical Society)

he explained the causes of the Dust Bowl in the 1930s. Malin's belief is that humans, when well versed in the laws of ecology, can master their environments. He saw the Dust Bowl as the result of recurrent climatic swings that humans could withstand with improved technologies and farming methods (Malin 1984). At nearly the same time, Walter Prescott Webb, whose work the *Great Plains* has greatly influenced environmental historians, had a very different view about people's abilities to adapt to living in the grasslands. In "The American West, Perpetual Mirage," he identified many themes that two to three decades later would be associated with a group of historians called the "New West." Webb especially chastised Westerners for their poor use and management of natural resources, water being the most extreme case. People's thinking about how to live in the grasslands was terribly out of sync with the realities of the ecosystems in the region, and Americans would pay a severe price for failing to change their ways. His piece, published in *Harper's Magazine*, hit a raw nerve, and many prominent writers and politicians wrote letters to the publisher harshly rebuking Webb (Webb 1957). In 2005, people throughout the grasslands still seem to gravitate around the polar extremes expressed by Malin and Webb right after the end of World War II. Those who feel pulled toward Malin have relied more on technological fixes and economic motivations to guide them. Those adhering to Webb's view look more toward what the grasslands themselves can teach about how to adapt to them.

Ian Frazier caught something else about the grasslands missing in Malin's and Webb's writing. Frazier, a New York–based writer, took an urban approach toward the region. His work, the *Great Plains*, reveals an ambivalence about the grasslands. He lived in Montana for a while and took the opportunity to explore his surroundings. His was an urban response to the grasslands. At once he stood in awe of the lonesome, open feel he experienced when crossing the plains. But he also took note of the impounded and depleted rivers; the wheat, alfalfa, and sorghum fields that had replaced the shortgrass plains of Montana, part of the breadbasket of the nation—a breadbasket feeding a very hungry urban nation. And even this "air shaft" of the nation itself appeared as a place of mostly city dwellers (Frazier 1989).

GRASSLAND PARK SYSTEMS

One of the interesting social phenomena arising out of an artistic depiction of the grasslands has been the struggle to create a grassland national park. Of all the biomes in North America, the tallgrass prairie is one of the last to be represented by a national park. Perhaps surprisingly, the tallgrass prairie biome was one of the earliest landscapes to inspire pleas for its preservation. Over a century before the end of World War II, George Catlin, a painter who marched alongside the army across the grasslands in the 1830s, made one, if not the first, argument for the creation of a national grassland preserve. In 1832, while reflecting upon the future of Indian peoples living throughout the prairies, he feared that the rapid advance of American culture threatened the existence of both peoples and their lands. He called for a "nation's park," one that would preserve the "wild and freshness" of nature (Catlin 1844).

Over a half century later, Walt Whitman in essence called for the same. On a jaunt through the grasslands in 1879, he was struck by "that vast something . . . which there is in these prairies, combining the real and ideal, and beautiful as dreams." On the same excursion, he depicted the grasslands as "America's Characteristic Landscape," a place many others at the time thought better represented by Yosemite, Niagara Falls, or the Upper Yellowstone. In making his case, he observed that the grasslands, "while less stunning at first sight, last longer, fill the esthetic sense fuller, precede all the rest" (Whitman 1963, 220–221). More than this, preserving such a landscape undeniably became linked in Whitman's mind with nurturing democracy. "I conceive of no flourishing and heroic elements of Democracy . . . without the Nature-element forming a main part—to be its health-element and beauty-element—to really underlie the whole politics, sanity, religion and art of the New World" (Whitman 1963, 294–295). The characteristic landscape of America, the Nature-element part that could breathe life into its democracy, would become the last landscape protected in national or state parks.

Perhaps Al Runte, a historian of national parks in the United States, has hit upon the reason why it has been so terribly difficult to create a national park in the grasslands. More than just scenic beauty and historic importance had to be attached to any place before Congress would incorporate it into a national park. The land itself, Runte maintains, must be of limited, or "worthless," economic value: if a mountain range had mining potential, if a forest were rich in trees suitable for lumbering, or the soil were fertile, then Americans preferred to see these lands placed into economic production rather than set aside for posterity. The resources of the grasslands have always had rich economic potential, and as we have already seen, Euro-Americans quickly converted them into commodities owned by private stakeholders.

There were some who thought about preserving the wild Southwest and intermountain grasslands. Notably, these grasslands are marked by few national parks, and only where National Forests were created are there any wilderness areas today. Only after World War II would there be an attempt to create a national park in the Great Basin. This one, supported by Nevada senator Alan

Bible, ran headlong into insurmountable opposition from ranchers and mining companies. Eventually, environmentalists were able to create a Great Basin National Park in 1986.

In other places, proponents favoring the preservation of wild grasslands took other routes besides creating national parks. The intense struggle to create the Konza Biological Research Station, a preserve that is neither a commercial enterprise nor a full-fledged public preserve, illustrates well the constraints people faced in creating grassland parks. When Lloyd Hulbert arrived at Kansas State Univeristy in 1955, he encountered a biology department already considering the merits of developing a scientific reserve for the study of tallgrass prairies. Discussions about the value of "biological stations," places where students and academics could acquire firsthand acquaintance with the ecosystems of a particular region, had been circulating among academic institutions for about a decade.

Establishing the first elements of a bluestem prairie research preserve took another decade of hard work by Hulbert with the cooperation of a growing cadre of supporters. The university purchased the first small tract of land from Elizabeth Landon, wife of former Kansas governor Alf Landon. The next big break came with acquisition of an additional 7,000-acre ranch adjacent to the Landon purchase. By 1975 the Nature Conservancy had begun its negotiations for the land and, behind the scenes, collaborated with Ted Barkley, a botanist at Kansas State University, and Hulbert, who worked with Katharine Ordway, heiress to the family fortune acquired through connections to the 3M Company. They convinced Ordway to underwrite the Nature Conservancy's purchase of the ranch for $3.6 million. In turn, the Conservancy entered into a lease arrangement with Kansas State University to manage Konza Prairie. As its first director, Hulbert oversaw an ambitious scientific research program that included investigating the effects of fire and grazing on the plant ecosystems; the nature of soil and water chemistry on the site; and wildlife behavior and habitats. The National Science Foundation established the Long-Term Ecological Research (LTER) program in 1980, and guided by Hulbert, Konza became one of the first six sites established in the nation, and the only one at the time dedicated to the study of grassland ecosystems. But this was not a park like Yellowstone National Park. While established as a preserve, it served a utilitarian function as a scientific research site, and as such, it provided an "economic" return.

Creating a national park would require another route. In the 1930s, Dr. V. E. Shelford, an ecologist at the University of Illinois, with the endorsement of the Ecological Society of America sent a proposal for preserving the tallgrass prairie to the National Park Service. Shelford noted two prime areas where this biome still existed: one was throughout the Flint Hills of Kansas, and the other was in portions of the Osage Hills in northeastern Oklahoma. By 1960, one site in particular drew the attention of the National Park Service—nearly 60,000 acres located in Pottawatomie County just to the east of Manhattan, Kansas, and the Blue River. During the Eisenhower administration, the Pottawatomie site had considerable support from the Manhattan area. Then secretary of the interior Frederick Seaton and his brother Edward, newspaper owner in Manhattan, Kansas, endorsed a prairie park in Pottawatomie County. The state legislature even appropriated $100,000 for purchasing the land if Congress would pay the remaining purchase costs. Public hearings were held, and most residents and the political leadership of Manhattan stood solidly behind creating the park. However, the 1961 Pottawatomie park bill never made it out of the congressional subcommittee considering it.

One event more than any other heralded the demise of this proposal. In December 1961, then secretary of the interior Stewart Udall and National Park Service director Conrad Wirth were observing the proposed land for the prairie park from their helicopters. Their pilots landed on Twin Mound, and there to greet them was a shotgun-wielding Carl Bellinger. Bellinger, a rancher, leased the land where the helicopters landed, and after a brief and less than cordial visit with the secretary of the interior, the rancher ordered Udall to get off "his" land. In fact, Udall not only left the site, but he and Park Director Wirth never returned.

As symbolic as the Bellinger-Udall confrontation is, two other factors accounted for the demise of the Pottawatomie Prairie Park idea. First, the building of Tuttle Creek Dam and Reservoir had generated extensive local distrust in the federal government. The farm families, who once lived throughout the Blue River Valley, actively opposed the building of the dam by the Army Corps of Engineers. They lost the political battle and their farms after nearly a decade of bitter protest during both the Truman and Eisenhower administrations. Second and more important, ranchers and their organizations actively opposed the park. This is the main reason that Professor Lloyd Hulbert emphasized over and over again that the Konza Prairie Botanical Station had to be a scientific field laboratory and not a prairie park. Hulbert knew that a scientific station had economic potential, as its research findings could have value for ranchers and farmers. A park, as far as many ranchers were concerned, would take agricultural lands out of production and mark the end of their livelihoods (Baldridge 1993).

Hulbert, who passionately championed a prairie park, clearly understood how most ranchers viewed the creation of a national park. When he had tried to enlist the support of state politicians for a park in 1975, he received this response from state senator Don Christy, a rancher advocate from Scott City, Kansas: "If you wish to see all facets of the prairies as they once were except for the few wild animals which can be seen in the zoo, all you have to do is drive through the tallgrass system now. . . . Consequently, I see no justification for removing that kind of acreage from the food-producing capabilities for this nation. Thanking you for writing, but do not count on my support for this piece of federal or state legislation" (Christy 1975). Over twenty more years would pass before President Clinton signed the bill creating the Tallgrass Prairie Preserve, a dream realized by the tireless campaigning of dedicated supporters such as Hulbert. And even with this, by 2005 a mere 140 acres stood open to the public while nearly 10,000 acres of the rest remained bound in a thirty-year lease with a Texas cattleman.

In protecting wild grasslands from development, the Nature Conservancy rather than the National Park Service has been the most effective institution in setting aside acreage. The push to create a national tallgrass park in Oklahoma fell by the wayside, but in 1989, the Nature Conservancy took advantage of the opportunity to acquire the 29,000-acre Barnard Ranch in Osage County, Oklahoma. Through fire management and the reintroduction of bison, managers of the preserve have worked to restore this southern end of the Flint Hills. Elsewhere throughout the grasslands, the Conservancy has worked to purchase holdings representing a variety of its ecosystems. For example, it bought the Niobrara Valley Preserve in northern Nebraska, around 60,000 acres of riparian woodlands and upland prairies. There bison have been reintroduced along with controlled burning to restore these ecosystems. Recently the Conservancy purchased a 16,800-acre site in western Kansas in order to preserve a shortgrass system that at one time blanketed the Great Plains. The Conservancy also has a large site in northern Montana and a wetlands preserve in North Dakota.

In contrast, most state governments in the grasslands have been slow to develop any kind of extensive preserves. One exception to this is Custer State Park in South Dakota. This 71,000-acre park is home to one of the largest bison herds in the United States and hosts busloads of tourist on "safari." The park was once Custer State Forest and became Custer State Park in 1919. James Philip and his Cheyenne wife Sarah Larrabee were primarily responsible for the bison herd in the park. Around 1900 he and Sarah had acquired around 100 head of bison and grazed them on 3,500 acres of public land leased from the federal government. He worked diligently to interest the state of South Dakota in creating a preserve for his herd. In 1914, the state purchased thirty-six bison from his son Scotty to begin a herd in the park. In the 1960s the herd had reached nearly 3,000 animals, and a regimen of culling was begun to keep the number to around 1,400 animals. Today these bison are managed carefully to enhance their genetic diversity, and these animals serve as a source for stocking other herds across the nation (Wood 2000).

URBANIZED GRASSLANDS

Much of the support for grassland parks has come from the urban centers in the grasslands, which is rapidly becoming the most urbanized region of the United States. Four states in the middle of the grasslands are excellent examples of this trend. In Kansas, nearly 20 percent of the state population resides in one highly urbanized county in the northeast portion of the state. Johnson County is not only highly urbanized, but it also has one of the highest per capita incomes in the entire United States. The county contains several corporate home offices, whose glass and steel buildings are set in highly manicured corporate landscapes bearing no resemblance to the former open grasslands of Kansas. While the "rural" population increased slightly between 1990 and 2000 from 765,003 to 768,337 people, the population living on farms fell sharply from 108,083 to 89,758. This movement highlights a trend becoming ever more commonplace—the movement into the countryside of people who make their livelihood in ways other than farming. Often, they simply seek an escape from city life and stake out a homesite on small acreages often referred to as "ranchettes." At the same time they still work, shop, and seek entertainment in cities.

As in Kansas, similar demographics in 2000 marked Nebraska with around one-half of the population in the state living in three counties: Douglas, Lancaster, and Sarpy. Two of these counties, Douglas and Lancaster, are home to Lincoln, the capital and site of the University of Nebraska, and Omaha, the largest city in the state. Sarpy is largely a suburb to the south of Omaha and is also the location of Offutt Air Force Base. In much of the rest of the state rural depopulation has been the general trend.

Many environmentalists, farm advocates, and locally elected officials fear this trend but largely for differing reasons. Environmentalists take alarm at the loss of wildlife habitat, groundwater recharge zones, scenic beauty, and further fragmentation of the grassland biome; farm advocates lament the passing of midsize farms, family-owned operations of 500 to 2,000 acres. Locally elected officials have lately realized that ranchettes more often than not cost counties money. For example, in Bandera County, Texas, for every $1 that farmlands returned to the treasury in the 1980s and early 1990s, the county expended $.26 in services. Ranchettes, on the other hand, demanded $1.10 in services for every $1 they generated in tax revenues. By 2002, many officials have become concerned about this growing development throughout the grasslands, as around two acres of farmland each day in the United States were being converted to residential developments. In Texas alone, from 1982 to 1992, the state lost 1.4 million acres of farmland to residential developments. Consequently, while "rural" populations might be growing in some portions of the grasslands, in recent years this trend has been more an urban phenomenon than an increase in farms.

Similar urban demographics are reflected in South and North Dakota. In the nonmetro counties of North Dakota, only one county has shown any population growth over the last two decades, and all of the other counties have been showing population declines in the same period. In fact, the population of the state declined altogether from 1990 to 2000. By the year 2000, only 6 percent of the population in the state lived on farms, so year by year fewer people experience a direct working relationship with the land itself. The same general trend has held true for South Dakota also. By 2000 only 7 percent of the population there lived on farms, while 51 percent of the population were in urban areas; by contrast, in 1940 over 75 percent of the people lived in the rural areas of the state.

In Colorado, the population in every grassland county in the state has shrunk. In 1927, Michael Creed Hinderlider confidently predicted to the chamber of commerce in Rocky Ford that irrigation and attendant economic development would create a vast urban corridor stretching from Pueblo eastward to the state line (Sherow 1989). By 2005 Hinderlider's prediction had clearly fallen far short of the mark. Over the last fifteen years, within the Arkansas River Valley east of the Front Range, the only counties showing any population growth were those with the major cities of Colorado Springs, Pueblo, and to a lesser extent, Trinidad. All of the other counties had population declines between 2000 and 2004, and most over the last fifteen years have undergone continual decreases. Kiowa, a county on the eastern border, has fewer than 1,500 residents. Yet overall, the population in the state increased by 30 percent from 1990 to 2000, and again by another 7 percent by 2005. From 1990 to 2004, the urban population growth in El Paso County alone, where Colorado Springs is, nearly doubled the *total* 2004 population of the predominantly rural counties in the rest of the Arkansas River Valley; that is exclusive of Teller and Pueblo counties, where explosive urban growth has also occurred over the last fifteen years.

CONTEMPORARY GRASSLAND HYDROLOGY

One overriding factor has caused great consternation for all city managers, businesspeople, and politicians in the post–World War II years, and that is securing enough water resources to supply continued growth. World War II spectacularly stimulated the growth of defense industries in many cities throughout the region, and this in turn led to significant population booms in each of these urban areas. Population growth continued unabated as defense contracts and facilities continued to play a key if not a dominant role in this trend. Some of the main cities that experienced this type of growth were San Antonio, Texas; Wichita, Kansas; Denver and Colorado Springs, Colorado; and Albuquerque, New Mexico. Las Vegas added gambling and adult entertainment to its post–World War II military economy, and this led to an exponential growth now threatening to deplete minor rivers throughout the Great Basin desert grasslands as its city leaders strive to supply a projected growth rate of 100,000 new residents each year (see Table 3).

As in Las Vegas, urban population growth throughout the grasslands produced an insatiable demand for a reliable water supply. Beginning during World War II, cities like Wichita, Kansas; Denver and Colorado Springs, Colorado; Albuquerque, New Mexico; and San Antonio, Texas, grew exponentially as a result of military bases and industrial production. The city managers had to develop new water supplies to meet residential and industrial needs, and this often meant constructing elaborate systems and spearheading changes to state water laws. These cities continued to grow after the war, and their ever-mounting needs for water often pitted them initially against farm interests and in time, with environmentalists, too. The depletion of groundwater supplies, such as the Equus Beds near Wichita or the Edwards Aquifer near San Antonio, has highlighted this problem. Competition for surface water supplies has made irrigation throughout the Upper Arkansas and Canadian River valleys less economically viable compared to the returns possible from urban uses in Albuquerque or Colorado Springs. At the same time, the federal government has built storage reservoirs to help meet increasing urban demands, although not all water developments were built to supply cities; many were constructed to protect cities from water. For example, a great flood in the Kansas River Valley in 1951 spurred the Army Corps of Engineers to construct several reservoirs in the

TABLE 3
Population Rounded to the Nearest 10,000

Year	San Antonio	Wichita	Denver	Colorado Springs	Albuquerque	Las Vegas
1940	254,000 (36)	115,000 (74)	322,000 (24)	37,000	35,000	8,500
1950	408,000 (25)	168,000 (58)	416,000 (24)	45,000	97,000	25,000
2000	1,145,000 (9)	344,000 (50)	555,000 (24)	360,000 (48)	449,000 (35)	478,000 (32)

Sources: For 1940 and 1950, see Campbell Gibson. *Population of the 100 Largest Cities and Other Urban Places in the United States, 1790–1990*, Population Division Working Paper No. 27. Washington, DC: U.S. Bureau of Population, 1998; for 2000, see Proximity, Community and Regional Analysis. "The 100 Largest U.S. Cities Based on Census 2000 Population," http://proximityone.com/plc100.htm (accessed September 2005).

Note: Number in parentheses indicates rank among the largest 100 cities in the United States.

watershed mainly to protect industrial interests over a hundred miles away in Kansas City, Kansas, and Missouri.

Development of the water resources for, or to protect, burgeoning urban populations was not always a smooth process. In 1954, Elmer T. Peterson published a best-selling book titled *Big Dam Foolishness.* He castigated the programs sponsored by the Army Corps of Engineers and the Bureau of Reclamation. In 1944, the corps and the bureau had joined forces to share the spoils of dam building throughout the Missouri River Basin. The agreement engineered by General Lewis A. Pick of the corps and W. Glenn Sloan, an assistant bureau director in Montana, divided up projects so as to accommodate the irrigation proposals of the bureau as well as the flood control and navigation goals of the corps. More than this, the agreement allowed state politicians throughout the region to kill any attempts at creating a Missouri River Valley Authority modeled on the Tennessee River Valley Authority. Significantly, it gave states the power to review all proposed projects, made irrigation the main objective for any project west of the 98th meridian, charged the corps with regulating "surplus" water in any project, and assigned the bureau the responsibility of marketing the hydroelectricity generated by the dams. As a result, the 1944 Water Act became the model for *all* water development in grasslands. This means of dam building, especially the projects detailed in Section 9 of the 1944 act, became the target of "big dam foolishness" critics.

Three fights illustrate the nature of the contests. In two the opponents failed to achieve their objectives, and in the other the antagonists won their fight. One of the most beautiful and agriculturally productive river valleys in Kansas was flooded by a Pick-Sloan project. Up to 1951, the residents had fought fairly successfully to prevent the construction of Tuttle Creek Dam in the Blue River Valley. But in June 1951, unseasonable downpours overflowed the tributaries to the Kansas River, which produced extreme flooding all along its course to the Missouri River. Every city in the valley suffered exceptional property damage, especially in the industrial river bottoms of Kansas City. The floodwaters also drowned the protests the of the Blue Valley residents, and by 1963 the functioning dam had covered their way of life with its waters.

The Mandans, Arikaras, and Hidatsas who lived on the Fort Berthold Reservation experienced similar treatment from the corps. The Garrison Dam project ruined their ranching economy, and the dam directors denied them any fishing rights in the reservoir. The business council president, George Gillette, who had no alternative but to sign the contract with the corps, did so, while making it clear that "the members of the tribal council sign this contract with heavy hearts." During the signing ceremony, a photographer caught a teary-eyed Gillette among the happy faces of the corps and dam champions (Lawson 1982).

In another notable encounter over a Pick-Sloan project, farm families and landowners organized the United Family Farmers (UFF) in the early 1970s. This grassroots group challenged the building of the Oahe Unit, a bureau project designed to tap the Oahe reservoir to irrigate around 750,000 acres in South Dakota. The UFF won control of the Oahe Conservancy Sub-District in an effort to stop the project because of its high potential to ruin farmland. Environmentalists joined the effort, and they demonstrated how the project would drain wetlands, channelize the James River, and increase stream pollution. The contest led to the defeat of Senator George McGovern in his reelection campaign of 1976 and the liquidation of the project in 1977 (Carrels, 1999).

In the years after World War II, the hydrological landscape in the grasslands, whether surface or groundwater, has suffered severe degradation with the advent of chemical farming techniques, changes in livestock production, and continued problems with soil erosion. As a result of chemical applications, researchers often found large traces of herbicides such as atrazine and metolachlor in surface water. According to analyses done in 1997 and 1998, reports from the Environmental Protection Agency revealed atrazine, a suspected carcinogen and a gastrointestinal, liver, reproductive, and skin toxicant, along with metolachlor, which is suspected as a carcinogen, appearing in significant quantities in streams and lakes. Pesticides have been also been found repeatedly in surface water affecting, for example, 30 percent of the bodies of water in Nebraska. The emergence of huge feedlots and poorly regulated grazing has resulted in the buildup of ammonia and/or pathogens affecting 56 percent of the water bodies in South Dakota and 65 percent in Kansas. Non-point pollution from these enterprises has often resulted in massive fish kills.

Consequently, environmentalists have frequently listed Kansas as having some of the worst surface water quality in the nation. Their case is supported by the endangered species list in the state. As of July 2005, fifty-nine species of mammals, birds, reptiles, amphibians, fish, and invertebrates have been listed in the state as either threatened or endangered. Representing just over half of the total were fourteen aquatic invertebrates, mostly mussels; sixteen fishes; and two frogs. Mussels are the most telling, as they do not move from a location once fixed, and they feed by filtering the stream flow for their nutrients, which means they also ingest any of the pollutants in the current. The building of dams has also destroyed the habitats of mussels by reducing downstream flows, and now mussels have become rarer in most Kansas streams below dams.

More than anything else, center-pivot irrigation has transformed the hydrology of the grasslands. After World War II, farmers began employing more efficient pumps powered by natural gas and gasoline to lift water from deep underground sources. Still, considerable care had to be taken in leveling fields, providing good drainage, and monitoring pipes and siphons. In 1949, Frank Zybach invented a machine that eliminated much of the labor required in this system. He attached aluminum pipes fitted with water sprays to A-frame tow-

Frank Zyback's center-pivot irrigation system dramatically transformed agriculture throughout the grasslands. (U.S. Department of Agriculture)

The Ogallala Aquifer.

ers riding on tandem wheels. This apparatus extended out like the radius line of a circle, and its center was a swivel where ground pipes tapped the water pumped to the surface. The pressurized water flowing into the pipes both powered the wheels, rotating the arm in a circle, and provided the supply to the sprays. Today these machines are carefully regulated with computerized software and ground moisture sensors, and one machine normally covers about 133 acres. From the air, the land covered by these sprays appears as a green circle in a square plot.

Center pivots made it possible for farmers to tap the Ogallala Aquifer (see map on p. 137). Essentially, the aquifer is the collection of Ice Age runoffs entombed by the accumulation of soil and sand. The total "drainable" aquifer holds enough water to fill Lake Huron, third largest of the Great Lakes. The saturated thickness of the aquifer varies, so that some irrigators have very deep sources beneath their lands while others do not. For example, some farmers around Lubbock, Texas, have completely pumped all of the water below their ground and have been obliged to return to dryland farming. By contrast, in the mid-1980s irrigators in Nebraska, where the saturated thickness of the aquifer is at its greatest, had used less than 1 percent of the water that was available in the aquifer prior to the inauguration of pumping. In general, however, this water, essentially "fossil" water because of its exceptionally restricted recharge rates, is a limited resource. By one estimate, given 1980 pumping rates, center-pivot systems will be unable to supply 80 percent of the area presently irrigated through this system by the year 2020 (White and Kromm 1992).

Grassland irrigators and their state legislators have taken steps to regulate the use of the aquifer through laws and institutions. In 1927, for example, New Mexico became the first of the grassland states to bring groundwater (defined as a public rather than private resource) under state control; the state engineer regulates the depletion rates within legally defined groundwater basins. Elsewhere, management systems vary from state to state, but all states from Nebraska to Texas that are fortunate enough to have this underground water endowment have institutions and regulations to manage its exploitation (Opie 1993).

CHANGES IN FARMING AND RURAL DEMOGRAPHICS

The economic effects of intensified capitalization of farming in the grasslands has had consequences far broader than changes associated with irrigation. From the end of World War II to the present, federal farm policy has emphasized three objectives: "maintain farm income, control production, and provide relatively cheap food for consumers." The federal government has pursued these objectives through farm legislation designed to implement price supports for crops, to control crop acreage under production, to enhance and/or regulate international agricultural imports and exports, and to establish what has been, in essence, a living income for farmers. More than this, however, has been the federal government's policy of abetting large-scale, corporate, industrialized farm practices at the expense of smaller scale, family-owned operations. Industrial farming techniques may best be described as specialized monocropping, the concentration of capital and land in single operations, and the nearly complete reliance on corporate seed, fertilizer, and pesticide companies for sources of information and stock (Gardner 2002; Wessel 1998).

The secretary of agriculture during President Eisenhower's administration was Ezra Taft Benson, and his recommendation to a group of farmers is the best summary of federal aspirations for farming in the grasslands. In what surely appeared to him a helpful suggestion for his farmer audience, Benson told the crowd to "get big or get out." During the post–World War II period, powerful forces have worked to underscore Benson's advice. For example, the Committee for Economic Development, an organization of many of the most powerful business leaders in the nation, played an influential role in shaping agricultural policy in the postwar years. Initially chaired by David Rockefeller, the committee advocated removing around 2 million people from farms and continuing to hold steady or reduce the number of farmers in the future (Ritchie and Ristau 1986).

In numbers alone, this policy has certainly shown remarkable success with the rapid decline of farmers throughout the grasslands. In nearly all of these states, census takers recorded the highest number of farms prior to full implementation of New Deal agricultural policies. Then the big fallout occurred during the Eisenhower, Kennedy, and Johnson administrations. The "get big or get out" mind-set worked its wonders with states such as Washington, which lost 44 percent of its farms from 1950 to 1970 while the amount of land in production fell by only 8 percent. In the middle of the Central Grasslands, the number of farms in Nebraska dropped by 33 percent while the amount of land in production declined by a mere 1 percent. From the end of the war to the latest numbers in 2004, Nebraska lost over 64,700 farms—equivalent to a 57 percent reduction.

One of the major developments in the "get big" mode of doing business has been the growth of the containment beef and hog industry. For years, large stockyards, slaughterhouses, meatpacking plants, and railroad tracks were commonplace in many grassland cities such as Wichita, Kansas City, Omaha, or Fort Worth. In some grassland pastures—for example the Flint Hills—small feedlots were built for fattening cattle before shipping the animals to facilities

A photograph taken within an Iowa Beef Processors (IBP) plant, one of the massive butchering and packaging plants in the grasslands. (AP Photo/Nati Harnik)

in the cities. By the 1960s, severe ecological problems associated especially with stream pollution, along with changing environmental values, drove feedlots out of the Flint Hills, and stockyards and slaughterhouses out of the cities. Today, feedlots are a thing of the past in the Flint Hills, and the "Stockyards" of Fort Worth is a national historic district with prospering restaurants and entertainment. This change became feasible with the development of a different type of containment operation placed in the mightily depopulated areas of the grasslands. In western Kansas, Oklahoma, and Nebraska companies such as Cargill, ConAgra, and IBP, Inc. (Iowa Beef Processors, Inc., which Tyson Fresh Meats, Inc. purchased in 2003) built huge factories. In 1997, the IBP plant in Holcomb, Kansas, had a slaughter capacity of over 6,000 animals per day, and the capacity of all plants in Kansas amounted to 22 percent of the national total, ranking first in the nation for cattle slaughtered per day. These corporations have flourished with the close proximity of abundant alfalfa and corn production based on center-pivot irrigation; with a plentiful, inexpensive, nonunion work force of mostly Mexican and Asian extraction; with little environmental opposition; and with the ready subsidies provided by local governments eager for economic development.

ENVIRONMENTAL PROBLEMS ASSOCIATED WITH MODERN AGRICULTURE

Massive animal containment facilities and beef production plants, while economic mainstays of the grassland economy, have created several ecological problems. Prior to 1970, the small-scale feedlots often were the source of deadly, massive fish kills in many streams. Grazing practices contributed to high bacteriological counts in many streams and groundwater sources throughout the grasslands. High concentrations of e-coli bacteria indicated water sources too polluted for human consumption. The practice of routinely inoculating cattle with antibiotics contributed to lowering the effectiveness of antibiotics in humans. The large-scale hog containment facilities have spurred local, and in some cases statewide, protests and political controversy. For example, the residents of Great Bend, Kansas, uniting with Sierra Club activists, opposed the plans of Murphy Farms to build a huge hog containment and processing plant. The concerns, based on observing what had happened in other communities near these hog farms, revolved around expressions of xenophobia; fears of groundwater, air, and soil pollution; concerns with falling property values of holdings near the operation; and charges of animal cruelty. One other aspect of cattle production, the detection of mad cow disease (bovine spongiform encephalopathy) has raised serious public concerns. This condition arises when cattle are fed on feed mixed with rendered animal remains. By mid-2005, the USDA had confirmed three cases of the disease in American herds, and cases have been detected in Canadian herds as well. The appearance of this disease, associated with its deadly consequences for humans who eat meat from contaminated animals, has led to heightened fears about the health safety of American cattle and has had detrimental market consequences for American beef exports.

Along with concerns about beef production, many scholars and critics see the unabated application of chemicals in farming as a serious and mounting problem in the post–World War II years. The U.S. Department of Agriculture (USDA) began tracking chemical applications in 1990 as a result of legislation passed in 1989, and the first reports summarized data gathered for the 1991

TABLE 4
Nitrogen and Phosphate Applications in Kansas
Winter Wheat Production, 1991 and 2004

Year	Acres applied	Nitrogen	Lbs. per acre	Acres applied	Phosphate	Lbs. per acre
1991	9.8 million	569.9 million lbs.	58.0	5.8 million	192.4 million lbs.	33.0
2004	9.0 million	788.6 million lbs.	87.6	6.2 million	281.8 million lbs.	45.5

Source: United States Department of Agriculture. *Agricultural Chemical Usage, 2004 Field Crops Summary.* Herndon, VA: National Agricultural Statistics Service, May 2005.

TABLE 5
Nitrogen, Phosphate, and Insecticide Applications in Nebraska
Corn Production, 1991 and 2004

Year	Acres applied	Nitrogen	Lbs. per acre	Acres applied	Phosphate	Lbs. per acre	Acres applied	Insecticides	Lbs. per acre
1991	8.2 million	1,094.6 million lbs.	135.0	5.2 million	191.7 million lbs.	36	4.3 million	6.8 million lbs.	1.56
2003	7.7 million	980.2 million lbs.	127.3	6.2 million	217.0 million lbs.	35	2.9 million	1.5 million lbs.	0.51

Sources: United States Department of Agriculture. *Agricultural Chemical Usage, 1991 Field Crops Summary.* Herndon, VA: National Agricultural Statistics Service, 1992; and United States Department of Agriculture. *Nebraska, Biotechnology Varieties, Chemical Usage.* Lincoln, NE: Agricultural Statistics Service, 2004.

season. A comparison of nitrogen and phosphate applications to winter-wheat fields in Kansas shows a continuing and growing reliance on these chemical fertilizers (see Table 4).

A different trend holds true when corn production in Nebraska is tabulated for the same period (see Table 5).

Farmers have slightly reduced their nitrogen applications, and another report reflects this trend as the amount used for a bushel of corn fell from 1.6 pounds per bushel in 1965 to less than 1 pound per bushel by 2004. Yet when phosphate use is analyzed, the amount applied per acre has declined a little while farmers spread it on more acres, which resulted in a greater demand for phosphate in 2003 than in 1991. When insecticides are examined, farmers across the United States applied .13 pounds per acre in 2000, and a report from Tufts University showed that insecticide use on corn crops in the country varied little from the percentage used from 1994 to 2002. In comparison, Nebraska farmers were certainly spraying more than this national average in both 1991 and 2003.

One other major environmental problem confronting farmers has been soil erosion. Despite the work of agronomists in the U.S. Department of Agriculture, and soil scientist and extension work in the land-grant university system, soil erosion has remained a vexing problem. In 1986, Robert Papendick and others reported that measured soil loss by the Soil Conservation Service significantly exceeded its recommended tolerance level. An inch of topsoil normally takes 500 years to accumulate, yet in the Palouse, the wheat lands of western Washington State, some farmers' fields in the years leading up to 1980 lost an inch of topsoil every eighteen months. In three years' time, on an acre where such erosion occurred, the amassing of 1,500 years of soil building, some 300 tons' worth, flew with the winds and flowed in the streams (Papendick et al. 1985). Moreover, soil material, along with its chemical contents, has washed steadily into the river system of the grasslands and has resulted in a growing problem becoming ever more manifest in the Gulf of Mexico. By 2000 there was an area in the Gulf the size of Massachusetts where every oxygen-dependent life-form had died. The Mississippi River had been washing into the Gulf ever larger concentrations of nitrogen and phosphorous, two chemicals that resulted in algae and phytoplankton blooms. This fluorescence has resulted in oxygen deprivation in the deeper waters and kills fishes, crabs, and shrimp. The source of the nitrogen and phosphorous has been from soil erosion sweeping into the Mississippi River. Not only has this occurred in the Gulf, but now water storage reservoirs in the grasslands have encountered similar problems of eutrophication along with attendant massive fish kills.

While much of the emphasis on environmental problems in the grasslands has been focused on land use or wildlife, the human population throughout the region has also encountered health problems associated with ecological changes. Some of these problems have resulted from water pollution, and others have arisen with increased chemical farming techniques. For example, the enchantingly named town of Pretty Prairie, Kansas, has a water purity problem faced by many communities in the farmed areas of the grasslands. The city water supply contains a high level of nitrates that have percolated into it as a result of fertilizer applications on the surrounding fields. High nitrate levels in drinking water can result in "blue baby" syndrome, or methemoglobinemia, a condition that hinders a baby's blood system in delivering oxygen to its body. Only 600 people with fewer than a handful of babies live in the town, and correcting the problem will be costly. The city government has offered to give bottled water to the affected families while the Environmental Protection Agency demands expensive remedies for the entire water system. By 2004 mediation techniques remained undetermined, and nitrate levels still hovered above federal limits.

Another health liability becoming increasingly serious with the advent of chemical farming has been the rise in cancer among farmers in the grasslands. In

Kansas, for example, a study of over 6,800 farmers who contracted cancer between 1980 and 1990 illustrated their higher than average risk in developing the disease. A number of cancers appeared more common in this population than in the population at large. According to the study, farmers were generally at a reduced risk of lung cancer but at a higher one for lip, skin, prostate, brain, non-Hodgkin's lymphoma, multiple myeloma, and leukemia. The data pointed toward environmental factors associated with modern farming techniques, especially the prolific use of chemical pesticides. In others states, reports have illustrated similar trends (Blair et al., 1993; Blair and Zahm, 1993; Frey, 1995).

While chemical applications have come under increasing criticism ever since the publication of Rachel Carson's *Silent Spring,* another fear among some environmentalists has been the rise in planting genetically modified (GM) crops. Genetically modified corn research began in earnest in the 1980s, and the federal government approved its planting in the early 1990s. The first large-scale planting of it began in 1996 but accounted for only 1.4 percent of the nation's crop, but by 2002 GM corn accounted for more than 24 percent of the total. In the grassland states of Kansas, Nebraska, and South Dakota, GM planting of corn has risen dramatically from 2001 to 2004 (see Table 6) (Ervin, et al. 2000).

Genetically modified corn has been a mixed blessing to corn producers in the grasslands. Certainly, better yields have occurred along with some slight reductions in chemical applications. In the last two decades, significant academic support has been given to genetically modified crops. An early example of this is *Agricultural Biotechnology and the Environment* by Sheldon Krimsky and Roger Wrubel of the Department of Urban and Environmental Policy at Tufts University (Krimsky and Wrubel 1996). But other researchers such as Fred Ackerman and Timothy Wise, both professors in the Global Development and Environmental Institute also at Tufts University, warned that genes from GM corn

TABLE 6
2001–2004 Corn Crops: Total Acres vs. Genetically Modified Acres in Kansas, Nebraska, and South Dakota (in 1,000s)

State	2001 acres	2001 GM acres	2002 acres	2002 GM acres	2003 acres	2003 GM acres	2004 acres	2004 GM acres
Kansas	3,450	1,311 (38%)	3,150	1,355 (43%)	2,900	1,363 (47%)	3,250	1,755 (54%)
Nebraska	8,100	2,754 (34%)	8,400	3,864 (46%)	8,000	4,160 (52%)	8,300	4,980 (60%)
South Dakota	3,800	1,786 (47%)	4,100	2,706 (66%)	4,500	3,375 (75%)	4,500	3,555 (79%)

Source: Pew Initiative on Food and Biotechnology, Factsheet. "Genetically Modified Crops in the United States," http://pewagbiotech.org/resources/factsheets/display.php3?FactsheetID=2 (accessed September 2005).

could be transferred to wild plants of similar DNA thereby producing weed plants immune to both herbicides and insects. Or such modified corn plants could result in "allergic reactions and plant toxicity." Americans have also had difficulty in marketing GM corn. Exports to the European Union have fallen from $305 million in 1996 to a mere $2.7 million in 2002 due to the reluctance of Europeans to purchase GM corn from the United States (Ackerman et al. 2003).

TRADITIONAL CONSERVATION IN THE GRASSLANDS

The vast majority of conservation techniques implemented and promoted by the U.S. Department of Agriculture have rested squarely on economic efficiency and the sustainable reuse of resources. The Soil Conservation Service, renamed the Natural Resources Conservation Service (NRCS) in 1994, launched a series of programs and policies to preserve soil and water resources across the grasslands. The agency has conducted extensive mapping of soils in every county and has made these easily accessible. It has subsidized conservation techniques such as farm impoundment ponds for controlling soil erosion, innovations in plowing for soil retention, and the effort to preserve wetlands throughout the grasslands. Critics have readily pointed out that on the one hand, NRCS policies have inordinately benefited large farmers at the expense of the small ones, while on the other hand, the agency has had too little effect in encouraging sensible conservation practices because of overriding market considerations in farming.

One of the programs touted for its innovative approach toward soil conservation has been the Conservation Reserve Program (CRP). The U.S. Department of Agriculture initiated this program in 1986, and it has paid farmers to take fields out of crop production and to plant them with perennial grasses. Certainly many farmers have benefited economically from this, as they have opened their new grass fields to hunters who seek game birds. These fields provide excellent cover for these birds, especially pheasants. While on the surface the fields appear to resemble wild grassland ecosystems, they differ greatly. For example, the below-surface microbiological systems of a cornfield placed into CRP will still resemble that of a cornfield forty years after the grass is planted. Ecologists have also noted that CRP land is considerably less diverse in terms of flowering plants and insect populations, and it does little if anything to protect threatened species such as prairie dogs or sage grouse. Moreover, perennial grass species non-native to North America, such as crested wheatgrass, may be planted. Worse yet, according to several critics, up until 2002, indigenous grasslands could be, and were, plowed and then placed into CRP for the subsidy payments.

The work of land-grant universities has also been highly criticized for failing to lead the way to more ecologically based farming practices in the grasslands. Jim Hightower, the former, twice-elected agricultural commissioner of Texas, launched one of the first salvos at these venerable institutions. In his *Hard Tomatoes, Hard Times* (1978), he castigated land-grant agricultural colleges for their approach toward teaching farming, stating that corporate agriculture dominated both the academic culture and the ways farming itself was practiced. The teaching of economies of scale and efficiency, genetic engineering, and heedless applications of technology and chemical inputs, along with an ever-growing reliance on outside funding from agribusiness, Hightower believed, destroyed both family farm operations and the land itself on which agriculture had to rest. He advocated a curriculum structured more around sustainable practices even if this might reduce economic returns. The end results would be healthy farm operations and land.

Ed Marston, the former editor of *High Country News*, summed up the indictment of land-grant schools in this manner. These institutions have focused exclusively on "elevated efficiency and scale of production" and in so doing have created a class of students who leave the farm to pursue agricultural research and agribusiness—"insurers and accountants and bankers and chemical and machinery salesmen." Moreover, this education more often than not enabled graduates "to move easily away from their rural communities and families into urban lives and jobs." For those who "are concerned with land and resources in a different way—as a land-use planner in a rural county, as a citizen concerned about grazing or logging or as owner of a recreational rafting firm—you are likely to be on your own," ignored by academic researchers and extension agents (Marston 1994, 6). The hide-bound nature of these schools, Marston maintained, seldom encouraged faculty to think new ideas, leaving this inquiring to others outside the land-grant tradition (Marston 1994).

In the last two decades movement toward teaching some forms of alternative or sustainable agriculture has appeared in several land-grant institutions. Washington State University and its college of agriculture serve as examples of how difficult steering a different course can be. In 1989 a new dean of agriculture requested funding from the state legislature for two innovations: one, a lab for examining the safety of pesticides, and the other, a "Center for Sustaining Agriculture and Natural Resources." The dean had farmers and environmentalists surveyed and their concerns noted, and surprisingly to many, the two groups shared nearly identical concerns. The dean mistakenly believed this unified support would translate into legislative funding for his initiative; however, only the lab for gauging the safety of pesticides received any backing. Nonetheless, the university central administration believed in the center and began funding it. The mission of this ongoing center has been "to develop and foster agriculture and natural resource management approaches that are economically viable, environmentally sound, and socially acceptable." The center also has striven to forge ties between the university, "growers, industry, environmental groups, agencies, and the people of Washington." The organizers of this center drew heavily upon the ideas of writers such as Aldo Leopold and Rachel Carson, who both stressed the interconnectedness of biotic and abiotic systems in agriculture, and more largely speaking, in human environments.

ALTERNATIVE AGRICULTURE

Perhaps Aldo Leopold's thinking is one of the best expressions of an alternative approach to traditional agriculture. Leopold had serious concerns about farmers' relationships to the land they farmed, and he began clarifying his thinking on the subject soon after the end of World War II. He challenged farmers to redefine their views of farming, and he criticized the New Deal embrace of mechanistic, large-scale practices. He asked how farmers could easily recognize their social obligations in funding roads, public schools, their churches, and the local sport programs but fail to understand that they needed to form a beneficial and reciprocal relationship with the land itself. Leopold also chastised farmers for implementing only those conservation practices that "yielded an immediate and visible economic gain for themselves." He wanted farmers to consider using practices beneficial to the entire "community" even if it meant a lowered individual economic return, as "a land-use ethic based wholly on economic self-interest" endangered the entire community. In Leopold's thinking, a community included more than just people; it also embraced the biosystems of forests, wetlands, streams, fields, and grasslands. This led him toward espousing an early version of his famous land ethic: "A thing is right only when it tends to preserve the integrity, stability, and beauty of the community, and the community includes the soil, waters, fauna, and flora, as well as people." He pleaded with farmers to change their priorities, and to "cease being intimidated by the argument that a right action is impossible because it does not yield maximum profits, or that a wrong action is to be condoned because it pays." In short, according to Leopold, farmers had to develop an *ecological conscience* (Leopold 1948, 112).

Much of Leopold's thinking would receive little public support until the 1970s with the rise of the environmental movement. Garth Youngberg and Charles Benbrook, two economists who worked in the Department of Agriculture, published important papers in 1979 and 1980, and in these they outlined the nature of organic farming. Organic farming differs in an important respect from "sustainable" farming, which will embrace synthetic fertilizers, pesticides, herbicides, and large-scale mechanized farming techniques under certain conditions. Organic farming, as Youngberg and others have understood it, relies heavily on highly diversified crop rotations and maximizes the use of manure, legumes, organic wastes, and biological pest control. When the report Youngberg had worked on became public it was poorly received both within the USDA and by other experts in the field. In disgust, or perhaps under some degree of pressure, Youngberg quit shortly afterward during the Reagan administration. However, Congress incorporated some of the ideas of Youngberg and his colleagues into the 1985 farm bill, and more were adopted in the 1990 bill, but both bills were hardly an endorsement of organic farming as whole.

Ironically, USDA conservation policy and organic farming can work together to destroy wild grasslands. Harold Miller, who began farming in the Milk River Valley of Montana in 1998, has conscientiously operated a certified organic farm. No chemical fertilizers enrich the soil, and synthetic pesticides and herbicides have not been applied. All of this fits the best practices of organic farming, but to achieve this Miller plowed under wild grasslands rather than recondition farmland already in production. Miller understood that any farmland doused with pesticides, herbicides, or synthetic fertilizers required at least three years before the chemical residues in the soils fell to a level that would allow his crops to be certified organic. Moreover, the USDA also helped in making possible this decision to plow rangeland because of the Conservation Reserve Program, which allowed farmers such as Miller to plant the chemical-laden fields into grass; this replanting creates an ecosystem dissimilar to wild grasslands, and up until 2002, farmers could plow wild grasslands, harvest them for a few years, and then place them into the CRP. At the same time, the USDA subsidized cropping better than it did ranching, so the incentive lay with plowing indigenous prairies rather than having cattle graze them. By 2004, Montana practitioners of organic farming had received fine returns on their production while increasing their acreage by a rate of 20 percent a year, the fastest growing rate of any grassland state. Concurrently, federal policies and economic incentives reduced wild grasslands by nearly 7 million acres nationally. Unquestionably, Miller's farm and others such as his retain soil fertility, produce healthy foods, and suffer little effects from wind or water erosion. On the other hand, traditional economics still drive many of the decisions made by farmers such as

Miller, and many environmental groups, such as the Nature Conservancy, have reacted in a spirited effort to preserve large parcels of what remains of the wild grasslands (Garrett-Davis 2004).

ENVIRONMENTALISM IN THE GRASSLANDS

Urban centers have been the source of the environmental movement in the grasslands, a movement that has taken root slowly in the region when compared to the rest of the country. This is illustrated well in the fight to protect Cheyenne Bottoms in Kansas, Courtney White's efforts to restore grasslands in New Mexico, and Frank and Deborah Popper's proposal to create a buffalo commons in the Great Plains. With a few exceptions, this urban-based movement has often found itself in conflict with agricultural values and aspirations, and this explains the weakness of environmental objectives given the importance of the farm economy throughout the grasslands. Nonetheless, the movement has grown and achieved some tangible results in the last fifty to sixty years such as the United Family Farmers' opposition to the Oahe project in South Dakota. Despite these victories, traditional environmental tactics of litigation and wild land preservation have waned in strength alongside the economic and ecological diminution of farming. While environmentalists and agriculturalists have been at loggerheads, between them a new form of environmentalism has emerged.

An example of a notable victory in the preservation of wild land and animals came with the protection of Cheyenne Bottoms, listed as a Wetland of International Importance. It derives its significance as one of if not the most crucial stop for migrating birds along the Central Flyway of North America. Over 320 bird species have been sighted there, and millions rest and eat there every year and have been doing so for millennia. As reported by the Nature Conservancy, over 40 percent of the wetlands in Kansas were destroyed between 1955 and 1978. Consequently, this sixty-four-square-mile (166 km^2) wetland became ever more necessary for the migratory fowl flying from the southern tip of Chile to the northern reaches of the Arctic.

Beginning in the 1980s, several ecologists and environmental activists, notably from the Audubon Society, observed a declining rate of water flowing into the Bottoms. According to Kansas water law, the Bottoms had prior appropriation water rights attached to it dating as early as 1949. In the 1960s, farmers had greatly expanded their use of center-pivot irrigation on farmlands surrounding the Bottoms, and this had lowered the groundwater levels that served as a source for the wetland. Advocates for the Bottoms started litigation to curtail the amount of water farmers were pumping, and the chief engineer for the state, David Pope, eventually reduced the farmers' rates of withdrawal. Pope had motivations other than environmental sympathy; Kansas water law places all beneficial uses on an equal plane whereas other grassland states have a hierarchy of beneficial uses. Consequently, the rights assigned to the Bottoms had equal standing with irrigation rights, so Pope limited irrigation pumping until he was satisfied that the rights supplying the wetlands, rights predating those for the center pivots, were fully supplied. In comparison, had a similar situation occurred in a state such as Colorado, the outcome for the wetland would have been disastrous.

Kansas irrigators begrudgingly acquiesced with the chief engineer's dictate, but other fights in the grasslands resulted in more virulent disputes. Cattle grazing on federal lands has always been a source of fierce contention. Environmentalists often accuse ranchers of destroying federal rangeland and not paying a fair price for either the grass their cattle eat or for any damage done to the pastures. In one notable case, the Laney family in New Mexico had a permit to graze cattle on 227 acres in the Gila National Forest. In the 1990s environmentalists showed that the family's 800 animals had destroyed riparian woodlands in the wilderness areas of the forest. Environmentalists forced the issue in court where a hotly waged contest resulted in the Forest Service forcing Laney's cattle off the rangeland.

Sometimes, though, ranchers and environmentalists get along together. Take the Quivira Coalition in New Mexico. Courtney White, an archaeologist with the National Park Service and a member of the Sierra Club, embarked on a mission to link environmentalists with ranchers to restore grasslands. White had little use for the wise-use movement of the 1980s and 1990s, but less tolerance for the litigious tactics employed by environmentalists. Influenced in part by Allan Savory's *Holistic Resource Management* (1988), White hoped to implement, as Tony Davis reported in the *High Country News*, a new "environmental ethic, one that restores the land, not by leaving it alone, but through the thoughtful use of human hands." White and the Quivira Coalition have had some limited successes in restoring rangeland through collaborative efforts; however, some economists fear the effort will fall short as long as profit margins for cattle remain about 2 percent. Moreover, the economics of ranching in the late twentieth century make it more attractive to sell grasslands for housing and commercial development than to keep it for grazing. One economist predicts that if post–World War II trends hold, as much as 45 percent of current rangelands will be developed within the next century (Davis 2005).

A BUFFALO COMMONS IN THE MAKING?

One of the most discussed and contested proposals for preserving wild grasslands has come not from any homegrown environmental movement but from two academics in New Jersey. In 1983, ninety years after Frederick Jackson Turner delivered his address on the significance of the American frontier, Frank Popper recognized the reemergence of the frontier in many areas of the grasslands—if a frontier is defined only in terms of six residents or fewer per square mile. Popper pointed to the accelerating depopulation throughout the grasslands, and he argued that native grasslands must be preserved for the future benefit of Americans. Popper saw an area of ecological, economic, and social collapse even though the amount of land in farm production of one sort or another had remained fairly constant in the post–World War II years. Without question, however, environmental problems and depopulation have been mounting in many areas, and Popper thought he had a cure for this. He called for the federal government to subsidize taking land out of production and cattle grazing, returning bison and other former animal inhabitants to the grasslands, and re-seeding many

Frank and Deborah Popper, originators of the "buffalo commons" idea. (Courtesy of Frank and Deborah Popper)

farmlands with wild grasses. In essence, the grasslands would become that immense park envisioned by George Catlin 150 years before.

Needless to say, Popper's ideas, and those of Deborah Popper, his wife and a respected academic in her own right whose views he shares, met with a great amount of hostility throughout the grasslands. Unperturbed, the Poppers took their message to communities, universities, and cities throughout the grasslands, and in time they began to win adherents. For example, Governor Mike Hayden of Kansas, one of their harsh critics during the late 1980s, had become one of their friends by the early 2000s. Governor Hayden still had doubts about turning the grasslands, especially the High Plains, into one massive "Buffalo Commons," but he came to recognize the value and legitimacy of their observations. Others, too, with even more biting critiques of the social, economic, and environmental health of the grasslands, have questioned whether some states should remain in existence. Jon Margolis, an East Coast journalist, asked, "Is North Dakota necessary?" Perhaps it might be better, mused Margolis, to have the depopulated areas of the Great Plains states revert to territorial status. He asked, how long will senators from states such as Illinois rest content sharing power with a senator from a state with a total population less than one-quarter that of Chicago, or more to the point, having Chicagoans massively subsidize the dwindling number of North Dakota farmers if senators from the grasslands fail to support social programs for city dwellers in the pursuit of tax reductions (Margolis 1995)?

FUTURE DIRECTIONS

The crystal ball presents a murky picture predicting the fate of the grasslands. It is not clear what will happen when the water of the Ogallala Aquifer finally is either too expensive to pump or is simply depleted. There are no ready answers as to the continued viability of animal containment operations. While wetland protections are written into law, these prairie jewels continue to disappear. Will unabated numbers of ranchettes so fragment wild grasslands that no hope remains of maintaining this unique biome? Many fear the family farm is a thing of the past, too dead for resuscitation. While bison are making a reappearance in the grasslands, does their presence foretell the re-creation of a grassland similar to the one before the arrival of the plow? And the fate of people in the grasslands—will they be relegated to ever-larger cities while the hinterlands become largely devoid of human presence? Perhaps, one might think, the "mirage" that Walter Prescott Webb described has lifted, revealing the futility of trying to sustain an American lifestyle in the grasslands. Is American culture on a crash course with the ecological and economic realities of living in the grasslands?

Certainly, this is the thinking of Wes Jackson, director of the Land Institute in Salina, Kansas. Jackson was on his way to a well-respected, highly paid academic career when he felt the calling to return to his native state, Kansas, to launch a unique effort to restore the human community along with wild grassland ecosystems. Recognized by *Life* magazine as one who in all likelihood will number among the 100 most important Americans of the twenty-first century—recipient of a MacArthur Genius Fellowship, a Pew Conservation Scholar Award, and the Right Livelihood Award—Jackson's work at the Land Institute has received considerable national and international attention. His labors, and the work of those alongside him, have been devoted to the development of perennial polyculture farming. For some time, American farmers have relied on annual monocultures of such domestic grasses as wheat, oats, rye, and corn supplemented with such crops as soybeans, sorghum, and domesticated sunflowers. Wild prairie grasses and forbs, in contrast, will reseed themselves or send new shoots annually. Many of these plants produce seeds with far better nutritional value than domestic grasses. The problem with wild grasses has always been in gathering enough of their tiny seeds to furnish adequate nourishment given the inordinate amount of time and energy required to make the harvest. Pound per acre, corn and wheat outproduce eastern grama grass (*Tripsacum dactyloides*), one of the wild grasses showing a potential for harvest.

If hybrids of prairie plants could be propagated, so Jackson reasoned, then a variety of mutually beneficial species could be sown together in one field, harvested, and left to proliferate for the following season. In the wake of such practices would come a substantial reduction of machinery, soil erosion and depletion, water pollution, crop-destroying insects, and synthetic fertilizers, pesticides, and herbicides. Farms would tend toward smaller sizes, which would mean a repopulation of the American farmscape and a revitalization of small, rural community life. By mimicking wild grassland ecosystems, such farms would mirror the nature of these systems. This method is the complete opposite of the one that relies on the elimination and replacement of prairie ecosystems with domestic monocrops and a "get big or get out" attitude. This practice, Jackson has maintained, has led to severe ecological problems such as non-point stream pollution, soil nutritional depletion, evolution of pesticide resistant insects, groundwater depletion, and severe economic dislocations associated with the ever-rising capital costs associated with industrial, mechanized agriculture. These problems have resulted in increasingly large operations in a depopulated farmscape dotted by abandoned houses and withering communities in an ecologically debased environment (Sherow 2005).

Wes Jackson, director of the Land Institute near Salina, Kansas. (The Land Institute)

The region of the United States where wild grasslands once flourished is entering a new, uncharted world of existence. The Little Ice that so shaped many of its features between 1400 and around 1870 is now a thing of the past. In the last 250 years the concentration of carbon dioxide in the atmosphere has increased by 30 percent, a level unknown in the last 20,000 years. The vast majority of atmospheric scientists have illustrated that human activity has been responsible for this increase, which in turn, has led to a rapid heating of the entire planet (Nijhuis 2005). Given this, the relationship between climate, people, animals, and plants that once formed the wild grasslands has disappeared, and a new, not necessarily better, system of people, animals, and plants has come to take their places. Remnants of the old wild grasslands still exist in many locales, but every year their area is reduced and their future existence held in question. Are the American adaptations to, and practices for, living in the grasslands, whether wild or domestic, sustainable, and if not, what will take their place? Are the wild grasslands themselves, along with their dwindling populations of prairie dogs and prairie chickens, worth keeping? What happens when pumping the Ogallala Aquifer is no longer feasible? Much is at risk if, as Walt Whitman once asserted, the wild grasslands are "America's characteristic landscape." Whitman earnestly believed that going to the wild grasslands, with their clear streams and abundant wildlife, breathed life and health into the "flourishing and heroic elements" of American democracy. Sometimes the fate of the grasslands is beyond the ability of humans to control as it was at the end of the Pleistocene, or during the altithermal. Humans, sentient beings at that, will shape much of the future of the grasslands if truly they are the keystone species. Choosing paths to the future will become clearer when past treks, and where they led, are better understood. The future of the grasslands hangs in the balance.

References

Ackerman, Frank A., Timothy A. Wise, Kevin P. Gallagher, Luke Ney, and Regina Flores. 2003. *Environmental Impacts of US–Mexico Corn Trade under NAFTA*. Working Paper No. 03–06. Medford, MA: Global Development and Environmental Institute, Tufts University.

Baldridge, George. 1993. "Pottawatomie County Says No to Prairie Preservation." *Kansas History* 16 (Spring): 94–107.

Bedichek, Roy. 1947, rev. ed. 1961. *Adventures with a Texas Naturalist.* Austin: University of Texas Press.

Blair, Aaron, et al. 1993. "Cancer and Other Causes of Death among Male and Female Farmers from Twenty-Three States." *American Journal of Industrial Medicine.* 23:729–42.

Blair, Aaron, and Shelia Hoar Zahm. 1993. "Patterns of Pesticide Use among Farmers: Implications for Epidemiologic Research." *Epidemiology* 4 (No. 1): 55–62.

Carrels, Peter. 1999. *Uphill against Water: The Great Dakota Water War.* Lincoln: University of Nebraska Press.

Catlin, George. 1844. Letter No. 31. *Letters and Notes on the Manners, Customs, and Conditions of North American Indians.* London: D. Bogue, reprinted New York: Dover Publications, 1973.

Cristy, Don to Lloyd C. Hulbert. 1975. File 1, Box 5, Konza Prairie Research Natural Area Collection, 1897–1995. Special Collections, Hale Library. Kansas State University, Manhattan, Kansas.

Davis, Tony. 2005. "Rangeland Revival: The Quivira Coalition Prophesies a New Era of Peace and Prosperity on the West's Rangelands, but Is the Group Bold Enough to Make That Vision Real?" *High Country News* (5 September): 6–9, 12, and 19.

Ervin, David E., Sandra S. Batie, Rick Welsh, Chantal L. Carpenter, Jacqueline I. Fern, Nessa J. Richman, and Mary A. Schulz. 2000. *Transgenic Crops: An Environmental Assessment.* Washington DC: Henry A. Wallace Center for Agricultural and Environmental Policy.

Frazier, Ian. 1989. *Great Plains.* New York: Farrar.

Frey, R. Scott. 1995. *Cancer Morbidity among Kansas Farmers, Report of Progress 724.* Manhattan: Agricultural Experiment Station, Kansas State University.

Gardner, B. 2002. *American Agriculture in the Twentieth Century.* Cambridge, MA: Harvard University Press.

Garrett-Davis, Josh. 2004. "Prairie Conundrum." *High Country News* (2 August): 8–14 and 19.

Heat-Moon, William Least. 1991. *PrairyErth (a deep map).* Boston: Houghton Mifflin Company.

Hightower, Jim. 1978. *Hard Tomatoes, Hard Times: The Original Hightower Report, unexpurgated, of the Agribusiness Accountability Project on the Failure of America's Land Grant College Complex and Selected Additional Views of Problems and Prospects of American Agriculture in the Late Seventies.* Cambridge, MA: Schenkman Publishing Company.

Krimsky, Sheldon, and Roger P. Wrubel. 1996. *Agricultural Biotechnology and the Environment: Science, Policy, and Social Issues.* Urbana: University of Illinois Press.

Lawson, Michael L. 1982. *Dammed Indians: The Pick-Sloan Plan and the Missouri River Sioux, 1944–1980.* Norman: University of Oklahoma Press.

Leopold, Aldo. 1948. "The Ecological Conscience." *Journal of Soil and Water Conservation* (3 July): 109–112.

Malin, James C. 1984. *History and Ecology: Studies of the Grassland.* Edited by Robert P. Swierenga. Lincoln: University of Nebraska Press.

Margolis, Jon. 1995. "The Reopening of the Frontier. *New York Times Magazine* (15 October).

Marston, Ed. 1994. "Land-Grant Universities: Their Roots Loosen as the West Changes beneath Them." *High Country News* (14 November): 6–7.

Nijhuis, Michelle. 2005. "What Happened to Winter?" *High Country News* (18 April): 8–11, 13, and 19.

Opie, John. 1993. *Ogallala: Water for a Dry Land: A Historical Study in the Possibilities for American Sustainable Agriculture.* Lincoln: University of Nebraska Press.

Papendick, Robert I., et al. 1985. "Regional Effects of Soil Erosion on Crop Productivity: The Palouse Areas of the Pacific Northwest." In *Soil Erosion and Crop Productivity,* edited by R. F. Follett and B. A. Stewart. Madison, WI: American Society of Agronomy, Crop Science Society of America, and Soil Science Society of America.

Ritchie, Mark, and Kevin Ristau. 1986. *Political History of U.S. Farm Policy.* St. Paul: Minnesota Agriculture Commission.

Samson, Fred B., and Fritz L. Knopf. 1996. *Prairie Conservation: Preserving North America's Most Endangered Ecosystem.* Washington, DC: Island Press.

Savory, Allan. 1988. *Holistic Resource Management.* Washington, DC: Island Press.

Sherow, James E. 1989. "The Chimerical Vision: Michael Creed Hinderlider and Progressive Engineering in Colorado." *Essays and Monographs in Colorado History* (Essays Number 9): 37–59.

Sherow, James E. 2005. "Wes Jackson: Kansas Ecostar." In *Famous Kansans,* edited by Virgil Dean. Lawrence: University Press of Kansas.

Webb, Walter Prescott. 1957. "The American West, Perpetual Mirage." *Harper's Magazine* 214 (May): 25–31.

Wessel, Thomas R. 1998. "Agricultural Policy since 1945." In *The Rural West since World War II,* edited by R. Douglas Hurt. Lawrence: University Press of Kansas, 76–98.

White, Stephen, and David Kromm, eds. 1992. *Groundwater Exploitation in the High Plains.* Lawrence: University Press of Kansas.

Whitman, Walt. 1963. *Prose Work 1892,* Vol. I, *Specimen Days.* Edited by Floyd Stovall. New York: New York University Press.

Wood, Judith Hebbring. 2000. "The Origin of Public Bison Herds in the United States." *Wicazo Sa Review* 15 (Spring): 157–182.

5

CASE STUDIES

CASE STUDY: STATE MAKING AND UNMAKING (AND MAKING IT BACK AT ALL): FOLLOWING ZEBULON PIKE ACROSS THE PLAINS IN 1806

Jared Orsi

Somewhere between the Platte and the Arkansas, in a chilling October rain, Zebulon Pike searched for his party. Two days earlier, the U.S. Army expedition he commanded had lost its way, and Pike had set off with a partner to search for the Spanish trail that he hoped would help the Americans regain their bearings. Fending off wolves and bracing themselves against the worsening autumn weather, the two men could locate neither the trail nor the rest of their party. "Excruciating," Pike wrote in his diary on October 17, 1806, as he contemplated his dwindling ammunition and the four hundred miles that lay between him and the "first civilized inhabitant" (Pike 1966). Although only two weeks before he had blustered his way through a tense encounter with the Pawnees on the Republican River by boasting of the vast reach of American power, now, lost on the plains, Pike became sensible to his remoteness from the nation's center.

Pike's anxiety stemmed more from his own ecological status in the West than from the intrinsic remoteness of the Great Plains. Indeed, the plains were not all that remote. Since leaving St. Louis the previous July, everywhere he had gone, it seemed, he encountered evidence of the Atlantic world economy. Among the Osages, near the modern Kansas-Missouri state line, he encountered

Note: The author wishes to thank Mark Fiege and James Sherow for reading early drafts of this essay and for their many helpful insights and suggestions along the way. The research for the essay was generously funded by grants from the Charles Redd Center for Western Studies, the College of Liberal Arts and Department of History at Colorado State University, the Colorado Endowment for the Humanities, and the Santa Fe Trails Association.

Zebulon M. Pike was an early explorer of the Old Southwest for whom Pike's Peak, Colorado, was named. (Library of Congress)

traders and abandoned trading posts of the mercantile houses of Manuel Lisa and Auguste Chouteau, traders whose operations reached from St. Louis and New Orleans to the Great Lakes to the Atlantic seaboard to the European continent and back again. A few weeks later, Pike met his first Pawnee, who wore a scarlet coat and medals of George Washington and the Spanish king. He also found Spanish mules, horses, bridles, and blankets among the Pawnees, and the Spaniards themselves had trekked from Santa Fe, arriving at the Pawnees' Republican River village a few weeks before Pike himself. Lieutenant Facundo Melgares had brought manufactured goods and other gifts to seal an alliance between the Pawnees and His Catholic Majesty.

These incursions of people and their artifacts onto the plains worried Pike's superiors. There was much wealth to be tapped in the grasslands, and the St. Louis traders and Spanish ambassadors seemed to be getting the upper hand. As both groups sought to direct its flow through their hands, they not only threatened to divert capital out of American control, but they also presented a slap in the face to the young American state, which claimed sovereignty to license trade and treat with Indians on the plains and in the rest of the recently acquired Louisiana Territory. Pike, therefore, was sent west to arrest unlicensed traders, sway Indians into allegiance to the American state, and bring back ecological and geographical knowledge to help the American state catalog and administer its recent real estate acquisition (Scott 1998). Far from being an errand into the wilderness, then, Pike's mission derived from just how crowded the West was already getting.

Ecologically speaking, however, Pike's problems were all too well grounded. He and his men were essentially invasive organisms in a grasslands ecosystem to which they were not well adapted. For many other species, including some human beings, there was plenty of food, fuel, water, and shelter on the plains. Securing the basics for keeping their bodies alive, however, was frequently beyond the ability of Pike and his comrades. At home, the institutions of their society provided the physical and social infrastructure that usually delivered a reliable supply of what they needed to keep their bodies warm, fed, hydrated, and protected from physical harm; but on the plains, Pike's command had to learn how to acquire its daily needs on its own. And that was hard. The single institutional lifeline supporting them in this regard was the party itself. Being separated from his men, then, made it even harder, if not nearly impossible, for Pike to sustain himself. Thus, Pike and his men frequently found themselves relying on Indians and Spaniards, peoples for whom meeting organic needs was less of a challenge. In doing so, Pike was reduced merely to staying alive day to day, a far cry from the grander purpose of his expedition, which was to determine the means whereby the grasslands could be incorporated into the service of the young and expanding nation-state. Pike's anxiety about isolation notwithstanding, the plains were filling up, and the ambitious American nation-state aimed to supervise the crowding. As the rain pelted Pike that bleak October day, it was this larger concern and his "fear of the failure of the national objects," as much as his own personal safety, that made him think of the distance that separated him from the world that reliably met his organic needs.

A Young Nation and a Young Soldier Facing West

Pike found himself lost on the plains in October 1806 because he was an archetype citizen of the new republic. (Pike's journey is illustrated in the map on p. 163.) Born two-and-a-half years after the signing of the Declaration of Independence, Pike was of a generation that took seriously the republican ideology that linked individual freedom and civic virtue. Citizens of the new republic aspired simultaneously both to selfless pursuit of the common good and Lockean pursuit of individual life, liberty, and (especially) property. And it was these founding principles of his nation that brought him to the plains. Although he does not seem to have been consumed by the pursuit of wealth, he did, throughout his life, exhibit a keen sense of the sanctity of private property, a point he repeatedly sought to impress on Indians, whom he disparaged as thieves. After one clash with Pawnees who stole a belonging of one of his men, Pike, with great ceremony, demanded the return of the item, proclaiming that "it was not the value of the article but the act [of theft] we despise." Never, he wrote, would he suffer "a thing of ever so little value to be taken without liberty" (Pike 1966, 332–333). Wealth may not have animated him, but achievement did.

The teetotaling Pike, who sought out and embraced the extreme physical deprivation occasioned by army exploration, never seems to have doubted the nation's promise that through self-discipline and hard work, any citizen, even one of modest birth such as himself, could rise in station. He was obsessed with self-betterment. He read constantly, whether teaching himself foreign languages and mathematics or scrawling moral and ethical notes in the margins of books like Dodsley's *Economy of Human Life* (Niles 1820). This ambition, however, was always dressed in the garb of republican virtue and the rhetoric of a citizen's service to country. "Preserve your honour free from blemish," he wrote to his wife Clara, with the instructions that the words be passed on to his son in the event of the father's death, and "Be always ready to die for your country."

As Pike entered adulthood, his idealized republic was in trouble. It was often difficult for a nation founded in rebellion and lofty promises of individual freedom to control its citizens. Almost as soon as the Revolution was over, disgruntled citizens with an amazing array of major and minor grievances began

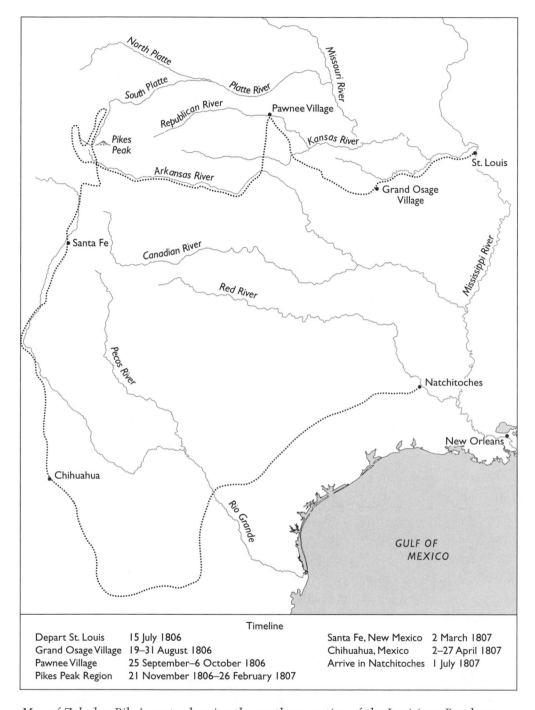

Timeline

Depart St. Louis	15 July 1806	Santa Fe, New Mexico	2 March 1807
Grand Osage Village	19–31 August 1806	Chihuahua, Mexico	2–27 April 1807
Pawnee Village	25 September–6 October 1806	Arrive in Natchitoches	1 July 1807
Pikes Peak Region	21 November 1806–26 February 1807		

Map of Zebulon Pike's route showing the southern section of the Louisiana Purchase.

falling prey to one rebellious or secessionist scheme after another. In 1786, high taxes and the post-Revolutionary recession incited 2,000 farmers to march on the federal armory in Springfield, Massachusetts, in the infamous Shays's rebellion, an event made more terrifying by rumors that the Spanish had offered to open the port of New Orleans if western settlers would secede from the union. A few years later, western Pennsylvania farmers began attacking federal officers in protest of a tax on whiskey that Congress had imposed in 1791. Here, again, far away New Orleans played a role, as farmers near Pittsburgh who could no longer ship their crops to market down the Ohio and Mississippi to that Spanish city had come to depend on selling their corn as alcohol, which moved overland much more cheaply. In 1794, the rising violence and threat of secession prompted President Washington himself to lead a military party west to crush the so-called Whiskey Rebellion.

Although these and other schemes that followed in the 1790s never bore fruit, a pattern was emerging of aggrieved Westerners invoking the Revolution's rhetoric and tactics to protest their own weak position in the union and the nation's similarly weak position in the Atlantic world. It was exactly these kinds of people to whom Aaron Burr appealed between 1805 and 1807 as he launched the most elaborate and ambitious of rebellions, a scheme to capture the northern provinces of New Spain and possibly sever western American states from the union. Even the self-consciously virtuous Pike would become entangled, perhaps unwittingly, in Burr's plot. Although hard to believe looking back from the twenty-first century after a hundred years of world power status for the United States, the merging of thirteen colonies into a nation was an outrageous experiment in the eighteenth century, one that during Pike's life guaranteed no success.

Citizens were difficult enough to control, but national expansion incorporated an ever-increasing population of Spanish, British, French, Indian, and mixed-race peoples, whose loyalties to the aspiring republic were tenuous at best. Pike got a taste of this at the outset of his expedition when among the Osage Indians he arrested three unlicensed French traders who had paddled their way from St. Louis in armed canoes on a mission to collect debts for Manuel Lisa and brought gifts to court an Osage chief. The seemingly minor affair was symptomatic of a bigger problem the United States encountered in the West. Pike complained in a letter to his commander James Wilkinson that traders in the employ of Lisa and Chouteau were discouraging the Osages from cooperating fully with Pike. He grumbled that he could secure but a few of the horses that he wished to purchase from the Osages, whereas rumor had it that Chouteau or Lisa could have acquired many.

Traders like the Chouteau family had operated in recent decades under French, Spanish, French (again), and American rule. British power exerted from Canada and the Illinois country had never been far away. And Indians on the prairies and eastern plains like the Osages, Kaws, and Pawnees still considered

themselves "and were in fact" independent peoples. Moreover, the Spanish officials, as Louisiana governor William C. C. Claiborne repeatedly complained to
his superiors in Washington, were refusing to leave New Orleans and even going so far as to open land offices, purportedly to sell real estate in August 1805,
two years after the Louisiana Purchase (Rowland 1917). How was a nation that
had difficulty enough protecting Massachusetts arsenals from its own citizens
going to enforce the sovereignty it claimed over Louisiana and its diverse and
sometimes uncooperative populace that was so accustomed to adapting to frequent changes in officialdom without fully submitting to the new authority?

The ecosystems of the grasslands held part of the answer. The capitalist
economy of the early republic, and much of the Atlantic world at the time, depended on the conversion of stored solar energy in minerals and organisms into
capital. From this perspective, the sea of grass and the bodies of animals on the
plains constituted a vast reserve of lightly tapped energy (West 1998). A beaver
pelt or bison hide was but a store of energy that the organism had converted
into a body part and that its captor could convert into capital by bringing it to
market. This capital could then buy goods or services to do work, by which the
captor acquired the energy of other people or organisms (Sherow 2001). What if
the nation could capture the enormous energy reserve of the grasslands and divert it into the Atlantic world markets?

Brokering such an energy transfer from one ecosystem to another held two
significant promises for the young nation-state. First, it promised to direct capital into the hands of American citizens instead Indians or the Spanish royal
treasury. Second, setting up the rules and regulations that enabled people to
profit would cement their loyalties to the state. If surviving and profiting in the
West increasingly required some degree of attachment to and cooperation with
the state and its laws, licenses, and property rules that rationalized the markets,
plots like Aaron Burr's would never be able to mobilize enough support to succeed. The troublesome Westerners might be corralled if the very rules that enabled them to make money also restricted their misbehavior. Thus, becoming
the master of energy transfer would be an important step in the consolidation of
state power and defeat of rival powers, peoples, and factions. In all this, exploration was the first step.

Pike's expedition offered him the perfect opportunity to pursue both the
personal improvement and virtuous service to country that he so desired. The
1803 Louisiana Purchase added a vast territory to the United States, but just
how vast was unclear. Its reaches needed to be explored. In 1805, a twenty-six-
year-old Pike gladly accepted from the U.S. Army commander General James
Wilkinson an appointment to explore to the headwaters of the Mississippi
River. He returned the following spring only to set off again a little more than

two months later for a second journey, one that would take him across the Great Plains and beyond. "Should my country call for the sacrifice of that life which has been devoted to her service from early youth, most willingly shall she receive it," he wrote to Clara, his wife, from Kaskaskia in the Illinois country. "The choisest [sic] tears which are ever shed, are those which bedew the unburied head of the soldier" (Niles 1820, 319). In undertaking the dangerous but glorious task of exploration, a task which if successfully completed promised to advance both Pike's career and the nation's power and glory, Pike was truly in his element.

His instructions from Wilkinson reflected the state's desire to control the crowding and to broker the energy transfer, and then refracted these impulses through a mysterious blend of public and private ambition. Ostensibly, he had three general orders: find Indians, induce their loyalty to the American state, find the headwaters of the rivers. With the boundaries of Louisiana in dispute and the United States and Spain on the brink of war, Wilkinson had justified the expedition to the secretary of war as a mission to find military routes to Santa Fe. The rivers of the plains were the way to go he said, but travel along the river required the friendship of the Comanche and other plains peoples. Moreover, the Spanish were courting these same people with gifts, and the United States needed to step up its meager efforts in this regard.

Now clearly Wilkinson was nothing if not a schemer, and he had many additional motives for conceiving this expedition to the grasslands. Among other things, he cast an eye along the Platte and the Arkansas rivers as avenues for directing some of the stored energy and latent capital of the Great Plains and Southwest into his own hands. Shut out of the Osage and Missouri trade by the Chouteaus and Lisa, Wilkinson saw the rivers of the central plains as an open field by which he might personally profit (Cox 1930–1931). He was also in league with Burr, and Pike's expedition was likely intended, at the very least, to gather geographical and political information for their maneuvers and possibly to provoke an incident that would bring Spain and the United States to war and provide cover for the filibustering plot. It is not clear how much of this many-layered scheme Pike understood, but regardless, he was embarking on a project that aimed simultaneously to consolidate and undermine the American state. The republican civic virtue and the individual pursuit of property merged and blurred. It was a quintessentially American undertaking. On July 15, he was under way, leaving St. Louis for the Great Plains. It would not be long before he was forced to set aside a large part of both of the state making and unmaking projects in favor of simply making it at all in the grasslands ecosystem.

Making It in the Grasslands

By October Pike was wet and lost, and the "national objects" were in jeopardy. Brokering the flows of energy on the Plains "either for Wilkinson's benefit or the nation's" was simply beyond Pike's means. In the end, he gathered little scientific data; imposed no deep, lasting friendships with Plains Indians; and failed to decipher the geography of the rivers. Instead, he became ecologically dependent on the very people and organisms he was supposed to be incorporating into the nation-state, and he came up short at almost every opportunity to extend the rules and regulations of the United States into the West. Pike did not fail for lack of will or talent, but rather his efforts foundered on the grim realities of surviving as an outsider in the grasslands ecosystem; and so small a party in so short a time could not adapt to, much less capture and control, an ecosystem as vast and complex as the Great Plains.

Traveling west from St. Louis and the last frontier outposts across the modern state of Missouri, Pike's party followed the Osage River. His first task was to escort some Osage Indians to their villages near the modern Kansas-Missouri state line. The Osage villages constituted something of a human and geographical transition. Their homelands lay in the region where the fertile, wet midwestern prairies and arid Great Plains converged. Consequently, like the Pawnees, Omahas, Kaws, and Wichitas, they had crafted a mixed economy that combined horticulture on their well-watered village lands and seasonal hunting expeditions onto the grasslands to the west. For more than a century before Pike arrived, and especially in the second half of the eighteenth century, they had added trade to this repertoire. Since the 1760s, they had traded horses, furs, and hides, all of which they obtained from the plains, to St. Louis merchants, who brought manufactured goods from Europe and North America via New Orleans.

Pike's treatment of the unlicensed French traders he encountered among the Osages was an attempt to impose American intervention on this far-flung trade that connected plains energy and Atlantic world markets. This first intervention, however, proved only marginally successful. The St. Louis trade had become part of the Osages' ecology, and they depended on it to supplement their merger of the two ecological worlds they straddled. They were consequently less than eager to help Pike discipline the interlopers. After Pike interrogated the men, the behavior of the Osages seemed to change. The Osage chief would not provide much information about who the traders were, and the resident Osage translator would not translate for him. To Pike, it seemed that the Osages were under the influence of the St. Louis traders and were resistant both to his calls for ties to the United States and to his requests for help in meeting his more immediate needs for continuing on his journey through the grasslands. Without Osage support and unable to spare the horses or men necessary to escort the unlicensed traders back to St. Louis, Pike was able only to intimidate them and try to impress upon them the extent of American power. After detaining their leader a "sufficient time to alarm him," Pike then settled for ordering them to return to St. Louis on their own accord and dashed off a letter to American authorities there asking them to prosecute the men (Pike 1966). Whether they actually were prosecuted and what impression of American power any of this left upon them is unknown.

Pike found himself in the untenable position of trying to extend American power without the resources to do so largely because the Osage village was something of an ecological transition point for him as well. Turning northwest from the Osage villages and heading to meet the Pawnees on the Republican River, Pike moved from waterborne travel to overland travel. Wilkinson had been right to see the rivers of the Great Plains as strategic. Traveling along them, one could expect a reasonable supply of game, water, and fuel. At some times of the year, along some of them, the rivers' buoyant waters even supported the loads of boats, reducing the human and animal energy necessary to move through space.

Away from the rivers, however, the necessities of life were harder to come by. Travel was difficult and often dangerous. Since the rivers of the grasslands head on the high plains or in the Rocky Mountains and flow toward the Mississippi, traveling east-west on the plains was always easier than north-south. As his course veered north after leaving the Osages, Pike needed more energy than he had needed so far, as water had borne the weight of some of his load. So not only could he not spare the human and animal energy to escort the miscreant traders to St. Louis, he was at the same time struggling to bargain with the only marginally cooperative Osages for additional men and horses to accompany him to the Pawnees. Thus, at his first opportunity to extend American sovereignty over the plains trade, immediate ecological circumstance forced him to sacrifice his national objects. It would not be the last time.

Not surprisingly, travel grew somewhat difficult in the following weeks. Over Pike's protests, the expedition's Osage guides had led the party on a lengthy, indirect route to avoid the hunting grounds of enemy Kaws. Some of the Osages had deserted the party entirely. In late September, in the vicinity of the Solomon River, the men in the expedition labored overland on blistered feet in search of the village of the Pawnees. A week before, Pike had dispatched a scouting party to venture ahead, but since then, their progress had been slow. They were delayed two days when a rainstorm had drenched their baggage, and they had halted early another day for fear of not finding water ahead. Some of the men suffered severe headaches, and everyone was thirsty. Game was unpredictable. Sometimes the men feasted; sometimes they went hungry. As Pike's second-in-command, Lieutenant James B. Wilkinson, remembered it, the march was painful, the sun oppressive, and the terrain irregular and broken. Three days had passed since Pike had expected to meet the Pawnees. The delay was becoming a "serious matter of consideration" (Pike 1966). Lost among the hills and river valleys of north-central Kansas, where vegetation was sparse and sand was plentiful, it must have seemed to Pike and his comrades as if they were nearing the edge of the world.

When they finally found their way into the Pawnee village, a now rested and well-fed Pike began to inform his generous hosts of American power. Discovering that a few hundred Spaniards had visited the village only a few weeks before and had left a Spanish flag flying outside the lodge of the Pawnee chief, Pike demanded that the standard be lowered and replaced with the American flag he offered to them. The Pawnees, he said, could not have two fathers; they must choose where to direct their loyalties. He was met with an awkward silence. Like the Osages, the Pawnees had carved out their niche by mixing farming, hunting, and trading in the ecological borderlands that straddled the short-grass and tallgrass biomes. The Pawnees, too, depended on commerce with Spaniards, French, and British, and they were not about to circumscribe their resource base by submitting to the demands for exclusive allegiance with the United States as a result of demands made by a mere twenty-six-year-old American and his puny party. Initially the chiefs complied with the flag demand, but they did so with such obvious dissatisfaction that Pike relented and suggested a compromise. The Pawnees could keep the Spanish flag but refrain from flying it while the Americans were among them. As he had with the unlicensed traders among the Osages, Pike again retreated from initial boldness, and he did so here in yet another ecologically dependent context, anticipating his planned southward trek to the Arkansas River and his need for fresh horses and guides. He could not afford to anger the people on whom he relied for his very ability to continue his mission. Once again, Pike's imperial pretensions were beginning to exceed the reach of American power.

And traveling southwest from the Pawnee village to embark on his next instruction—finding the headwaters of the Arkansas—Pike passed beyond the reach of American power completely. As he moved through the central grasslands, Pike undertook the unlikely move of following the trail of his enemies—the very Spaniards who had crossed the plains from Santa Fe a few weeks before with orders to arrest him. Although like Pike's small party, the Spanish were also interlopers in the grasslands, the ecology of their travel was quite different.

If Pike struggled to live off the land, the Spaniards had brought their ecosystem along with them. As he followed their several-week-old path across the plains, Pike had his men count the remains of the Spanish campfires. By such crude but ingenious ecological sleuthing, Pike estimated that the party of Spaniards numbered 354 men, and he calculated from Pawnee accounts that they herded 2,000 head of livestock with them. Bringing their own food and energy supply on their journey not only allowed the Spanish a more secure ecological status on the plains, but the enormous herds of men and beasts left an unmistakable scar of campfires, beaten-down grass, and livestock manure on the grasslands, which, even weeks later, was frequently visible to Pike.

Sensibly enough, he decided to follow them, calculating "that they had good guides, and were on the best route for wood and water" (Pike 1966, 334). Ironically, Pike was now dependent on the very people who were after him, for the Spanish commander Facundo Melgares had orders similar to Pike's—to find and arrest undocumented strangers who might pose a threat to Spanish ability to control the politics and trade of the plains. Pike may have stepped beyond the furthest reach of American power, but he had not entered an empty or unconnected land. In the harsh grasslands ecosystem, which was so stingy in providing necessities for visitors, everyone needed fuel, game, and water. Since these could be found easily in only certain spots, there were actually few places that human strangers were likely to go. That the American and the Spanish paths would intertwine in the vast geography of the plains was not quite so accidental or unlikely as it might seem. Because of his ecological dependence on following his Spanish pursuers, losing their trail (for example, as when a herd of bison crossed the path, trampling evidence of the Spanish passage through the area and erasing any trace of the direction they had taken) seemed to an alarmed Pike a "mortifying stroke" that jeopardized the well-being of his party. It was just this sort of mortifying stroke, and the need to regain his important lifeline, that caused him to split off from the party, his other lifeline. By October 17, he appeared to have lost both.

That was the low point of his time in the grasslands. Unbeknown to Pike, only a few miles separated him from the main party, but not until the morning of the eighteenth did they reunite. They spent a couple of days looking for the Spanish trail but could not locate it. Little matter, however, for they now had found something better, the Arkansas River. Pike recounted the days along the Arkansas as a time of plenty. The men found game, and Pike found time to record one of the few scientific observations in his journal, a lengthy description of the prairie dog. When they had arrived, only a trickle flowed in the riverbed; in fact, Lieutenant James B. Wilkinson, who had commanded the party in Pike's absence, had camped right in the bed the first night, not realizing in the darkness that they had reached the river. But an autumn rain augmented the meager late-season flow, and the crew hastily fashioned two canoes, one from a small cotton-wood and another from the hides of bison and elk they killed, and prepared to depart in time to take advantage of the energy the rising water would provide.

Lieutenant Wilkinson, the son of the general, would lead five men in the boats down the Arkansas and back to the United States, carrying maps and other documents. He complained to his father by letter that he was "perhaps more illy equipd [sic] than any other Officer, who ever was on Command, in point of Stores, Ammunition, Boats & Men." And indeed, his detachment suffered numerous hardships on their descent of the river. Pike's upstream-bound portion of the party, however, found the Arkansas a fairly reliable lifeline that provided a steady supply of water, game, and feed for the horses for several more days. The banks of the river were "covered with animals." On November 4, he estimated a herd of bison to number 3,000 animals, and on the sixth, the party feasted sumptuously on the meat of a bison cow that Pike declared was "the equal to any meat I ever saw" (Pike 1966, 343).

As November wore on, however, and the party ventured further west, "the face of the country considerably changed." It grew more rugged, and fodder became less abundant. Several horses gave out and had to be abandoned, but Pike, undaunted by more than three months of ecological struggle, kept his mind on his promise to serve the nation's goals and determined to "spare no pains to accomplish every object." As the horses continued falling, the loads of the fallen animals were redistributed to the others, as were 900 pounds of bison meat that the party slaughtered, fearing that the herds might soon be thinning. Meanwhile, there was another alarming development: the sign of Indian war parties grew more frequent, and Pike's men advanced "with rather more caution than usual" (Pike 1966, 347).

"Voila un Savage," the cry rang out on the twenty-second. Pike's mixed-blood interpreter had sounded the alarm. Suddenly, Pawnees were everywhere. They ran toward the party from several directions and quickly surrounded it. Pike dismounted to greet them, and in a flash an Indian was astride his horse, riding it away. As the chaos settled and the Pawnees returned the American horses, Pike learned they were a war party returning home to their village after an unsuccessful search for Comanches. Here Pike encountered another stage in the Indian annual ecological cycle. On the way home to harvest their crops after the summer hunt on the plains, bands of warriors would detach from the homeward-bound group to raid on the plains and even as far away as New Mexico. By this mechanism, they acquired horses, slaves, and Spanish manufactured goods for use in the villages on the eastern edge of the plains or for trade to Europeans, Americans, and Indians from the Mississippi Valley.

There were sixty warriors in this particular party to Pike's sixteen men, and lacking other options, Pike presented the warriors with gifts of knives and other goods. But they demanded ammunition—one of the stores of energy Pike was able to bring from his own ecosystem for use in this new one—which Pike was unwilling to spare. As the Americans packed their horses and prepared to depart, pandemonium again erupted. The Pawnees isolated the Americans, surrounding each man with several braves and taking every item within their reach. Pike took out his pistol and threatened to kill the next man who touched the Americans' cargo. At this, the Pawnees began filing away, but not without making off with a sword, a tomahawk, a broad axe, five canteens, and numerous other small articles from the strange skirmish.

That evening, Pike reflected on the day's events with indignation. He was "mortified, that the smallness of my number obliged me thus to submit to the insults of a lawless banditti, it being the first time ever a savage took any thing from me with the least appearance of force" (Pike 1966, 349). For a third time on this trip across the plains, grasslands ecology had trumped American power. Here, Indians, for whom raiding was a regular part of their ecological base, flaunted the American law that Pike presumed to be imposing on Western ecology. To Pike, this was lawlessness, but it looked different from the Pawnees' perspective. Their attempt to acquire resources by raiding the Comanches had failed, and in this context Pike's party presented an opportunity for the dispirited party to compensate for this breakdown of their regular ecological strategies. For Pike the encounter was theft of the most sacred thing in the post-Revolutionary world that had produced him: property. And that truly rankled his Lockean sensibilities. The trouble was, Locke held no sway in the grasslands ecosystem—yet.

A few days earlier, Pike had famously spotted the mountain that would come to bear his name. Rising above the plains to an elevation of 14,110 feet, "Pikes Peak" signaled that he would soon be leaving the grasslands. To his dismay, he fared no better in the Rockies than he had in the grasslands, and he wandered lost among the peaks for several weeks before the Spanish army finally caught up with him, arrested him, and took him first to Santa Fe and then to the city of Chihuahua to answer to the Spanish commandant general, Nemesio Salcedo. From there he received a Spanish military escort across Texas and back to the United States. Both Salcedo's and Pike's political enemies back home insisted he was a spy for the Burr conspiracy, something Pike disavowed to his death in the War of 1812.

Pike's most famous legacy, however, does not stem from the unanswerable question of his involvement with Burr, but rather from the official report he issued on his return, in which he compared the Great Plains to the "sandy deserts of Africa" (Pike 1966). His army explorer successor Stephen Long would even more famously tar the region with the label "Great American Desert" after his 1820 expedition to the area. And so its reputation was established early on. Both Pike and Long suffered there, and it is perhaps not surprising that they would cast it as a barrens. This appellation has been credited (or blamed) for delaying American settlement of the region for decades. Not only, then, did Pike not bring the grasslands ecosystem under the control of American institutions, apparently he also helped dissuade others from attempting that project for years to come.

A Desert No More

Susan Shelby Magoffin, however, did not find it to be a desert. In the summer of 1846 she crossed the plains from St. Louis to Santa Fe, and covered much of the territory that Pike had crossed four decades earlier. She enjoyed a much more comfortable interaction with the region's ecology, and thus her adventure offers a marker of the progress of the national objects that Pike brought to the grasslands. The young bride of eight months, who celebrated her nineteenth birthday on the plains, rode in a carriage with her maidservant as part of a large trading caravan of 14 wagons, 20 men, dozens of mules, and some 200 oxen hauling goods to the northern part of the now independent nation of Mexico. She slept on a bed with sheets in a tent with a dressing bureau, combs, and pillows. Among her chief troubles at the beginning of the trip were howling wolves that kept her awake at night and rain that seeped through the tent and wet her Philadelphia-manufactured carpet.

She did not get lost. She was only occasionally wet, cold, or hungry "and other times she enjoyed wine and dessert with her meals." Neither Indians nor anyone else contested her Lockean rights to her property by trying to appropriate it into their own ecosystem. So great was her comfort that she would periodically get bored and leave her carriage to rough it a bit, walking alongside the caravan picking flowers until she tired. To be sure, her trip after sighting Pikes Peak was not easy. She would suffer a miscarriage, illness, and depression before her travels were over, but throughout the journey her ecological security was far greater than Pike's ever was, largely because she benefited from a system of connections that brought carpets from Philadelphia to the grasslands and brought furs and Mexican silver out of the West (Drumm 1982).

Much had happened during the years when the Great Plains evolved from Pike's North American Sahara into the site of what one historian has called Susan Magoffin's "extended honeymoon safari." The United States had indeed become the broker of the transfer of energy between the grasslands and the Atlantic World, and its institutions and laws had provided the commercial environment to induce many people to cooperate with the young nation-state—even Mexicans. One worry that dogged Magoffin's party was the threat of war breaking out between the two nation-states. While Susan was dallying on the plains, her merchant brother-in-law James, who had married a Mexican woman, was brokering a compromise by which a few people in Santa Fe who favored American rule would transfer the city to U.S. control.

For too long, in their opinion, Santa Fe had suffered as an outpost of unstable and frequently corrupt Mexican regimes that limited the frontiers people's market access and supplied them with few and expensive trade goods. If a relationship with the United States could convert their resources and goods into capital more profitably, why not join that ecopolitical system? So much illegal trade was already crossing the plains that joining the United States would in some ways merely give sanction to what was already economic and ecological fact: Santa Fe depended on shipping its natural wealth to market via St. Louis. Many in Santa Fe disagreed with this logic, but those who did not were numerous and powerful enough to hand the city over to the Magoffins' nation in the summer of 1846, initially without a fight. The United States was simply better at protecting and expanding property, and turning Western ecology into profits. Pike did not live to see it, but Locke had made it to the grasslands and beyond after all.

References

Cox, Isaac Joslin. 1930–1931. "Opening the Santa Fe Trail." *Missouri Historical Review* 25 (October–July): 30–60.

Drumm, Stella M., ed. 1982. *Down the Santa Fe Trail and into Mexico: The Diary of Susan Shelby Magoffin, 1846–47.* Lincoln: University of Nebraska Press.

Niles, John M. 1820. *The Life of Oliver Hazard Perry.* Hartford, CT: William S. Marsh.

Pike, Zebulon. 1966. *The Journals of Zebulon Montgomery Pike with Letters and Related Documents*, edited by Donald Jackson. Norman: University of Oklahoma Press.

Rowland, Dunbar, ed. 1917. *Official Letter Book of W.C.C. Claiborne, 1801–1816*, Vols. 1–6. Jackson, MS: State Department of Archives and History.

Scott, James C. 1998. *Seeing Like a State: How Certain Schemes to Improve the Human Condition Have Failed.* New Haven: Yale University Press.

Sherow, James E. 2001. "Water, Sun, and Cattle: The Chisholm Trail as an Ephermeral Ecosystem." In *Fluid Arguments: Five Centuries of Western Water Conflict*, edited by Char Miller. Tucson: University of Arizona Press.

West, Elliott. 1998. *Contested Plains: Indians, Goldseekers, and the Rush to Colorado.* Lawrence: University Press of Kansas.

CASE STUDY: WETLANDS OF THE AMERICAN GRASSLANDS

Margaret Aline Bickers

Mention "Great Plains" or "North American Grasslands" and the mental image that first springs to mind is probably one of long sweeps of flat or gently rolling land covered solely in grass. Those more familiar with the area may add thin stripes of cottonwood and willow along streams and rivers, but grassy uplands still predominate the mental picture. "Great Basin" conjures pictures of barren, isolated mountain ranges overlooking stretches of sagebrush desert, or perhaps the Great Salt Lake. Ask someone who has flown over the Dakotas and the Canadian prairie provinces in a wet year, or the High Plains after May and June's storms have drenched the area, or the western Great Basin during the late winter and spring of 2005, and the observer will add a significant feature to the picture—playa lakes and prairie potholes, small wetlands shining like scattered diamonds in the slanted light of early morning or late evening. A bird watcher may include Cheyenne Bottoms, the depression in central Kansas that attracts waterfowl and birders alike during the spring and fall. A biologist or geologist may mention the sandhills wetlands of Nebraska, or the riparian marshes and thermal springs found in the Great Basin. In short, the grasslands contain much more than grassy uplands.

These wetlands of the North American grasslands form vital ecological components of the grasslands. Less than half of the extensive wetlands that once dotted the area survive. These remnants range from the seemingly endless prairie potholes in the Dakotas, Minnesota, and Iowa to solitary marshes at Cheyenne Bottoms and Quivara in Kansas to the ephemeral shallows of High Plains playas and the sinks and springs of the Great Basin desert grasslands, and they do more than house mosquitoes and migrating ducks. They supply aquifers, shelter migrating shorebirds, absorb snowmelt, dissipate energy from floodwaters, provide oases in the midst of the sandhills, and support much of the intense variety of life on the grasslands. Since the last Ice Age, playas, sloughs (pronounced "sloos"), potholes, bottoms, river flats, sandhill lakes, desert sinks, thermal seeps, and rainwater basins have provided water, cover, and food for creatures living on the prairies and plains.

Wetland Hydrology

While many definitions of "wetland" exist, William Mitsch and James Gosselink provide a succinct definition: a wetland is a place that has water on or close to the surface, with soil that is different from that of the surrounding uplands and that supports "vegetation adapted to the wet conditions" (Mitsch and Gosselink 2000). All forms of wetland that dot the North American grasslands meet Mitsch and Gosselink's conditions, despite their varied hydrology and geologic and biologic features. From north to south, these wetlands are prairie potholes and sloughs, Nebraska's Sandhills lakes and rainwater basins, Cheyenne Bottoms in Kansas, and the playa lakes of the High Plains and Great Basin, as well as Great Basin springs and sinks. Riparian bottomlands extend across the plains in bands from the Assinibione River in Manitoba down the Red River toward the James, Missouri, Niobrara, Platte, Kansas (or Kaw), Arkansas, Cimarron, and Canadian rivers, and along the Great Basin's Humbolt, Carson, Sevier, and Walker rivers, as well as edging other smaller streams.

Each variety of grassland wetland, while sharing common features with the others, is uniquely formed. The prairie potholes of the northern plains are a geologically young landscape created by glaciers that covered North America until roughly 15,000 years ago. The undulating region of low topographic relief (mostly remnant ground moraines) and poor drainage that remained after the ice sheets melted covered roughly 20 million acres (eight million hectares), with individual basins as densely packed as 150 per square mile (58 per square kilometer) in what is now Iowa, Minnesota, and the Dakotas. Because the winter snow melts before the ground thaws in the northern prairies, the potholes collect water in early spring, filling before the seasonal summer rains begin. Clayey glacial soils drain poorly unless cracked by drying, so most potholes contribute little to groundwater. Because the individual depressions do not connect and the overall topographic relief of the region is low, water caught in potholes spills into streams and rivers only in exceptionally wet years. Some potholes maintain open water from spring thaw until winter freeze, but most dry out over the summer, leaving communities of grasses or bare white salt and alkali flats to mark their place (Van der Valk 1989).

South and west of the pothole prairies, the sandhills lakes of western Nebraska mark where the waters of the High Plains (Ogallala) and other aquifers come to the surface in a stabilized dune field. Like the pothole prairies, the sandhills lakes stem from the last ice age, marking where sand blown from the ends of retreating glaciers covered gravels washed off the rising Rocky Mountains (the Ogallala formation) that in turn overlie more impervious sandstones and shales. The highly porous sand allows for a rapid recharge of water levels after storms but also leads to equally rapid drainage of some wetlands into others. The region's semi-arid climate and highly seasonal precipitation make the sandhills lakes an important wildlife habitat within the short- to mid-grass prairies that cover the dunes (McIntosh 1996). Rainwater basins of south-central Nebraska appear similar to the playas of the High Plains and Great Basin but stem from different origins. Geologists theorize that they mark pre-glacial watercourses now buried under thick loess, or the remains of other ice induced features like the modern Todd Valley. The 400 basins remaining in the region feature a great diversity of plant and animal species and provide an important stopping place on the Central Flyway (Smith 2003).

To the south is one of the most important, large, open, freshwater wetlands in all of North America. Cheyenne Bottoms, designated in 1988 as a Wetland of International Importance, lies in a fault-depression, separated from the Arkansas River by a low bedrock divide that traps the flow of two creeks (see map on p. 178). Before large-scale irrigation began upstream, groundwater flowing into the creeks provided much of the water of the Bottoms, but precipitation is now the main water source despite diversions from the Arkansas River. Pollen studies suggest that the Bottoms existed as a wetland beginning 100,000 years ago. When the climate became drier, grasslands covered the Bottoms prior to the Wisconsonian glacial period 21,000 years ago, followed by a wooded parkland habitat until 10,000 years ago. Since then Cheyenne Bottoms has remained a shallow lake and marsh of varying size, expanding and contracting with precipitation, groundwater depletion, and evaporation (Zimmerman 1990). To the south of Cheyenne Bottoms are the playa lakes. Scientists have not decided on the factors that led to their formation except to agree that they are not potholes. Some playas may mark ancient fluviatile (watercourse) basins, but most owe their existence to a variety of factors, including wind deflation of loose soil, organic acids dissolving a carbonate layer below the topsoil and leading to the development of a clay-floored basin, or bison pawing out wallows that the wind enlarged. These broad, shallow basins depend on precipitation, especially late spring rains, for their water supply and frequently dry out by August or September. A few playas received water from springs in the Ogallala Aquifer, such as Wild Horse Lake and McDonald Lake in Amarillo, and the salt lakes northwest of Lubbock (all in Texas). Although spring-fed lakes are not technically playas, the modern decline of local groundwater has transformed these lakes into

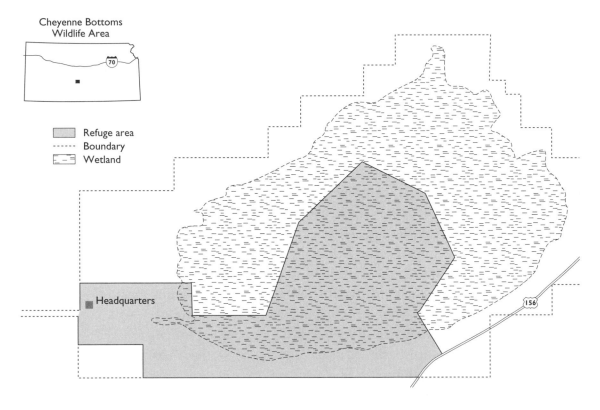

A pool at the Cheyenne Bottoms Wildlife Area near Great Bend, Kansas. (AP Photo/Charlie Riedel) The location of Cheyenne Bottoms is illustrated in the map, below.

playas by turning them into seasonal, precipitation-dependent water bodies. The same holds true for the series of salt lakes west of modern Lubbock, Texas, that once featured freshwater springs along their margins (Smith 2003; Steiert and Meinzer 1995).

The wetlands of the Great Basin formed over several geological eras in a land active with numerous fault systems tending roughly northeast-southwest. These faults are partly responsible for the small mountain ranges that subdivide the Great Basin into hundreds of smaller basins. During the wetter, cooler Pleistocene Era, 2 million years ago, these basins filled with water and eventually interconnected into large lakes. The mountains around these lakes continued to rise and erode, filling the lake basins with layers of sediment, and as the ice retreated and advanced, the lakes expanded and shrank, leaving terraces marking old shorelines. When the lakes finally dried out 10,000 years ago, they left behind hundreds of feet of water-rich sediment trapped by the surrounding mountains, just as the surface waters had been trapped. In addition, geothermal warm and hot springs appeared along the faults—400 have been recorded by the U.S. Geological Survey, with at least 75 thermal springs within Lake Lahontan's former expanse alone. Several rivers flow into the Basin, the largest of which are the Truckee, Carson, Humbolt, Bear, Sevier, and Walker. These rivers support riparian wetlands as well as supplying fresh water to Pyramid Lake, the Humbolt Sink, Walker Lake, the Great Salt Lake, and other smaller lakes. Runoff from the ranges within the Great Basin feed ephemeral playa lakes, many of which are saltier than seawater after thousands of years of leaching and evaporation (Chronic 1990; Harper et al. 1994; Houghton 1994). Marshes are common and often large within the Great Basin. Along the Quinn River in northern Nevada, a typical Great Basin stream, bluejoint grass and wild clovers grow along the banks in wet meadows, and rushes (tules) fill the places where the river slows and spreads into marshy flats. The Humbolt River was once known for the acres of giant wild rye grass and marshes that grew along its banks, and for the plentiful beaver that cut willows, cottonwoods, and other trees. The Smith Valley marshlands in the Walker River drainage in Nevada fill in wet years like 2004–2005; they attract waterfowl that nest among the tules and feed on young tules and desert nutgrass. Prior to upstream water diversions, Pyramid Lake and Walker Lake had freshwater marshes surrounding the central open waters.

Wetland Flora and Fauna

Sunlight, water, and microorganisms form the foundation for all grassland wetland food chains. Although much research remains to be done on the exact numbers and varieties of microbes, microinvertebrates, and other small species inhabiting plains wetlands, and how their role in the natural system differs from that of microorganisms in other types of wetlands, some aspects of pothole, playa, and marsh biology are certain. Algae, including blue-green algae and diatoms, can be found in all varieties of wetlands, as can the larvae of chironomids (midges). Bacteria, rotifers, and other creatures aid the decomposition of plant matter and other waste material while serving as food for other species like the chironomids. Waterfowl, amphibians, reptiles, and larger mammals consume plants, small animals, and insects. Although the exact species composition and food web varies greatly with temperature, latitude, and moisture, this basic series of consumption, reproduction, and decomposition can be found in all grassland wetlands.

Similar to all grassland wetlands, the vegetation associated with playa lakes varies with moisture availability and the season. Nearly all playa plants can survive periods of no moisture, either by use of sturdy seeds, buried tubers (like *Sattigaria*, or arrowhead), or other techniques. A scant few playas have cattails, while most have western wheatgrass, varieties of *Chenopodium* (goosefoot, lamb's quarters), buffalo grass, spikerush, rough pigweed, and saltmarsh aster. Beginning in spring, grasses dominate the bottom of the playa until enough moisture falls to produce standing water. At that point arrowhead and other species that thrive in wet soil will dominate, followed by annuals and grasses as the playa dries out. Species distribution normally occurs within zones in a playa basin, as plants like arrowhead and pondweed that prefer "wet feet" grow in the lowest part of the playa while those that tolerate more variable conditions thrive on the playa margins. As the dry season progresses, upland plants like buffalo grass and wheatgrass begin moving into the playa and may eventually cover the bottom completely once the soil dries out. Some playas retain bare soil in their central depression as a result of long inundations. The variety of flora described here in turn supports a wide variety of fauna. Bacteria and algae can be found in the water, along with fairy and brine shrimp, snails and several varieties of insect larvae, including mosquitoes and midges. Frogs and toads, salamanders, and horned frogs feed on the insects found in and around playas, as well as on each other. Migrating birds use playas, sometimes spending their winters there in mild years. Prior to intensive farming on the southern plains, coyotes, swift foxes, badgers, and weasels could be found, but muskrats apparently did not find enough permanent water and cattails to supply their needs. Bison and pronghorn grazed and watered in and around playas, and bison made use of dried out playas for dust wallows, eventually contributing to the popular notion that all playas were formed solely by bison.

Farther north, Grand Quivara and Cheyenne Bottoms also support multiple varieties of organisms from rotifers and bacteria to cattails, bald eagles, and snapping turtles. As John Zimmerman points out in *Cheyenne Bottoms: Wetland in Jeopardy*, bloodworms and midge pupae are the main food source for almost everything up the food chain. Bloodworms eat decaying plant matter and diatoms, and are in turn eaten by minnows, carp, birds, and other creatures. Herbivores and omnivores (like ducks) also feed on mats of duckweed and other water plants, while muskrats and white moths feed on cattails. When the margins of the marshes begin to dry in later summer, annuals like kochia, goosefoot, and smartweed move onto the exposed flats, providing seeds for other varieties of birds and small mammals. Floating green mats of duckweed and stands of cattails, bulrushes, and prairie cordgrass provide cover and sheltered nesting places for smaller fish and amphibians (Zimmerman 1990; Collins et al. 1994).

The sandhills wetlands have a variety of flora and fauna comparable to that of the playas and potholes. Wet meadows, areas where the groundwater comes within three feet (a meter) of the surface but does not emerge, feature stands of big bluestem and other "true prairie" plants. Sedges and prairie cordgrass (*Spartina* Sp.) thrive where the water table comes within eighteen inches (half a meter) of the soil surface. Buffalo grass and little bluestem favor drier uplands around sandhills basins and wet meadows. Some of the lakes are too alkaline to support vascular plants but do contain varieties of algae. Arrowhead, varieties of sunflower, and sedges thrive in longer lasting pools or along streams. Willows and cottonwoods mark streams and rivers emerging from the sandhills (Novacek 1989). Over half the lakes in the sandhills support fish, but most suffer winterkill due to the shallowness of the basins. Fairy shrimp and fly larvae are common in all types of sandhills wetland, alkaline as well as freshwater. Frogs, toads, tiger salamanders, and bullsnakes thrive in the wetlands. Mammals associated with the wetlands include beaver and muskrat, raccoons and mink, as well as mice, shrews, and voles. Long-billed curlew, mallards, blue-winged teal, pintail, and northern shoveler ducks all nest in and around sandhills wetlands, while roughly 2 percent of the mallards using the Central Flyway nest in the Nebraska Sandhills. For thousands of years, other migratory waterfowl and songbirds have used the wetlands as stopovers during their migrations, including avocets, pelicans, bitterns, and several varieties of heron.

In the prairie potholes to the north, vegetation varies along with water availability. If water stays in the depression year-round, or at least for most of the year, bulrushes and cattails will grow along the surface line in the lowest part of the pothole, and algae and water weeds flourish below. Above these where the shallow marsh loses its water cover, whitetop, manna grasses, and common reed will grow, while sedges and annuals grow in the moist meadow areas. These zones vary within each pothole and across the region. Not every wetland will have all these botanical zones, and drought or sedimentation often changes the zones within the same pothole from one year to another. The salinity of a wetland also affects the plant communities. For example, higher concentrations of salts favor sago pondweed, a preferred food for tundra swans, canvasback, and several other ducks. Of course, wetlands provide a rich habitat for mammals, too. Muskrats (*Ondatra zibethicus*) are the predominant mammals of the pothole prairie wetlands, but they are not the only mammals to use the marshes and sloughs. A variety of shrews, voles, and mice make use of wetland plants for food and shelter, or hunt the insects and smaller reptiles that live around the marshes. Foxes, raccoons, weasels and mink (*Mustela* sp.), skunks, and white-tailed deer inhabit the area as well, while coyotes are replacing wolves as the largest predator. Muskrats can change a pothole from a cattail marsh into an open-water pond lake in two or three years by "eating out" semi-submerged plants like cattails and bulrushes, leaving only free-floating and fully submerged plants behind. Muskrats also utilize wetland plants for lodge building, while some birds and small mammals in turn use the lodges for nest sites or burrows. Muskrats prefer the central, deeper water parts of potholes and prairie lakes, and the smaller rodents make use of the greater species diversity found on the marsh edges and uplands. Whitetail deer occasionally use the potholes for winter cover and fawning, as well as grazing during the fall and winter. Raccoons, foxes, and coyotes are the most common predatory mammals, but

Prairie potholes. (Tom Bean/Corbis)

raptors like red-tailed hawks and harriers also frequent not only the potholes but other wetlands of the grasslands as well (Fritzell 1989).

Throughout the Great Basin, hot and cool spring wetlands, most often found along the faults beside the fault-block mountain ranges, support many kinds of life. Pupfish and minnows, many unique to the pools where they now live, survive in springs sometimes six times saltier than seawater. Rushes, reeds, and tall grasses surround the springs and playas, forming inviting habitat for birds, deer, pronghorn, jackrabbit, and the predators that feed on them. Salt-tolerant species like inland saltgrass, iodine bush, and pickleweed survive near salt springs as well as on salt playas like the Great Salt Lake. Great Basin playas tend to have higher salinity than High Plains playas, because plains playas lose most of their water through absorption, while basin playas generally dry out due to evaporation. Great Basin playas are integral to groundwater recharge, like their plains counterparts, and can change from dry to wet and back to dry as the water table rises and falls with precipitation and the seasons. As evaporation rates decline in winter, the water table can rise enough to rejuvenate a playa, just like rain showers or snowmelt do. The salinity of the playa also varies with evaporation and precipitation, which in turn affects the plant communities around wet playas. Dry playas support more xerophytic (drought-tolerant) plants like winterfat and shadscale.

Humans and Plains Wetlands

Humans, like the other mammals mentioned above, have made use of the grassland wetlands since the end of the last Ice Age. Humans need water as much as other animals, possibly more if food preparation is taken into consideration. Wetlands and watercourses attract birds and animals, which throughout their varied existence have attracted human hunters. People living in the grasslands have also made use of the large variety of fruits, tubers, and grasses found around wetlands, as compared to the surrounding uplands. Anglo-Americans have made some use of wetlands for watering cattle and irrigation, but more commonly have sought to eliminate potholes, playas, and wet meadows in order to farm more land. This large-scale destruction of Great Plains wetlands has had many long-term consequences, including reduced groundwater recharge and increased river flooding.

Paleoindian peoples made great use of the wetlands in the grasslands. Both the Clovis (10000–8600 BCE) and Folsom (8900–8100 BCE) type sites are in watercourses in New Mexico, while the Clary Ranch site in Nebraska sits on a tributary of the North Platte. Scattered finds across the land surface of the Nebraska Sandhills show that Paleoindians also made use of the sandhills wetlands. Finds are not as common in the pothole prairies, with some scholars suggesting that Native Americans moved into the northern prairies comparatively late (8500–6000 BCE), while also pointing out the paucity of sites of all periods found so far (Holliday 1997). Archaic (6000 BCE–1 CE) peoples and later historic Indian peoples also made extensive use of wetlands. Those peoples exploited plants such as beaked sedge, tall mangrass, buttercup, arrowhead, and cattails. They ate arrowhead and cattail tubers either raw or cooked to help provide starch in the diet. Farther south, along the Big Blue River in northeastern Kansas, the remains of charred bulrush seeds suggests that people ate them, along with the seeds of wetland margin plants like goosefoot and smartweed. Wetlands with seasonal flows, like Cheyenne Bottoms, or along the edges of Great Plains rivers like the Republican and Arkansas provided easy-to-catch fish as the waters receded in late summer. Waterfowl could be caught with snares, or later (5000 BCE) with bow and arrows. As would later Anglo-American plains dwellers, Indian peoples used wood from riparian areas for building lodges, making tools, and burning as fuel. Farther south, in the modern Oklahoma Panhandle, a spring and series of playas provided traders, hunters, and other travelers with a "temporary shelter and kill opportunities" as they crossed the plains going to and from the flint quarries on the Canadian River. South of the Canadian River, playas and springs provided the only surface water where hunters found bison, elk, pronghorn, and migrating fowl gathered. When droughts dried most of the playas and reduced spring flows, plains residents dug temporary wells in the spring-fed wetlands to reach water, especially in the southern plains where the water table remained fairly high. Indian peoples also made use of the variety of plants growing in these same plains wetlands (Kuhnlein and Turner 1991; Wedel 1986; Ballenger 1999). Paleoindians and Native Americans made use of the Great Basin wetlands much as did their counterparts on the Great Plains. After the megafauna disappeared at the end of the last ice age, desert dwellers came to depend on smaller game like jackrabbit and occasionally pronghorn and mountain sheep. Later Basin peoples like the Piute, Shoshone, and Bannock traveled the area seasonally, hunting animals and waterfowl, and gathering foods like pinyon nuts and acorns. Lakes provided trout and cattail shoots in spring.

Later, Europeans and New Mexicans, as did Indian peoples, depended on playa lakes as they crossed the southern plains. Francisco Vásquez de Coronado and his party probably followed the chain of saline lakes located northwest to southeast from New Mexico toward Yellowhouse Canyon and modern Lubbock, Texas. Later, *ciboleros* (bison hunters) and *Comancheros* (traders) who ventured onto the southern plains in search of meat, hides, and slaves also depended on playas and springs for water. In fact, it has been tentatively suggested that the name for this region, the Llano Estacado (staked plains), may be a corruption of Llano Estancado, "plain of many pools" (Morris 1997). Cheyenne Bottoms and the Arkansas River wetlands received attention from Coronado's men and later travelers like Zebulon Pike. Native Americans had been hunting the abundant waterfowl and animal life in the Bottoms and the riparian wetlands along the river valley for centuries, as well as gathering plants, berries, and fruit to augment their maize, squash, beans, and sunflower crops. Coronado's chronicler, Juan Jaramillo, commented in 1541 that the Bottoms was some of the best land the group had seen thus far. Pike, unimpressed with the Arkansas River's lack of flow in October 1806, did note a "low and swampy" area north of the main valley. Following parties of Anglo-Americans agreed with Pike's assessment of the river valley and the Bottoms, eventually settling around the basin but not taking up homestead acres or farming the basin itself (Harvey 2001).

In contrast, the sandhills wetlands remained relatively unknown to Europeans longer than did the playas, potholes, or eastern bottoms. French traders and trappers probably knew of the sandhills wetlands through their Native American trading customers and associates. Anglo-American settlers, on the other hand, avoided the sandhills wetlands in general, while bands of Lakota and Cheyenne hunted and traveled through the area. Although some Anglo settlers attempted to farm near the few streams running through the area before the 1890s, the sandy soil, erratic rain, and hilly nature of the region discouraged settlement in favor of grazing and ranching. This remained the pattern until the development of irrigation, and especially of center-pivot irrigating equipment in the 1960s. Even today, ranching remains the primary land-use pattern, a pattern that has helped to preserve the sandhills wetlands to a greater extent than other types of grassland wetlands (Sandoz 1966; Kepfield 1993).

Euro-Americans began settling the Great Basin in the 1850s. The best known are the Latter Day Saints (Mormons), who began irrigating along the tributaries of the Great Salt Lake. Non-Mormons ("gentiles") moved into the area from the west and south, often drawn by the prospect of making money from the mineral rushes in the area. The miners and prospectors who flooded Nevada looking for the next gold and silver lodes needed supplies, and Mormon and "gentile" ranchers and farmers soon obliged them, grazing cattle and growing truck crops with water diverted from the few year-round rivers. Although some artesian wells were dug, most of the basin-bottom wells needed pumps and windmills to bring water to the surface. As cities like Las Vegas and Reno, Nevada; Salt Lake City, Utah; and Los Angeles grew in size and political clout, river waters were diverted from the Owens Valley, Carson, Truckee, and other rivers to supply their needs. Irrigators along the Humbolt diverted river water in order to grow hay for winter livestock feed, which along with irrigated grains account for 90 percent of the state's farm production. Away from the streams, groundwater pumping has led to water table declines, drying springs and creeks in some parts of the Great Basin, while increased erosion and stream cutting have lowered the water table in other areas, drying wet playas and springs. This in turn changes the local plant communities from wet meadows to broad meadows and eventually to sagebrush and rabbit brush flats as the groundwater sinks (Chambers and Miller 2004).

In general, Euro-American practices worked to destroy wetlands. The pothole prairie wetlands serve as an example of how culture and commodity agriculture could combine to change a physical environment. While Indian peoples traditionally made great use of marshes and wetlands, Euro-American settlers of the grasslands and Great Basin generally took a dim view of these "useless bogs." Northern Europeans and their American descendents tended to view swamps and marshes as wet, dank, useless swaths of land (the name "Great Dismal Swamp" sums it up) to be drained or avoided. The best thing one could do to a swamp was to find a way to drain the waters, level the land, and make a "productive" field out of it. To help this process along, the United States government passed the Swamp Land Act of 1850, giving swamp- and marshlands to the individual states in an attempt to help speed drainage and development (Lewis 2001). Settlers in the pothole prairie wetland regions of Iowa, Minnesota, and the Dakotas had the blessing of tradition and the federal and state government to drain the potholes. In the northern prairies, "wild hay" from summer-dried potholes could prove nourishing for farm animals in the winter. As Laura Ingalls Wilder told it in *Little Town on the Prairie* and *The Long Winter*, the hay harvest was only a step away from cultivation and civilization as epitomized by wheat, maize, and small-grain farming. By the late 1800s, the advent of tiling and ditching to drain wet meadows and potholes resulted in major reduction to the northern wetlands as farmers replaced them with uniform fields of wheat, oats, and maize. During the twentieth century farmers increased the size of farms with larger machinery, and this also contributed to wetland destruction. Wider farm implements could not navigate around the soft soil, and this motivated farmers to drain and fill the irregularities.

In large part because of their need for cattle-watering places, southern ranchers viewed playas more favorably than northern homesteaders looked at potholes. Amarillo, Texas, was founded near a spring-fed playa (Wild Horse Lake, now Amarillo Lake) because it served to water cattle before they were shipped to market by rail. Early irrigation farmers also viewed playas as useful features because they could put a pump in the playa bottom and capture unused irrigation water ("tailwater") flowing off a field. Even with the newer center-pivot systems, deepened playas ("tailwater pits") served as catch basins for runoffs. However, these uses for playa wetlands have not prevented their destruction. Many playas have filled in because of erosion, either with aeolian (windblown) sediments or soil washed in from the surrounding plowed fields. Other playas have been converted into stormwater runoff catchments (Lawrence Lake in Amarillo, Texas), or been built over by developers (Carson 1996). Certainly draining and ditching are not the greatest dangers faced by Great Basin wetlands. Water diversions, groundwater pumping, erosion due to overgrazing, chaining of brush, and mining all contribute to the disappearance of wetlands. While some protection has come from the Endangered Species Act and the 1985 Food Safety Act, restrictions on swampbusting, groundwater depletion, and stream exports still threaten many Great Basin wetlands. The desert climate of the Great Basin makes all surviving marshes, playas, sinks, and springs that much more important to the plants and animals of the Great Basin grasslands.

Over the years, some people have begun to question the deliberate or accidental destruction of grassland wetlands. Wildlife hunting and conservation groups like Ducks Unlimited and the Izaac Walton League were among the first to lobby for wetland preservation. The membership argued successfully that without the pothole prairies for duck and geese nesting, there would be no waterfowl to hunt. This eventually led to congressional passage of the 1929 Migratory Bird Conservation Act. The saline marshes that now form the Quivara National Wildlife Refuge southwest of Cheyenne Bottoms were preserved because most of them were owned by private hunting clubs prior to the formation of the refuge in 1955. When local farmers threatened to drain Cheyenne Bottoms in 1925–1928, opponents included businessmen in the towns of Hoisington and Great Bend who profited from visiting hunters, as well as the hunters and conservationists themselves. Birders and hunters had given more value to the wetlands as marsh than as cropland in the eyes of local wetland supporters.

Scientific research into the hydrologic function of wetlands has also won converts to preservation and restoration. Pothole prairies once captured up to 80 percent of the spring snowmelt and runoff in the region, and this greatly reduced the amount of water that eventually flowed into Devil's Lake, the upper Mississippi River, and the Red River [of the North]. Since many of the wetlands now drain into the streams and rivers, more problems with downstream flooding have occurred, with 1998 being only the most memorable example. Plowing the potholes did not cause the floods that spring, but the reduction in meltwater storage and the faster runoff rate probably contributed to the size and duration of the flood. The sandhills wetlands and High Plains playas both contribute to the recharge the Ogallala Aquifer, the body of fossil water that underlies much of the southwestern Great Plains grasslands. Rain and snowmelt percolate quickly through the sandy soil and recharge the portion of the aquifer under the sandhills, while the playas hold water that soaks into their bottoms and thence into the groundwater. In addition, the Randall clay that forms the base of High Plains playas cracks deeply as it dries. Precipitation runs through these cracks, augmenting the groundwater before water saturates the clays and they swell closed. Playas are the only current recharge source of the southern portion of the Ogallala Aquifer. Riparian and other wetlands serve as filters for water flowing into and through streams, removing sediment and chemicals by slowing overland flow and supporting plant growth that feeds on suspended nutrients like phosphorus and nitrogen.

Wetland conservation and restoration efforts, like those under way at Hackberry Flat near Fredrick, Oklahoma, often encounter difficulties despite increasing evidence about the importance and value of wetlands as wetlands. Conservationists are faced with social and technical difficulties when trying to protect or repair grassland wetlands. Many wetlands are on private property, and landowners often have deep suspicions about perceived outsiders "trying to tell them what to do with their property," especially governmental outsiders. Because wetlands are taxed the same as the income-producing cropland around them, landowners have little incentive to preserve wetlands that are costing them money but generating little or no income. Creditors may insist on drainage for maximum yield and income, even if the farmer would rather preserve the wetland. The culture of farmers also plays a role in reducing interest in wetland preservation, especially if the community traditionally calls a "good" farmer one who has straight, even rows of crops and an efficient, orderly farm. Wetlands are not conducive to straight rows or order. The technical difficulty of wetland restoration and preservation also causes problems. After land has been intensively farmed for many years, putting the water back will not automatically restore the wetland. "Weeds" establish themselves much more quickly than native wetland plants thereby reducing species variety and habitat redevelopment. Sedimentation from surrounding cropland can reduce the size of wetlands, while deep plowing can break up the soils; both problems are often associated with playa restoration and preservation. Wetlands that depend on groundwater for much of their moisture, like Cheyenne Bottoms and the Nebraska Sandhills wetlands, must compete with irrigators, towns, and others (Seabloom and Van der Valk, 2003; Duram, 1995).

Progress in preservation legislation is being made, if slowly. The federal government passed the 1985 Food Security Act, which included penalties for destroying or altering wetlands after that year. This so-called "swampbuster"

legislation penalized those who plowed up or drained wetlands after 1985 by reducing or canceling agricultural subsidy payments. The 1991 *North Carolina
Wildlife Federation et al. v. Tulloch* decision defined discharging material into a
wetland, something that requires a federal permit, as including any material related to digging or draining. This greatly broadened the scope of activities in and
around wetlands requiring authorization from the Army Corps of Engineers or
the Environmental Protection Agency. The doctrine of "no net loss," suggesting
that for wetlands drained or altered, an equal area must be restored or created,
became federal policy in 1993, but like the earlier "swampbuster" law it seems
to have only slowed wetland loss, not eliminated it.

The outlook for the future of Great Plains wetlands appears mixed. Educational programs like the "Playas: Jewels of the Plains" photo exhibit and the
1983 publicity that rallied support for Cheyenne Bottoms did help to persuade
the general public that rather than being mosquito-infested nuisances, wetlands
serve important ecological functions. On the negative side, property owners
with wetlands remain reluctant to pay taxes on and preserve wetlands that produce no apparent benefits. Landowners can become actively hostile when faced
with legislators and environmentalists who seem uninterested in the farmers'
and ranchers' economic difficulties, as happened in South Dakota in 2000. That
spring the state attorney general announced that all surface water belonged to
the state and was open to public use. Property owners were not pleased when
they found hunters and fisherman literally in their backyards on what had been
dry land before that year's heavy snows and rain inundated the eastern section
of the state. Continuing depletion of the groundwater that feeds the sandhills
wetlands and the destruction of riparian wetlands because of upstream water diversion also contribute to the reduction of plains wetlands. On a positive note,
projects like the preservation of Cheyenne Bottoms and the restoration of Hackberry Flat in Oklahoma, a pluvial (rainwater) wetland area covering 7,120 acres
(2,881 hectares), have helped to restore some of the drained and farmed land.
Groups involved in the restoration include the Oklahoma City Sportsman's
Club, the Williams Companies, Ducks Unlimited, Phillips Petroleum, the Bureau of Reclamation, the U.S. Fish and Wildlife Department, and the North
American Wetlands Conservation Council, among others.

The wetlands of the North American plains, despite the loss of over half
their original area, remain vital features of grassland life. From the potholes of
the Canadian prairies to the Smoky Hill riparian marshes and High Plains
playas, plains wetlands provide food and shelter for migratory waterfowl and a
variety of reptiles, amphibians, and mammals. For thousands of years, they
have served human communities as rich hunting grounds and watering sources
for their domestic animals. They also contribute to flood control, water quality

preservation, and groundwater recharge while helping preserve botanical diversity across the region. The wetlands were valued by American Indian peoples but commonly denigrated by the Euro-Americans who settled the plains; however, the general public is slowly regaining an appreciation for the benefits of preserving and restoring grassland wetlands. In recent years they have changed their attitudes, recognizing the critical functions of wetlands in preserving wildlife and shaping the grasslands. Perhaps Hugh Prince is correct in asserting that wetlands have been, are, and will continue to be the sine qua non of the grasslands (Prince, 1997).

References

Ballenger, Jesse A. M. 1999. "Late Paleoindian Land Use in the Oklahoma Panhandle: Geoff Creek and Nall Playa." *Plains Anthropologist* 44 (May): 189–207.

Carson, Paul. 1996. *Empire Builder in the Texas Panhandle: William Henry Bush.* College Station: Texas A&M Press.

Chambers, Jeanne C., and Jerry R. Miller, eds. 2004. *Great Basin Riparian Ecosystems: Ecology, Management and Restoration.* Washington, DC: Island Press.

Chronic, Halka. 1990. *Roadside Geology of Utah.* Missoula, MT: Mountain Press.

Collins, Joseph T., Suzanne L. Collins, and Bob Gress. 1994. *Kansas Wetlands: A Wildlife Treasury.* Lawrence: University Press of Kansas.

Duram, Leslie Aileen. 1995. "Water Regulation Decisions in Central Kansas Affecting Cheyenne Bottoms Wetland and Neighboring Farmers." *Great Plains Research* 5 (Feburary): 5–19.

Fritzell, Erik K. 1989. "Mammals in Prairie Wetlands." In *Northern Prairie Wetlands*, edited by Arnold G. Van der Valk. Ames: Iowa State University Press.

Harper, Kimball T., Larry L. St. Clair, Kaye H. Thorne, and Wilford M. Hess, eds. 1994. *Natural History of the Colorado Plateau and Great Basin.* Niwot: University Press of Colorado.

Harvey, Douglas S. 2001. "Creating a 'Sea of Galilee': The Rescue of Cheyenne Bottoms Wildlife Area, 1927–1930." *Kansas History* 24 (Spring): 2–17.

Holliday, Vance T. 1997. *Paleoindian Geoarchaeology of the Southern High Plains.* Austin: University of Texas Press.

Houghton, Samuel G. 1994. *A Trace of Desert Waters: The Great Basin Story.* Reno: University of Nevada Press.

Kepfield, Sam S. 1993. "The 'Liquid Gold' Rush: Groundwater Irrigation and Law in Nebraska, 1900–1993." *Great Plains Quarterly* 13 (Fall): 237–250.

Kuhnlein, Harriett V., and Nancy J. Turner. 1991. *Traditional Plant Foods of Canadian Indigenous Peoples: Nutrition, Botany and Use.* New York: Gordon and Breach.

Lewis, Jr., William M. 2001. *Wetlands Explained: Wetland Science, Policy and Politics in America.* New York: Oxford University Press.

McIntosh, Charles Barron. 1996. *The Nebraska Sand Hills: The Human Landscape.* Lincoln: University of Nebraska Press.

Mitsch, William J., and James G. Gosselink. 2000. *Wetlands*, 3rd ed. New York: John Wiley.

Morris, John Miller. 1997. *El Llano Estacado.* Austin: Texas State Historical Society.

Novacek, Jean M. 1989. "The Water and Wetland Resources of the Nebraska Sandhills." In *Northern Prairie Wetlands*, edited by Arnold G. Van der Valk. Ames: Iowa State University Press.

Prince, Hugh. 1997. *Wetlands of the American Midwest: A Historical Geography of Changing Attitudes.* Chicago: University of Chicago Press.

Sandoz, Mari. 1966. *Love Song to the Plains.* Lincoln: University of Nebraska Press.

Seabloom, Eric W., and Arnold G. Van der Valk. 2003. "Plant Diversity, Composition and Invasion of Restored and Natural Prairie Pothole Wetlands: Implications for Restoration." *Wetlands* 23 (March): 1–12.

Smith, Loren M. 2003. *Playas of the Great Plains.* Austin: University of Texas Press.

Steiert, Jim, and Weyman Meinzer. 1995. *Playas: Jewels of the Plains.* Lubbock: Texas Tech University Press.

Van der Valk, Arnold G., ed. 1989. *Northern Prairie Wetlands.* Ames: Iowa State University Press.

Wedel, Waldo. 1986. *Central Plains Prehistory.* Lincoln: University of Nebraska Press.

Zimmerman, John L. 1990. *Cheyenne Bottoms: Wetlands in Jeopardy.* Lawrence: University Press of Kansas.

CASE STUDY: JOHN WESLEY POWELL AND THE ROCKY MOUNTAIN WEST

James E. Sherow

John Wesley Powell had a comprehensive vision for living in the West. He hoped to avoid ecological, social, and economic failures throughout the Rocky Mountain region by warning people that the laws and techniques they had used to populate the subhumid and humid lands of the nation would fail in the West. Powell saw a symbiotic relationship between the Rocky Mountains and the grasslands, and he believed this had to be preserved to avert human-created and human-borne catastrophes. He understood his role as a mission to redeem the nation from its ill-begotten ways.

The best touchstone for understanding Powell's undertaking is his *Report on the Lands of the Arid Regions of the United States,* 1878, delivered to Carl Schurz, at that time secretary of the interior. Powell saw conditions in the Rocky Mountains requiring a different manner of settlement from that practiced east of the Missouri River. For nearly all of this region, he advocated eliminating the Homestead Act, preemption laws, the Timber Culture Act, and desert land provisions. Legislation such as the Desert Land Act, 1877, Powell anticipated, would make land dispersal subject to unbridled speculation. Competent, educated officials, Powell declared, should classify the land as grazing, timber, irrigation, coal, or mineral before anyone could gain access to it. The grid system of land surveying, Powell further believed, was completely inappropriate for establishing land parcels; rather, land units should be classified and recorded by drainage basin. People occupying a given basin would be held responsible for devising the conservation measures for water—the most important resource for successful settlements. Once the land was classified, a person could acquire up to 80 acres within an irrigation district, and a rancher could acquire 2,560 acres, or four sections of grazing land. Powell emphasized that most of the land would remain pasturage, while only a small portion could ever be irrigated. The mountains themselves would remain publicly owned and scientifically managed for their timber resources and the conservation of their water resources for downstream uses, especially water intended for irrigated agriculture (Powell 1962).

Those bent on the rapid economic development of the region quickly saw Powell's report as an impediment, and they were correct in their fears. Powell's ideas necessitated an orderly, planned occupation of the land according to its classification and use, along with cooperative planning for the exploitation of its natural resources. Powell's was not a path to quick riches; however, it was a direction toward a thoughtful, carefully planned system for human life in this

Portrait of John Wesley Powell. (Library of Congress)

region, an existence retaining the ecological base on which the well-being of its inhabitants depended. As Donald Worster aptly summarized: "[Powell] stood . . . at the center of a change that began late in the last century and is still inching forward today, away from a careless, unplanned exploitation of nature and toward a more thoughtful, scientifically informed ethic of conservation. . . . A utilitarian at core, he nonetheless appreciated the land for its wild beauty as well as for its economic potential. Conservationists and environmentalists would rightly look back on him as one of their founding giants" (Worster 2001, 573). In the context of Worster's insights, Powell's thinking gives us a historical gauge by which to interpret the environmental history of the Rocky Mountains today.

Powell's Life

Raised in a family environment of strong religious and reformist convictions, young Powell developed a sturdy moral streak that would guide him all his life. His father, Joseph, and mother, Mary, valued education highly, and they strongly encouraged John to take advantage of opportunities to learn as they arose. Both moral reform and education would shape his life's work, giving him a religious sense of right and wrong rooted in scientific rationalism. Work on farms gave him a love of the great outdoors and led to his early interests in botany. A tour of the South in 1860 internalized and solidified his hatred of slavery.

His first opportunity to join his sense of righteous indignation with science came with the onset of the Civil War. Motivated to preserve the Union and to destroy the evil of slavery, this remarkable young man immediately enlisted in the army. His work in engineering and artillery demanded an understanding of mathematics and physics. This blending of science and morality led the young, twenty-eight-year-old man to the battlefield of Shiloh, where in April 1862, a rifle ball wound necessitated the amputation of his right arm. His wife of less than one year, Emma, nursed him back to health, and in time Powell returned to serve for the duration of the war.

After the war, the lure of exploring the outdoors led the fledgling geology professor at Illinois State Normal University out of the classroom to the Rocky Mountains. Seeking ever-greater support for his expeditions, Powell secured funding from the Smithsonian Institution and the Chicago Academy of Science to supplement the substantial support already provided him by the Illinois Natural History Society. From May through August 1869, Powell guided four boats with eleven men, three of whom abandoned the expedition shortly before it concluded, on the Green River toward the first descent down the Colorado to its confluence with the Virginia River in present-day eastern Nevada (Powell's journey is illustrated in the map below). Press coverage and Powell's own reports of this adventure established him as one of the preeminent explorers in the United States. With his ready reputation, he garnered further support from the Department of Interior for several additional surveys of the Rocky Mountains during the 1870s (Stegner 1954).

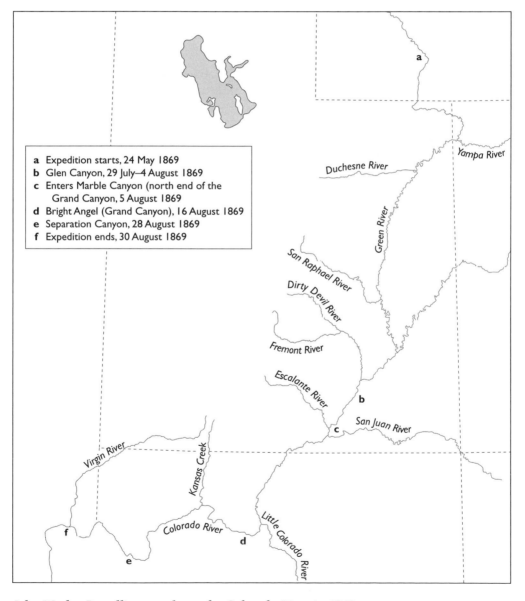

a Expedition starts, 24 May 1869
b Glen Canyon, 29 July–4 August 1869
c Enters Marble Canyon (north end of the
   Grand Canyon, 5 August 1869
d Bright Angel (Grand Canyon), 16 August 1869
e Separation Canyon, 28 August 1869
f Expedition ends, 30 August 1869

John Wesley Powell's route down the Colorado River in 1869.

He spent the last three decades of his life creating and directing some of the most important national scientific bureaus of his day. His efforts were instrumental in initiating the United States Geological Survey in 1879, and he became its second director in 1881. His treks through the West provided Powell firsthand experiences with Indian peoples, which led to his fascination with and study of them. Given this keen interest, he would complete his bureaucratic life serving as the director of the Bureau of Ethnology in the Smithsonian Institution until his death in September 1902.

Geology of the Rocky Mountains

As a pathbreaking ethnologist, naturalist, and geographer, Powell had an unquenchable curiosity about the West, and he took up a firsthand study of its resources. He knew that the mountains and canyons of the West were the result of ancient tectonic forces that had lifted and folded the crust of the earth, but exactly how these forces worked is still the matter of academic debate today. Most geologists will agree, however, that approximately 75 million years ago what would become the present-day Rocky Mountains was an elevated, level landmass bordered by an inland sea covering what humans would recognize today as the grasslands east of the Front Range. Millions of years before, the land above the seas had once been towering mountains, but over the course of time these had been eroded and flattened considerably, and covered by inland seas. Dinosaurs roamed the land, and giant creatures swam in the seas, their fossilized skeletal remains found, studied, and cataloged millions of years later by humans (Kearey and Vine, 1996).

The crust of the Earth is divided into huge solid plates floating atop a superhot, incredibly deep, molten layer of rock called the asthenosphere, Greek for "fragile or soft area," as opposed to the upper layer, the lithosphere, "rock or solid area." On colliding, these plates folded, thrust upward, or bent inward to form surface features over geologic time. According to one theory, such forces began working on two particular plates, the Farallon Plate underlying the Pacific Ocean drifting eastward, and the North American Plate, pushing westward, butting against it and forcing it underneath. This collision heaved, bent, and fractured the landmass that would eventually form the Rocky Mountains over a 45-million-year period. In the last 30 million years winds, rains, ice, and volcanism have scoured and shaped this stony uplifted mass, creating a setting for mountainous forests, meadows, rivers, and thermal basins (Bird, 1988).

Today, the Rocky Mountain System (illustrated on p. 197) is divided into the Southern Rocky Mountains, which include the ancient, extinct volcanos known today as the Spanish Peaks and ranges such as the Sangre de Cristos, Front Range, and Indian Peaks along with Rocky Mountain National Park. The Middle Rocky Mountains include the active thermal basins of Yellowstone National Park as well as the towering Teton and Wind River Ranges. The Northern Rocky Mountains include Glacier National Park. To the west of these are the

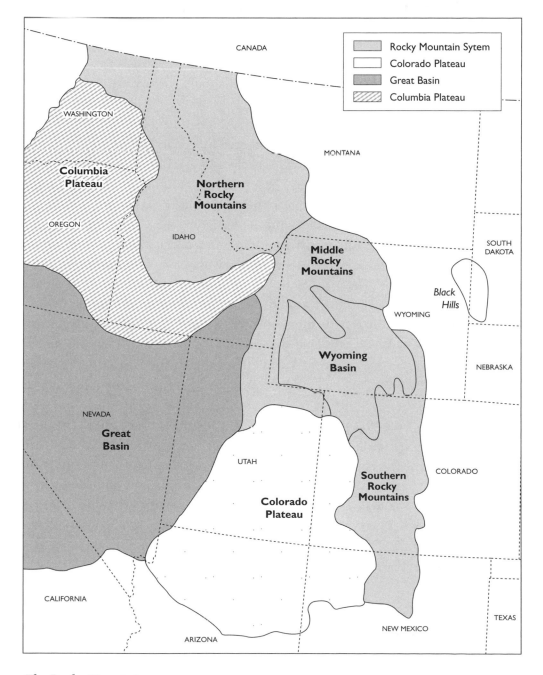

The Rocky Mountain ranges.

great plateaus such as the Colorado with its stunning canyon lands and to the north the great Columbia Plateau. Between these plateaus and the Sierra Nevada Mountains to the west are the immense desert grassland basins.

Learning the Geography of the Rocky Mountains

As one of the founders of the National Geographic Society, Powell knew the importance of understanding the lay of the land if he were to devise a system for its development and conservation. His own explorations through the region immeasurably contributed to this knowledge, but he also drew on the work of explorers who preceded him. Looking to American explorers besides Lewis and Clark, he drew on the knowledge gathered by mountain men such as Manuel Lisa, who traveled the upper reaches of the Missouri River in his effort to create a fortune based on the fur trade. The Chouteau family followed in his wake and controlled a fur trading empire reaching up into every tributary of the Missouri River. These men, and hundreds more like them, were "expectant capitalists," ones whom Powell could identify with in terms of their love for exploration and adventure in the great outdoors, which they blended with an unquenchable acquisitive impulse.

The federal government also joined in the spirit of exploration in mapping and establishing an American empire tapping the rich resources of the mountains. These expeditions included some of the best scientific minds and artists of the time. Colonel John James Abert organized the Army's Corps of Topographical Engineers in 1838, and through this command professional explorers and scientists—personified most aptly by Charles J. Frémont—mapped and made note of resources and peoples. The War Department, under the direction of Secretary Jefferson Davis, charted railroad surveys that eventually became the main transcontinental lines in the decades following the Civil War. It was the Union Pacific Railroad trains that would eventually take Powell to the Green River where he began his trip down the Colorado River. These mountain men, fur traders, and army explorers gave Powell a map of the Rocky Mountains and some of its geology. He would have been lost without their remarkable achievements (Goetzmann 1966).

Pre-contact Mountain Biomes

Powell recognized, in an oblique sort of way, the importance of Indian peoples in shaping the mountain biomes. "To preserve the forests they must be protected from fire," Powell wrote. His following sentence read: "This will be largely accomplished by removing the Indians." What Powell had observed were Indian fire practices, and for thousands of years these same practices had had a profound effect on shaping the ecosystems of the mountains. Fires enhanced grassland meadows, cleared underbrush through the forests, promoted certain food sources such as camas or acorns, controlled insects, and favored the growth of some trees to the exclusion of others. Broadcast fires constrained the extent of forestlands in the mountains and left large, mature trees in their wakes (Pyne 1982).

While certainly Indian burning practices significantly shaped mountain biomes, differences in topography, climate, and geography also shaped a multiplicity of life communities. For example, the Sierra Nevada Ranges act as a wall, catching most of the moisture-rich, Pacific Ocean air currents coursing eastward across the continent. As a result, the Great Basin is left parched throughout. Moving across the Basin and the Colorado Plateau, these westerlies again become more humid only to have their waters caught by the high ridgeline of the Continental Divide. And again, a "rain shadow" extends eastward across the Great Plains. Western slopes are wetter than eastern ones; general temperatures become colder from south to north; forests go from primarily juniper in the south to towering ponderosa pines to the north. Throughout the mountains are "parks," high-altitude basins of well-watered grasslands.

Within these varied zones a rich variety of wildlife has lived. In the dry reaches of the shrublands, greasewood dominates the landscape. In this semidesert environment live small mammals such as white-tailed jackrabbits, white-tailed prairie dogs, along with birds such as the Gambel's quail and the ubiquitous striped whipsnake and the sagebrush lizard. Above along the sides of the mountains are piñon pine-juniper woodlands. Here have thrived mule deer, the piñon mouse, and the piñon jay that flourishes by feeding on piñon nuts. Farther up begin the Douglas fir and ponderosa pine forests. Today, these stands are often more dense than when Indian peoples occupied them as a result of fire suppression and the decline of grazing in the grassland meadows that thrived in the wake of burns. In these woods that humans have valued highly for building habitation have lived a rich abundance of wildlife. Still present, but in diminished numbers, are large mammals such as black bears, mule deer, elk, big horn sheep, and mountain lions. Grazing animals have prospered in the meadows dominated by grasses such as Idaho fescue, blue grama, June grass, and needle and thread. Hundreds of species of wildflowers, such as the beautiful wild iris or the elegant Colorado blue columbine, grace these parks and woods. As one ascends or travels to the northern latitudes, aspen and lodgepole pines become more commonplace, and higher up still the forest gives way to tundra where introduced mountain goats gingerly scale steep, rocky slopes, and the small pika scurries among boulders and scree (Mutel and Emerick 1984).

Fur Companies, Indian Peoples, and Miners

Powell certainly viewed Indian peoples and the fur trade as an impediment to the full exploitation of the resources within the mountains. The fur trade, especially its destruction of beavers, revealed itself as a short-lived, highly extractive enterprise. But in many ways the fur trade economy established itself as the prototypical economic model for the mountains. As Bernard DeVoto viewed it in his famous essay, "The West against Itself," the wealth of the fur trade "went east into other hands and stayed here. The absentee owners acted on a simple principle: get the money out. And theirs was an economy of liquidation. They cleaned up and by 1840 they had cleaned the West out. A century later, beaver has not yet come back" (DeVoto 1947, 1). To understand the fur trade as a prototype is to grasp William Goetzmann's definition of mountain men as "expectant capitalists" for whom acquiring wealth was a form of religion. Given the inevitable depletion of fur-bearing animals, these men kept their eyes open for any and all opportunities—land speculation, transportation, farming, or especially mining (Goetzmann 1966). Those who made a transition from the fur trade to other endeavors assumed, like Powell, a need to remove Indian peoples in order to build railroads. In turn, such transportation links would complement the full development of industries such as mining and allow the forests to "increase in extent and value." Like furs, timber and ores would be shipped east.

In retrospect, it is odd that Powell thought forestry would be improved with the development of mining that in turn would flourish with railroad connections. Powell, like many others of his time, placed great faith in the "thousands of hardy, skilful [sic] men" whose mining districts would remain within the public domain. He did set off one form of mining from open exploitation—coal mining. To protect energy supplies, he believed coal mining districts should be set aside by governmentally sponsored geological surveys. In general, Powell clearly recognized how gold rushes set into motion the economic development of the mountains in a manner far different from that of the fur trade. Mining required urban centers with transportation, communication, and retail networks linking them to East and West Coast markets. He also noted how these urban mining centers greatly stimulated the rapid development of irrigated agriculture, its produce necessary for feeding thousands of hungry miners. He saw all of this as a good thing.

Prior to Powell's report, the federal government stimulated mining through the passage of the Hardrock Mining Act of 1872. This law, which is still the basis for mining the public domain, made it easy to establish a claim with little to no obligation to the federal treasury. The federal government had to also deal with the fallout with Indian peoples as a result of gold rushes such as the one to the Black Hills. Sometimes there was open conflict and brief triumph for the Native Americans, as when the Lakota and Cheyennes dealt the Seventh Cavalry a severe defeat at the Battle of Greasy Grass (Battle of the Little Big Horn) in June 1876, but more often Indian peoples suffered the most and found themselves quickly impoverished and confined to reservations.

While mining was bad for Indian peoples, it also did little good for the forests on the public domain. Hydraulic mining techniques polluted and filled streams with debris. Refining gold ore sent volumes of arsenic into the air and streams. Deforestation was always a problem around any hard rock mine. For example, the Comstock Mine in Nevada consumed over 600 million board feet of timber for structural supports in the tunnels. Moreover, the federal government allowed mining companies to cut freely on the public domain whatever timber was needed for their operations.

The transitory nature of mining also produced its own urban problems. Around the turn of the twentieth century, hard rock mining around Leadville, Colorado, placed high levels of lead into the drinking water of the city and arsenic in the air. In Butte, Montana, some tried to argue that the sulphur in the smoke from the Anaconda Copper mines was a cure for the flu and that arsenic was good for women's complexions despite the exceptionally high rate of respiratory related illnesses and deaths. Even though environmental problems associated with mining were commonly understood by foresters such as Gifford Pinchot, only after World War II would mining engineers give much attention to the pollution caused by mining. In 1909, for example, the young, very successful engineer Herbert Hoover told a lecture audience that "no question of public utility enters [into mining operations], so that all mining projects have by necessity to be from the first weighed from a profit point of view alone" (Smith 1987, 88). After World War II, middle-class Americans became increasingly enthralled by outdoor recreation such as hiking, mountain climbing and skiing, trout fishing, river rafting, and outdoor camping. These recreational interests came into sharp conflict with the economics and traditions of mining. Even those companies, such as AMAX, that added environmental mitigation to their administration lost out when they wanted to reopen mines around Crested Butte, Colorado, in the 1980s. Crested Butte had become a popular ski resort and the city had tied its future to tourism, which appeared to be a far more stable form of income than the ups and downs of mining. Moreover, contemporary mining techniques, such as the open pit operations of the Kennecott Mine in Utah, have considerable undesirable environmental consequences regardless of the effort put into mitigation.

Forests and Parks

Oddly, Powell seemed to see little need for the "conservation" of forests as advocated by either Gifford Pinchot or by George Bird Grinnell or Enos Mills, boosters of national parks. "No limitation to the use of the forests need be made," Powell wrote, when these lands were properly protected from Indian-set fires. Certainly, Gifford Pinchot would come to see the matter quite differently. As he succinctly stated his position: "The job was not to stop the ax, but to regulate its use" (Pinchot 1947, 29).

Pinchot was not the first to see danger in the unrestricted cutting of the forest on the public lands throughout the Rocky Mountain West. In many respects, a book published in 1864 first raised American awareness about the need for forest conservation. In his *Man and Nature,* George Perkins Marsh recommended farming trees with scientific methods. He had students who internalized these warnings and recommendations, and they later became important governmental officials when Powell was conducting his great expeditions. Among these, Charles S. Sargent helped to found the American Forestry Congress in 1882, and John W. Noble, secretary of the interior in President Harrison's administration, worked to create forest preserves for 13 million acres in 15 different preserves. By 1905, Gifford Pinchot, with President Theodore Roosevelt's support, enlarged this effort with a staff of over 1,500 employees who managed more than 150 million acres of forest reserves now administered by his Forest Service, which was housed in the Department of Agriculture. Pinchot was a hands-on manager who believed that "[f]orestry is Tree Farming... [W]hat the forester gets is a crop.... Farmer and forester alike get a lot of other products on the side. Good farming yields also such things as butter, eggs, apples, calves. Good Forestry, in addition to lumber, firewood, and other produce, yields such services as regulation of stream flow, protection against erosion, and some influence on climate" (Pinchot 1947, 31). Unlike Powell, who thought forest preservation could be left to individual initiative, Pinchot believed this could be achieved only through governmental oversight by scientifically trained foresters (Miller 2001).

Powell lacked interest in creating forest preserves and gave little regard to preserving the wild. Perhaps it seemed nearly impossible to him that something as vital and powerful as the Colorado River could ever be tamed by a dam, or a topography as vast and sublime as the Grand Canyon might be filled with water or mine tailings. When in the upper portion of the Grand Canyon, Powell marveled at walls "more than a mile in height—a vertical distance difficult to appreciate. Stand on the south steps of the Treasury building in Washington and look down Pennsylvania Avenue to the Capitol; measure this distance overhead, and imagine cliffs to extend to that altitude, and you will understand what is meant" (Powell 1964, 251). At the end of his *Canyons of the Colorado* (1895) he rhapsodized: "The glories and the beauties of form, color, and sound unite in the Grand Canyon—forms unrivaled even by mountains . . . It is a region more difficult to traverse than the Alps or the Himalayas, but if strength and courage are sufficient for the task, by a year's toil a concept of sublimity can be obtained never again to be equaled on the hither side of Paradise" (Powell 1964, 397) Understandably, he had thoroughly underestimated the technological might of humans and the fragility of the wild.

Photography, however, displayed to the American public an incredibly beautiful western landscape, and Powell was, in part, responsible for this. His photographer during the 1871 canyon expedition gave the country a view of this nearly inaccessible realm. At the same time, others such as Timothy O'Sullivan and William H. Jackson added immeasurably to this photographic record by depicting other portions of the Rockies. Inspired in part by such pictures, President Theodore Roosevelt wanted to preserve the beauty of the canyon, and in 1908 he invoked the Antiquities Act to set aside over 800,000 acres encompassing the Grand Canyon as a "national monument" within national forestlands.

Creating national parks, however, required more than a desire to preserve the stunning beauty of western lands. The history of Yellowstone, Glacier, and Rocky Mountain National Parks provides other good examples of the emphasis on a utilitarian rather than preservationist approach toward national parks. Congress set aside Yellowstone Park in 1872 but did little to protect it until public pressure led Congress to charge the army with patrolling the park, a responsibility the U.S. Calvary undertook in 1883. Not until passage of a National Park Service bill, first introduced in 1911 and subsequently enacted five years later, would a bureau be created to oversee a growing number of parks, most of which were located in the western mountains. The organic act has always had at its heart a dual tension: "[which] is [1.] to conserve the scenery and the natural and historic objects and the wildlife therein and [2.] to provide for the enjoyment of the same in such manner and by such means as will leave them unimpaired for the enjoyment of future generations." On one hand the service is charged with enhancing public enjoyment and use of the parks; on the other it is instructed to preserve their wild ecosystems. Stephen Mather, the first director of the National Parks System, saw no problem in enhancing tourism to the point of staging bear shows and building a golf park, while his successor tried hard to have the winter Olympics staged in the park. Moreover, the service pushed out of the park the Indian peoples who had once managed the same land especially through their use of fire, which the park service rangers sought to eliminate. Ecologists began to understand the value of fire to forest health by the 1940s, and in the early 1970s the park service began introducing controlled, small-scale burns. In 1988, all of this changed with massive fires in Yellowstone that before winter snuffed the remaining embers had charred 36 percent of the park. The following year showed the value of such burns with the emergence of rich grassland meadows, purity of streams, and animal wildlife abundance (Sellars 1993).

Similarly, Glacier and Rocky Mountain National Parks came into being. George Bird Grinnell, the editor of *Field and Stream*, and Louis W. Hill, president of the Great Northern Railroad Company, collaborated to create Glacier National Park in 1910. Once congressmen could see that the grasslands were too sparse to support cattle, that the timberlands were largely inaccessible to loggers, and the climate was too harsh for agriculture, they were satisfied that the area could be preserved for its scenic beauty and was suitable for tourism. Even before the creation of the park, the federal government removed the Blackfoot Indians, who were largely responsible for shaping the ecosystems of the park. To the south, Enos Mills played the key role in creating Rocky Mountain National Park. As an innkeeper near the foot of Long's Peak and a self-taught naturalist and journalist, he highly valued the beauty and ecosystems of the Rocky Mountains. Colorado congressman Edward Taylor thought the area of "no value for anything but scenery," but Congress allowed the Reclamation Service to retain access to the park just in case some resources of value might be later found. Indeed, by 1959, the Bureau of Reclamation had completed a $160 million transmountain water diversion project that included a thirteen-mile tunnel bored right through the heart of the park (Runte 1979; Spence 1996, 1997).

While the national parks would not be protected from all types of human development, whether for tourism or natural resource uses, there would come into existence those places called wilderness areas where human effects on the land would be held to a minimum. The work of Howard Zahniser, director of the Wilderness Society, along with the legacy of Bob Marshall, the original founder of the society, led to passage of the Wilderness Act of 1964 (Harvey 2005). Following this was congressional passage of the Endangered Species Act in 1973, which called for the identification and protection of "critical habitat" where an endangered species lived. Most of the wilderness lands would be carved out of national forests and Bureau of Land Management tracts, whereas critical habitat designations often came into conflict with developers of natural resources. Interest in wilderness preservation and endangered species protection would lead to the reintroduction of wildlife such as wolves in Yellowstone Park in 1995. This move, of course, worried anxious nearby ranchers who feared for the safety of their sheep and cattle herds. Careful management by the federal government and cooperative work with local ranchers have alleviated many, but certainly not all, of the problems that have arisen with these introductions. In addition to concerns about the newly returned wildlife, problems with appropriate access to wilderness areas have arisen. For example, should snowmobiles, all-terrain vehicles, or mountain bikes be allowed near or in these set-aside lands? Lately in Montana, the representatives from the Mountain Wilderness Association and the Montana Snowmobile Association amicably negotiated and resolved some of these issues as they apply to over 2.5 million acres of forestlands (King 2006).

The West's Reservoirs

In his report, Powell, like the Reclamation Service to follow, foresaw the vast value of the mountains as the reservoirs for irrigation and urban development. In fact, as far as Powell was concerned, the basis for all wealth in the region lay in its water and in cooperative planning for its development and use. The Big Thompson project that delivered water from the west side of the continental divide, through Rocky Mountain National Park, to the eastern slope and irrigated fields beyond was one system that partially resembled Powell's vision. The western side of the Rockies collects much more snow and water than the eastern slope does, but most of the intense urban growth and nearly all of the irrigated agriculture occur on the eastern side and Great Plains. In years of complicated negotiations among people living on the western slope, those in the rapidly growing urban areas of the Front Range, and the irrigation farmers along the semi-arid High Plains, a complex system of water development and use was devised (Tyler 2003).

One other place in the intermountain West had nearly unimaginable potential as an agricultural cornucopia—the unlikely place that Spanish and Indian peoples knew as the Valley of Death. It was a basin well below the western end of the Grand Canyon, and its soils could produce abundant crops if the flow of the Colorado River could be diverted onto those parched lands. Charles R. Rockwood took on this mission, and by 1905 the Imperial Land Company had over 67,000 acres irrigated in high-yielding, profitable truck crops. About the same time, the Colorado River made an unexpected shift: river flows increased dramatically; they scoured the basin and began shifting course, ripping apart diversion dams and headgates. A huge flood was produced that once again filled large portions of the basin creating what is called today the Salton Sea. Edward Harriman directed the Southern Pacific Railroad Company in spending millions of dollars trying to tame the river, and by the time that was accomplished all involved in either farming or transporting crops knew that they lacked the resources to control the river. They all turned to the Reclamation Service seeking a solution (deBuys and Myers 1999).

What Powell had not anticipated was the monumental main-stem dams to be built on the Colorado River. The Reclamation Service, however, did, and it advised building a huge dam plugging the Black Canyon of the Colorado, which was considerably upstream from the Imperial Valley. The dam could control periodic floods and regularize the flows of the river to the irrigation works. The problem was that such a dam would affect flows in the Colorado River and its tributaries in seven states altogether. By the time the service had become serious about such a project, World War I was over, *Kansas v. Colorado* had been decided, and Delph Carpenter, a renowned water lawyer and state politician, had already attempted negotiating an interstate compact between Colorado and Kansas. Later Carpenter championed an interstate compact for the Colorado River Basin, as the state of Colorado would be one of the parties affected by the dam. Eventually, representatives from all of the affected states met in Santa Fe, New Mexico, and with Secretary of Commerce Herbert Hoover as chair of the committee, the delegates hammered out the terms of what became the Colorado River Compact of 1922. This agreement, despite its many flaws—most notable the miscalculation of the average flow volume of the river—still governs the dispersal of water throughout the region, and the Hoover Dam, begun in 1931 and completed in 1936, physically regulates the flows of the river (Reisner 1986).

With this dam the folks in the Bureau of Reclamation (the service was renamed in June 1923) became excited about extending dam building throughout the Colorado River Valley. This led to one notable dispute among many. The proposal to build a dam on the Yampa River, a tributary to the Colorado, became highly controversial, as the site was within the boundaries of Dinosaur National Monument. After World War II, the environmental movement and outdoor tourism had gained considerable strength, and a combination of environmentalists, such as David Brower of the Sierra Club and Howard Zahniser of the Wilderness Society, river raft outfitters, and officers within the Army Corps of Engineers such as Ulysses S. Grant III, defeated the bureau's plans in the early 1950s. Still, this did not stop the building of other dams such as Glen Canyon, which, as pointed out by environmentalists, lowered the water temperature in what had once been the warmest river in North America. This temperature reduction killed aquatic species, and water outflows scoured the riverbed downstream from the dam

(Harvey 1994). Lately, more attention has been given to regulating discharges in an effort to rebuild sandbars throughout the canyon in hopes of reestablishing wildlife habitats. Regrettably, much of this effort has been severely hampered by an extended drought and the lowering of the water level in Lake Mead, which stores the reservoir water feeding the Colorado River below the Hoover Dam.

Beyond the environmental concerns raised by dams such as Glen Canyon and Hoover, the legal division of the Colorado River, based on inaccurate estimates of stream flow, has led to many interstate conflicts and disputes. The state of Arizona contends with California over the flows that support urban growth in Phoenix and Tucson. The cities of Los Angeles and Las Vegas wrangle over these same flows, while all upstream states look nervously downstream and wonder whether their citizens will be prohibited from using the water flowing by their fields and cities. At stake is the livelihood of 30 million people and a $1.2 trillion dollar economy based in Los Angeles, Las Vegas, Phoenix, and Denver. Sometimes city representatives have negotiated unique agreements to deal with their situations, such as the one completed in February 2006, which allows farmers within the Colorado River Basin to lease water to cities such as Los Angeles, Phoenix, or Las Vegas (Jenkins 2006).

The Urbanized Mountains

Of all the developments Powell foresaw, he had little means to predict the vast urban nature of the Rocky Mountain West, yet it has always been the most urbanized region of the entire nation. Mining has lain at the core of this development throughout the mountains as amply illustrated by Duane Smith's *Rocky Mountain Mining Camps* (1974). Mining camps such as Central City, Colorado, or Virginia City, Nevada, became instant small cities with the discovery of gold and other metals. As in all such cases, these places prospered with an extractive economy subject to boom and bust cycles as the natural resource base became exhausted. Moreover, their environments were highly toxic as a by-product of refining ores and dumping tailings into creeks and rivers. These mining camps depended on larger urban centers that served as depots for imported goods and the collection points for shipping out the extractive wealth of the mountains. For example, cities such as Denver and San Francisco served these economic ends for the interior mining camps. Other cities such as Cheyenne, Wyoming, abetted the extractive economy of the region by becoming a shipping depot for cattlemen and miners. A few cities, such as Colorado Springs, Colorado, became early health spas, which complemented their mining service economies.

Mining may have produced fantastic wealth for some, but it also proved highly mercurial as the history of Butte, Montana, certainly reveals. Early gold discoveries in the mid-1860s had led to the establishment of Butte, but copper mining boomed in the town from around 3,000 in the late 1870s to nearly 11,000 by 1890. In another ten years over 30,000 people lived in the city with a couple thousand miles of tunnels underneath. Known throughout the West as the "richest hill on earth," wealth did not buy health, as miners and citizens both lived in a toxic environment. By the end of World War II, deep shaft mining proved uneconomical and the Anaconda Company turned to open pit mining; by doing so, Anaconda and its successor, the Atlantic Richfield Company, created the most polluted superfund site in the nation—the Berkeley Pit, which covers more than 5,000 acres. Acidic heavy metals in its water form a corrosive brew that dissolves many other metals on contact and kills any wildlife unfortunate enough to drink from it. Today, Butte has made the transition to tourism and struggles to maintain a population that is still 10,000 fewer than during its mining heyday in the 1920s. Oddly enough, the Berkeley Pit is one of the advertised tourist attractions of the city (Peirce 1972).

Berkeley Pit near Butte, Montana. (Library of Congress)

By the end of World War II, the West, and the planet, had been transformed by the role of the federal government. In the little village of Los Alamos tucked away in the rugged mountains near Santa Fe, New Mexico, the federal government had funded the creation of the atomic bomb. The bomb and the military industrial complex both became permanent features of the region. By the end of the war Denver, whose economy had basically stalled, hummed with the infusion of federal dollars funneled into the Rocky Mountain Arsenal, Lowry Field, and the Martin Marietta Company. Rocky Flats, to the northwest of the city and owned by Dow Chemical, produced nuclear warheads along with environmentalists' claims of plutonium pollution in the surrounding countryside. Other companies and federal agencies flocked into the region such as IBM and Ball Brothers in Boulder, and the Air Force Academy in Colorado Springs. All of this brought with it massive urban growth and continued pressure for the development of additional water resources to quench its thirst. Generally, this new population growth had a higher income level than those previously associated with the extractive economy, and they had more leisure time, which often was spent outdoors, where people wanted to experience an unpolluted and vibrant environment (Hevly and Findlay 1998; Fernlund 1998).

Certain towns rapidly became known for their "quality of life" amenities such as Taos, New Mexico; Missoula and Bozeman, Montana; Jackson Hole, Wyoming; and most noteworthy of all, Aspen, Colorado. The economies of these communities became centered around tourism and outdoor life. The cities became attractive places for wealthy individuals to live, and soon, long-time residents found themselves forced out by mounting taxes and higher costs of living. Cities such as Aspen are now populated by "equity exiles," people living only part-time in the city and who can afford houses in different parts of the nation. Few people can sustain a lifestyle such as that in Aspen, where, for example, an average house cost $3.65 million in 2004 (Lichtenstein 2004). Urban growth has also spilled into the surrounding forestlands. In 1955, Auguste Spectorsky called this phenomenon "exurbia." Private timberlands are most subject to this growth, and environmentalists have worked to create conservation easements to protect such areas from exurban development, as was the case for the Blackfoot River in Montana, the setting for Norman Maclean's novel, *A River Runs through It* (1976). In other places, such as the foothills to the west of Denver, exurban growth proceeds inexorably causing ecological degradation of plants and animal wildlife (Best 2005).

The urban development of the West continues, fueled by its proliferating middle class with a disposable income and an appetite for outdoor recreation, its draw of eastern capital, and its dreams of economic development with housing tracts in the "wilderness" that are to be preserved by state and federal gov-

Las Vegas is the fastest growing urban area in the desert grasslands. (Corbis)

ernments. What becomes starkly apparent is that there are many contravening forces at work. This trend was noted and predicted to worsen by Bernard De-Voto shortly after the end of War World II. DeVoto saw a western population caught up in the dream of "economic liberation" from eastern ownership through "continually increasing federal subsidies." He clearly saw what has been a continuing schizophrenic attitude toward the federal government. On the one hand, Westerners wanted the government to preserve natural resources for "sustained, permanent use"; at the same time they wanted to conduct "an assault on those resources with the simple objective of liquidating them." Following this scenario, some Westerners associated with the Sagebrush Rebellion or the Wise Use Movement and similar organizations have pressured the federal government to relax or relinquish its oversight or ownership of public lands. DeVoto initially called this a "land grab" with the objective of private development. And in many respects, the problems he predicted have arisen due to the ecological and social troubles associated with exurbia, and by what other critics call the "hollowed-out" resort communities where in some cases over 60 percent of the housing is occupied for only part of the year. Many environmentalists in 2006 would agree with what DeVoto wrote in 1947:

[The dream of economic development and local ownership and control] envisions the establishment of an economy on the natural resources of the West, developed and integrated to produce a steady, sustained, permanent yield. While the West moves to build that kind of economy, a part of the West is simultaneously moving to destroy the natural resources forever. That paradox is absolutely true to the Western mind and spirit. But the future of the West hinges on whether it can defend itself against itself. (DeVoto 1947, 13)

Some Westerners have sought means for resolving this paradox. For example, West Slope ranchers such as Dorothy and Norman Kehmeier have placed some of their extensive pastures into conservation easements in order to prevent unwanted development.

The West's Playground or the Nation's Energy Reserve?

The paradox of the West against itself has played out distressingly well in its national parks. The good Major Powell could never have imagined needing a permit to float on the Colorado River through the Grand Canyon, much less

T. J. Hittle's photograph captures the danger still inherent in rafting the Colorado River down the Grand Canyon in the twenty-first century. (Courtesy of T. J. Hittle)

having to wait approximately twelve or more years from the time of application to do it. After World War II, Georgie White found he could make money by taking people down the Colorado through the canyon in surplus army rafts. The popularity of rafting grew so that by 2005 over 26,000 people annually floated through the canyon. The river has become considerably more crowded now than in Powell's day as people are loving it to death.

One can shorten the wait by employing a licensed raft company, but that's considerably more expensive than outfitting oneself, with rafting companies charging anywhere from $150 to $300 for a day trip, and around $3,500 for a multiday trip starting at Lee's Ferry and ending in Lake Mead, the reservoir formed by Hoover Dam. Navigating the canyon in a privately organized group not only requires patience—in 2001, there were 7,200 people on the wait list to float the canyon—but it also requires considerable river skill. Handling a kayak or river raft is highly dangerous in several sections of the river and is no place for a novice.

Other recreationists have taken to the mountains themselves. Long before skiing became a recreational sport, residents used what were called "Norwegian snowshoes" for transportation, and even several postal employees used them to traverse deep snow. By 1930 expert skiers from Europe had begun skiing the Rockies, and in 1936 the first ski run in national forests was created in Sun Valley, Idaho. Skiing received an unintended boost as a result of World War II. The army decided to train an elite mountain division in Colorado for combat operations in the mountain ranges of Europe. Training of the 87th Mountain Regiment as one unit of the 10th Light Infantry Division began in early 1943 at Camp Hale near present-day Leadville, Colorado. Several of the young men trained here returned to Colorado after the war and helped create the ski tourism industry that has left an indelible mark on the Colorado economy. Friedl Pfeifer and Peter W. Seibert developed the ski slopes of Aspen and Vail. Gordon Wren managed Steamboat Springs, while Steven Knowlton promoted the entire industry as the first director of Colorado Ski Country USA. By 1973, over 200 ski areas had been carved out of national forests alone. All of this was made possible by the startling and amazing growth of American middle-class leisure time. As the mining economy in towns such as Park City, Utah, faltered, tourism based on skiing rejuvenated its prospects. All of this development has created social conflict, as locals often resent those they call "trust fund" children. Other conflicts arose as some radical environmentalists associated ski resorts with environmental degradation and destruction of public forests—in one case, bombing Vail Resort. Undoubtedly, the ski industry has clogged highways leading to resorts and raised questions about appropriate uses and management of national forests while creating a prosperous tourist industry that many see bringing unequal benefits to those associated with it (Benson 1984; Coleman 2004).

Skiing is not the only questionable use of public lands in the mountains, and DeVoto would certainly recognize today the issues tied to the economics of energy development. Four years after the end of World War II, an obscure geologist wrote a startling article published in the journal *Science*. M. King Hubbert introduced to the world the concept of "peak oil," the time at which oil production in any given locale would begin to decline as a result of drained reserves. In 1949, he predicted that oil production in the United States would "peak" in 1970, an exceptionally accurate forecast (Hubbert 1949). The federal government has responded to the increasing demand for domestic oil production by allowing drilling on large tracts of Bureau of Land Management acreage throughout the Rocky Mountains. This stepped-up drilling has raised the ire of environmentalists who point to severe ecological degradation from the operation of over 7,000 wells for oil and gas on these lands. And yet, domestic oil production still continues to decline.

One promising source of oil lies tightly trapped in shale rock deposits over large portions of north-central Colorado and south-central Wyoming. By some estimates, there may be over seven times more oil reserve in these formations than is currently in Saudi Arabia, the source of around 15 percent of U.S. imports in 2005. This 15 percent amounted to 10 million barrels a day at the time, and the Saudis claimed they had enough proven reserves to continue pumping up to 15 million barrels per day for another fifty years. However, the cost of extracting shale oil is enormous, as the rock has to be heated to extreme temperatures to draw out the oil. In the 1980s, oil prices seemed high enough to support the costs of this type of technology, and an oil boom occurred in Colorado. Then oil prices sank, and quickly pulled down into its vortex were all the boomtowns and shale oil companies. On May 2, 1982, Exxon Corporation closed shop, and the residents of Rio Blanco County, Colorado, refer to this as "Black Sunday." Now with oil prices soaring, shale oil production appears profitable again to several companies, and they are back. So are environmentalists who see in this exceptionally intrusive form of mining the destruction of wildlife habitat and stream degradation from the dumping of brackish water, a waste product of shale oil drilling (Lay 2005).

Driving the race to develop the energy resources of the West is the amazing American appetite for oil, the basis of its economy. The construction of highways and interstates has made exurbia possible, ski resorts prosperous, and the exploitation of shale oil economical. In a bizarre feedback loop, the automobile may become, or already is, a major contributor to the destruction of this economy and highly mobile way of life. Many scientists already recognize global warming as the main cause of a prolonged drought affecting the mountain states in the last two decades. Over the last fifty years, the number of frost-free days in Aspen has increased by a month. In the next twenty years, an estimated increase of average temperatures by just one degree Fahrenheit will double the number of frost-free days. Some scientists predict a general rise of around three degrees Fahrenheit by 2050, and to many in the area the portents for snowpack and the ski industry are devastatingly clear. Many people see global warming as a real problem with potentially disastrous climatic consequences for the Rocky Mountains, and they have begun to act. Through organizations such as Cities for Climate Protection, in places such as Tucson, Arizona, and Salt Lake City, Utah, they are attempting to enact ordinances to reduce CO_2 emissions below those recommended in the Kyoto Protocol of 1997. Such occurrences and developments reveal a Rocky Mountain West very different from the one seen and experienced by Powell (Nijhuis 2006).

Powell's Unrealized Dream

Now what remains of John Wesley Powell's vision for the Rocky Mountain West? Cooperative, regional planning within watershed boundaries never materialized. The difficulties in planning water development and conservation within the Colorado River watershed certainly demonstrate this. The neat, mostly quadrilateral-shaped state boundaries bear no resemblance to the shape of their watersheds. The citizens of each state jealously guard their own water resources with laws that encourage use and economic development over ecological preservation. Even interstate water compacts such as the Colorado Compact of 1922 have not resolved pressing water allocation issues and may, in fact, only have heightened the intensity of some conflicts. Ranching and irrigated farming emerged without regard to centralized land use planning. The only place where centralized planning occurred at all was with the creation of the Forest Service, the National Park Service, and the Bureau of Land Management, and nearly none of this has been done within the cooperative framework envisioned by Powell.

Charles Wilkerson of the law school at the University of Colorado has harshly criticized an irrational, economically driven development of the natural resources in the Rocky Mountain West. He calls for a revised plan for living in the West, one modeled on Powell's notions of cooperative planning within reasonably established ecological boundaries. He hopes to discard into the ash bin of history what he calls the "lords of yesterday," those laws such as prior appropriation, the Hardrock Mining Act of 1872, Bureau of Land Management grazing policies, National Forest Service policies enhancing private lumber companies, the Bureau of Reclamation and its subsidies to agribusiness, and the prior appropriation doctrine, which has little use for streams other than economic development. Wilkerson certainly believes that public subsidies should end for mining companies, timber corporations, and large agricultural corporations (Wilkerson 1992). Cooperative planning for the preservation of the natural resources that will sustain ecosystems and economies must take place.

Powell wanted resource development, and that certainly has occurred. But he wanted neither big government nor centralized planning, nor did he want unbridled, unplanned economic plundering of the Rocky Mountain West.

> To a great extent, the redemption of the [arid Rocky Mountain West] will require extensive and comprehensive plans, for the execution of which aggregated capital or cooperative labor will be necessary. . . . For its accomplishment, a wise prevision, embodied in carefully considered legislation, is necessary. (Powell 1962, 8)

Perhaps Powell's vision is sometimes fulfilled as in those negotiations between the Forest Service personnel, the Montana Wilderness Association, and the Montana Snowmobile Association that resolved how to protect wild lands yet also provide recreational access to them. While Westerners have certainly learned to exploit the natural resources of the Rocky Mountains as Powell had once hoped, they have yet to devise workable, cooperative planning that will preserve and enhance those same resources for the life of both humans and other living entities. Powell's dream for that kind of Rocky Mountain West remains unrealized.

References

Benson, Jack A. 1984. "Skiing at Camp Hale: Mountain Troops during World World II." *Western Historical Quarterly* 15 (April): 163–174.

Best, Allen. 2005. "How Dense Can We Be?" *High Country News* 37 (June): 8–12, 14–15.

Bird, Peter. 1988. "Formation of the Rocky Mountains, Western United States: A Continuum Computer Model." *Science* 239 (March 25): 1501–1507.

Coleman, Annie Gilbert. 2004. *Ski Style: Sport and Culture in the Rockies.* Lawrence: University Press of Kansas.

deBuys, William. 1999. *Salt Dreams: Land & Water in Low-Down California.* Albuquerque: University of New Mexico Press.

DeVoto, Bernard. 1947. "The West against Itself." *Harper's Magazine* 194 (January): 1–13.

Fernlund, Kevin J., ed. 1998. *The Cold War American West, 1945–1989.* Albuquerque: University of New Mexico Press.

Goetzmann, William H. 1966. *Exploration and Empire: The Explorer and the Scientist in the Winning of the American West.* New York: Vintage Books.

Harvey, Mark. 1994. *A Symbol of Wilderness: Echo Park and the American Conservation Movement.* Albuquerque: University of New Mexico Press.

Harvey, Mark. 2005. *Wilderness Forever: Howard Zahniser and the Path to the Wilderness Act.* Seattle: University of Washington Press.

Hevly, Bruce, and John M. Findlay, eds. 1998. *The Atomic West.* Seattle: University of Washington Press.

Hubbert, M. King. 1949. "Energy from Fossil Fuels." *Science* 109 (February 4): 103-109.

Jenkins, Matt. 2006. "Colorado River States Reach Landmark Agreement." *High Country News* 38 (February): 3, 6.

Kearey, Philip, and Frederick J. Vine. 1996. *Global Tectonics.* Oxford: Blackwell Science.

King, Ray. 2006. "Snowy Middle Ground: Wilderness Advocates and Snowmobilers Come to Terms in Montana." *High Country News* 38 (March): 4–5.

Lay, Jennie. 2005. "Congress Bets on Oil Shale." *High Country News* 37 (December): 8–9.

Lichtenstein, Grace. 2004. "Part-Time Paradise." *High Country News* 36 (October): 7–10, 12.

Maclean, Norman. 1976. *A River Runs through It, and Other Stories.* Chicago: University of Chicago Press.

Miller, Char. 2001. *Gifford Pinchot and the Making of Modern Environmentalism.* Washington, DC: Island Press.

Mutel, Cornelia Fleischer, and John C. Emerick. 1984. *From Grassland to Glacier: The Natural History of Colorado.* Boulder, CO: Johnson Books.

Nijhuis, Michelle. 2006. "Save Our Snow." *High Country News* 38 (March): 8–12, 19.

Peirce, Neal R. 1972. *The Mountain States of America: People, Politics, and Power in the Eight Rocky Mountain States.* New York: W. W. Norton.

Pinchot, Gifford. 1947, 1987. *Breaking New Ground.* Introduction by George T. Frampton, Jr. Washington, DC: Island Press.

Powell, John Wesley. 1962. *Report on the Lands of the Arid Region of the United States, with a More Detailed Account of the Lands of Utah,* edited by Wallace Stegner. Cambridge, MA: Harvard University Press.

Powell, John Wesley. 1964. *Canyons of the Colorado.* New York: Argosy-Antiquarian Ltd.

Pyne, Stephen J. 1982. *Fire in America: A Cultural History of Wildland and Rural Fire.* Princeton, NJ: Princeton University Press.

Reisner, Marc. 1986. *Cadillac Desert: The American West and Its Disappearing Water.* New York: Viking Penguin.

Reisner, Marc, and Sarah Bates. 1990. *Overtapped Oasis: Reform or Revolution for Western Water.* Washington, DC: Island Press.

Runte, Alfred. 1979, 1987. *National Parks: The American Experience,* 2nd ed. Lincoln: University of Nebraska Press.

Sellars, Richard. 1993. "Manipulating Nature's Paradise: National Park Management under Stephen T. Mather, 1916–1929." *Montana: The Magazine of Western History* 43 (Spring 1993): 2–13.

Smith, Duane A. 1974. *Rocky Mountain Mining Camps: The Urban Frontier.* Lincoln: University of Nebraska Press.

Smith, Duane A. 1987. *Mining America: The Industry and the Environment, 1800–1980.* Lawrence: University Press of Kansas.

Smith, Duane A. 1992. *Rocky Mountain West: Colorado, Wyoming, and Montana, 1859–1815.* Albuquerque: University of New Mexico Press.

Spence, Mark David. 1996. "Crown of the Continent, Backbone of the World: The American Wilderness Ideal and Blackfeet Exclusion from Glacier National Park." *Environmental History* 1 (July): 29–49.

Spence, Mark David. 1999. *Dispossessing the Wilderness: Indian Removal and the Making of the National Parks.* New York: Oxford University Press.

Stegner, Wallace E. 1954. *Beyond the Hundredth Meridian: John Wesley Powell and the Second Opening of the West.* Boston: Houghton Mifflin.

Stegner, Wallace E. 1987. *The American West as Living Space.* Ann Arbor: University of Michigan Press.

Stevens, Joseph E. 1988. *Hoover Dam: An American Adventure.* Norman: University of Oklahoma Press.

Tyler, Daniel. 1992. *The Last Water Hole in the West: The Colorado-Big Thompson Project and the Northern Colorado Water Conservancy District.* Niwot: University Press of Colorado.

Tyler, Daniel. 2003. *Silver Fox of the Rockies: Delphus E. Carpenter and Western Water Compacts.* Norman: University of Oklahoma Press.

Wilkerson, Charles F. 1992. *Crossing the Next Meridian: Land, Water, and the Future of the West.* Washington, DC: Island Press.

Williams, Michael. 1988. *Americans and Their Forests: A Historical Geography.* New York: Cambridge University Press.

Worster, Donald. 2001. *A River Running West: The Life of John Wesley Powell.* New York: Oxford University Press.

DOCUMENTS

CORONADO'S REPORT TO THE KING OF SPAIN, SENT FROM TIGUEX ON OCTOBER 20, 1541

Near Mexico City, Francisco Vásquez de Coronado assembled and began an expedition seeking cities of gold located far to the north. This was in 1540, and before the spring of 1541, he and his company had wintered on the Rio Grande where they were perched to enter the grasslands of the present-day United States. While undoubtedly disappointed in not finding golden cities, Coronado had unquestionable admiration for the fecundity of the grasslands as depicted in his letter excerpted below.

After nine days' march I reached some plains, so vast that I did not find their limit anywhere that I went, although I traveled over them for more than 300 leagues. And I found such a quantity of cows [bison] in these, of the kind that I wrote Your Majesty about, which they have in this country, that it is impossible to number them, for while I was journeying through these plains, until I returned to where I first found them, there was not a day that I lost sight of them. And after seventeen days' march I came to a settlement of Indians who are called Querechos, who travel around with these cows, who do not plant, and who eat the raw flesh and drink the blood of the cows they kill, and they tan the skins of the cows, with which all the people of this country dress themselves here. They have little field tents made of the hides of the cows, tanned and greased, very well made, in which they live while they travel around near the cows, moving with these. They have dogs which they load, which carry their tents and poles and belongings. These people have the best figures of any that I have seen in the Indies. They could not give me any account of the country where the guides were taking me. I traveled five days more as the guides wished to lead me, until I reached some plains, with no more landmarks than as if we had been swallowed up in the sea, where they strayed about, because there was not a stone, nor a bit of rising ground, nor a tree, nor a shrub, nor anything to go by. There is much very fine pasture land, with good grass. And while we were lost in these plains, some horsemen who went off to hunt cows fell in with some Indians who also were out hunting, who are enemies of those that I had seen in the last settlement, and of another sort of people who are called Teyas; they have their bodies and faces all painted, are a large people like the others, of a very good build; they eat the raw flesh just like the Querechos, and live and travel round with the cows in the same way as these. I obtained from these an account of the country where the guides were taking me, which was not like what they had told me, because these made out that the houses there were not built of stones, with stories, as my guides had described it, but of straw and skins, and a small supply of corn there. . . .

The province of Quivira is 950 leagues from Mexico. Where I reached it, it is in the fortieth degree. The country itself is the best I have ever seen for producing all the products of Spain, for besides the land itself being very fat and black and being very well watered by the rivulets and springs and rivers, I found prunes like those of Spain [or I found everything they have in Spain] & nuts and very good sweet grapes and mulberries. . . . And what I am sure of is that there is not any gold nor any other metal in all that country, and the other things of which they had told me are nothing but little villages, and in many of these they do not plant anything and do not have any houses except of skins and sticks, and they wander around with the cows. . . .

Source: PBS—The West, "Coronado's Report to the King of Spain, Sent from Tiquex on October 20, 1541," http://www.pbs.org/weta/thewest/resources/archives/one/corona9.htm (accessed June 15, 2006).

GEORGE CATLIN, LETTER—NO. 33. FORT LEAVENWORTH, LOWER MISSOURI, 1844

Still today many scholars debate how purposefully Indian peoples once managed and shaped the grasslands of North America. George Catlin, known for his artistic rendering of Plains Indian peoples, gives in this letter excerpted below his understanding of Indians peoples' burning practices—how extensive and why.

I MENTIONED in a former epistle, that [Fort Leavenworth] is the extreme outpost on the Western Frontier, and built, like several others, in the heart of the Indian country. There is no finer tract of lands in North America, or, perhaps, in the world, than that vast space of prairie country, which lies in the vicinity of this post, embracing it on all sides. . . .

In this delightful Cantonment there are generally stationed six or seven companies of infantry, and ten or fifteen officers; several of whom have their wives and daughters with them, forming a very pleasant little community, who are almost continually together in social enjoyment of the peculiar amusements and pleasures of this wild country. Of these pastimes they have many, such as riding on horseback or in carriages over the beautiful green fields of the prairies, picking strawberries and wild plums—deer chasing—grouse shooting—horse-racing, and other amusements of the garrison, in which they are almost constantly engaged; enjoying life to a very high degree. . . .

The prairies burning form some of the most beautiful scenes that are to be witnessed in this country, and also some of the most sublime. Every acre of these vast prairies (being covered for hundreds and hundreds of miles, with a crop of grass, which dies and dries in the fall) burns over during the fall or early in the spring, leaving the ground of a black and doleful colour.

There are many modes by which the fire is communicated to them, both by white men and by Indians—par accident; and yet many more where it is voluntarily done for the purpose of getting a fresh crop of grass, for the grazing of their horses, and also for easier travelling during the next summer, when there will be no old grass to lie upon the prairies, entangling the feet of man and horse, as they are passing over them.

Over the elevated lands and prairie bluffs, where the grass is thin and short, the fire slowly creeps with a feeble flame, which one can easily step over; where the wild animals often rest in their lairs until the flames almost burn their noses, when they will reluctantly rise, and leap over it, and trot off amongst the cinders, where the fire has passed and left the ground as black as jet. These scenes at night become indescribably beautiful, when their flames are seen at many miles distance, creeping over the sides and tops of the bluffs, appearing to be sparkling and brilliant chains of liquid fire (the hills being lost to the view), hanging suspended in graceful festoons from the skies.

But there is yet another character of burning prairies, that requires another Letter, and a different pen to describe—the war, or hell of fires! Where the grass is seven or eight feet high, as is often the case for many miles together, on the Missouri bottoms; and the flames are driven forward by the hurricanes, which often sweep over the vast prairies of this denuded country. There are many of these meadows on the Missouri, the Platte, and the Arkansas, of many miles in breadth, which are perfectly level, with a waving grass, so high, that we are obliged to stand erect in our stirrups, in order to look over its waving tops, as we are riding through it. The fire in these, before such a wind, travels at an immense and frightful rate, and often destroys, on their fleetest horses, parties of Indians, who are so unlucky as to be overtaken by it; not that it travels as fast as a horse at full speed, but that the high grass is filled with wild pea-vines and other impediments, which render it necessary for the rider to guide his horse in the zig-zag paths of the deers and buffaloes, retarding his progress, until he is overtaken by the dense column of smoke that is swept before the fire—alarming the horse, which stops and stands terrified and immutable, till the burning grass which is wafted in the wind, falls about him, kindling up in a moment a thousand new fires, which are instantly wrapped in the swelling flood of smoke that is moving on like a black thunder-cloud, rolling on the earth, with its lightning's glare, and its thunder rumbling as it goes.

Source: Catlin, George. 1989. *North American Indians.* Edited and with an Introduction by Peter Matthiessen. New York: Penguin Books.

YELLOW WOLF AND THOMAS FITZPATRICK, "BIG TALK," 1847

By the end of the Mexican-American War, many Indian peoples in the Central and Southern Grasslands understood their way of life was rapidly becoming unsustainable. Some, such as Yellow Wolf, one of the Cheyenne chiefs, sought an alternative to a pastoral and barter-based economy. Thomas Fitzpatrick had this conversation with Yellow Wolf at Bent's Fort, and Fitzpatrick hoped to convince the Bureau of Indian Affairs to underwrite the transition costs of the Cheyennes' conversion to agriculture. Yellow Wolf never lived to see this change, as he was killed during the Sand Creek Massacre of November 1864.

I reminded [the Arapahos and Cheyennes present] of the great diminution and continual decrease of all game, and advised them to turn their attention to agriculture, it being the only means to save them from destruction. I pointed out and enumerated the many evils arising from the use of spiritous liquors, and advised them to abandon altogether so degrading and abominable a practice.

In reply to what I had said, one of the principal chiefs (Yellow Wolf [Southern Cheyenne]) spoke as follows: "My father, we are very poor and ignorant, even like the wolves in the prairie; we are not endowed with the wisdom of the white people. Father, this day we rejoice; we are no more poor and wretched; our great father has at length condescended to notice us, poor and wretched as we are; we now know we shall live and prosper, therefore we rejoice. My father we have not been warring against your people; why should we? on the contrary, if our great father wishes our aid, the Cheyenne warriors shall be ready at a moment's warning to assist in punishing those bad people, the Camanches [sic]." Here I interrupted him, saying that their great father had plenty of soldiers at his command—moreover, it was not his wish to embroil his red children in war with each other—on the contrary, he wished to see them unite in harmonious brotherhood. He continued—"Tell our great father that the Cheyennes are ready and willing to obey him in every thing; but, in settling down and raising corn, that is a thing we know nothing about, and if he will send some of his people to learn us, we will at once commence, and make every effort to live like the whites. We have long since noticed the decrease of the buffalo, and are well aware it cannot last much longer. . . ."

On the conclusion of the "big talk" with the Cheyennes, I addressed myself more particularly to the Aripohoes [sic], who were present, remarking that all they heard, applied equally to them as well as all other Indians who conducted in a peaceable and proper manner, and asked what they had to say in reply. They said, their ears were open and heard all, but could make no answer at present, inasmuch as they knew not the sentiments of their tribe. . . .

I do not wish to be understood as placing much confidence in the profession of the Indians of this country. . . . Many of them appear very desirous to commence raising corn, but I fear the effort will be found too laborious for them, unless they are encouraged and assisted. If the government wishes those Indians to settle down, they must give them some assistance, at least towards a beginning. A few dollars expended with those who are now willing to commence, might work some good, and be the means of inducing others to follow the example; and by the time the buffalo is all gone, those Indians will be prepared to live without them.

Source: U.S. Congress. Senate. 1847. Thomas Fitzpatrick, Indian Agent, Upper Platte and Arkansas to Thomas H. Harvey, Superintendent of Indian Affairs, St. Louis, Missouri, September 18, 1847. Vol. 1, S. Doc. 1, 13th Cong., 1st sess, pp. 238–249.

CATHERINE SAGER PRINGLE, "ACROSS THE PLAINS IN 1844, 1860"

Euro-American migrations across the grasslands often proved difficult and fraught with suffering. But more than Euro-Americans suffered as a result. Indian peoples, draft animals, and wildlife all felt the brunt of the "overlanders" on the Oregon Trail. A few insights into these travails come from the following excerpts from Pringle's published childhood experiences on the trail.

My father was one of the restless ones who are not content to remain in one place long at a time. Late in the fall of 1838 we emigrated from Ohio to Missouri. . . . In 1843, Dr. Whitman came to Missouri. The healthful climate induced my mother to favor moving to Oregon. Immigration was the theme all winter, and we decided to start for Oregon. Late in 1843 father sold his property and moved near St. Joseph, and in April, 1844, we started across the plains. The first encampments were a great pleasure to us children. We were five girls and two boys, ranging from the girl baby to be born on the way to the oldest boy, hardly old enough to be any help.

Starting on the Plains

We waited several days at the Missouri River. Many friends came that far to see the emigrants start on their long journey, and there was much sadness at the parting, and a sorrowful company crossed the Missouri that bright spring morning. The motion of the wagon made us all sick, and it was weeks before we got used to the seasick motion. Rain came down and required us to tie down the wagon covers, and so increased our sickness by confining the air we breathed.

Our cattle recrossed in the night and went back to their winter quarters. This caused delay in recovering them and a weary, forced march to rejoin the train. This was divided into companies, and we were in that commanded by William Shaw. Soon after starting Indians raided our camp one night and drove off a number of cattle. They were pursued, but never recovered.

Soon everything went smooth and our train made steady headway. The weather was fine and we enjoyed the journey pleasantly. There were several musical instruments among the emigrants, and these sounded clearly on the evening air when camp was made and merry talk and laughter resounded from almost every camp-fire.

Incidents of Travel

We had one wagon, two steady yoke of old cattle, and several of young and not well-broken ones. Father was no ox driver, and had trouble with these until one day he called on Captain Shaw for assistance. It was furnished by the good captain pelting the refractory steers with stones until they were glad to come to terms.

Reaching the buffalo country, our father would get some one to drive his team and start on the hunt, for he was enthusiastic in his love of such sport. He not only killed the great bison, but often brought home on his shoulder the timid antelope that had fallen at his unerring aim, and that are not often shot by ordinary marksmen. Soon after crossing South Platte the unwieldy oxen ran on a bank and overturned the wagon, greatly injuring our mother. She lay long insensible in the tent put up for the occasion.

August 1st we nooned in a beautiful grove on the north side of the Platte. We had by this time got used to climbing in and out of the wagon when in motion. When performing this feat that afternoon my dress caught on an axle helve and I was thrown under the wagon wheel, which passed over and badly crushed my limb before father could stop the team. He picked me up and saw the extent of the injury when the injured limb hung dangling in the air.

The Father Dying on the Plains

In a broken voice he exclaimed: "My dear child, your leg is broken all to pieces!" The news soon spread along the train and a halt was called. A surgeon was found and the limb set; then we pushed on the same night to Laramie, where we arrived soon after dark. This accident confined me to the wagon the remainder of the long journey.

After Laramie we entered the great American desert, which was hard on the teams. Sickness became common. Father and the boys were all sick, and we were dependent for a driver on the Dutch doctor who set my leg. He offered his services and was employed, but though an excellent surgeon, he knew little about driving oxen. Some of them often had to rise from their sick beds to wade streams and get the oxen safely across. One day four buffalo ran between our wagon and the one behind. Though feeble, father seized his gun and gave chase to them. This imprudent act prostrated him again, and it soon became apparent that his days were numbered. He was fully conscious of the fact, but could not be reconciled to the thought of leaving his large and helpless family in such precarious circumstances. The evening before his death we crossed Green River and camped on the bank. Looking where I lay helpless, he said: "Poor child! What will become of you?" Captain Shaw found him weeping bitterly. He said his last hour had come, and his heart was filled with anguish for his family. His wife was ill, the children small, and one likely to be a cripple. They had no relatives near, and a long journey lay before them. In piteous tones he begged the Captain to take charge of them and see them through. This he stoutly promised. Father was buried the next day on the banks of Green River. His coffin was made of two troughs dug out of the body of a tree, but next year emigrants found his bleaching bones, as the Indians had disinterred the remains.

Source: PBS—The West. "Catherine Sager Pringle, a Girl on the Oregon Trail," http://www.pbs.org/weta/thewest/resources/archives/two/sager1.htm (accessed June 15, 2006).

HOMESTEAD ACT, 1862

As Euro-American colonizers poured into the grasslands, a Republican-dominated Congress sought to stimulate the domestication of the wild grasslands through the rapid elimination of the public domain. The Homestead Act, which made the promise of "free" land to bona fide settlers was a key piece of legislation in this grand undertaking. The excerpts below give some of the key provisions in the act.

An Act to secure Homesteads to actual Settlers on the Public Domain.

Be it enacted by the Senate and House of Representatives of the United States of America in Congress assembled, That any person who is the head of a family, or who has arrived at the age of twenty-one years, and is a citizen of the United States, or who shall have filed his declaration of intention to become such, as required by the naturalization laws of the United States, and who has never borne arms against the United States Government or given aid and comfort to its enemies, shall, from and after the first January, eighteen hundred and sixty-three, be entitled to enter one quarter section or a less quantity of unappropriated public lands. . . .

SEC. 5. And be it further enacted, That if, at any time after the filing of the affidavit, as required in the second section of this act, and before the expiration of the five years aforesaid, it shall be proven, after due notice to the settler, to the satisfaction of the register of the land office, that the person having filed such affidavit shall have actually changed his or her residence, or abandoned the said land for more than six months at any time, then and in that event the land so entered shall revert to the government.

SEC. 6. And be it further enacted, That no individual shall be permitted to acquire title to more than one quarter section under the provisions of this act . . . .

SEC. 8. And be it further enacted, That nothing in this act shall be so construed as to prevent any person who has availed him or herself of the benefits of the first section of this act, from paying the minimum price, or the price to which the same may have graduated, for the quantity of land so entered at any time before the expiration of the five years, and obtaining a patent therefore from the government, as in other cases provided by law, on making proof of settlement and cultivation as provided by existing laws granting preemption rights.

Source: An Act to Secure Homesteads to Actual Settlers on the Public Domain, *U.S. Statues at Large*, 37th Cong., 2nd sess., May 20, 1862, pp. 392–393.

MORRILL ACT, 1862

As the Homestead Act attempted to make land accessible to the common person, the Morrill Act opened the doors of higher education to the same class of Americans. Heretofore most higher education was for an elite class who could afford attending private institutions. With this act, not only had the "industrial classes" of Euro-American males gained entry to higher education, but so had women and minorities. Along with the passage of the Hatch Act in 1887, which created experiment stations in land-grant institutions, and the passage of the Smith-Lever Act in 1914, which underwrote the extension service, the Morrill Act began a fundamental transformation of American higher education, which had a dramatic effect on domesticating the grasslands.

An Act Donating Public Lands to the several States and Territories which may provide Colleges for the Benefit of Agriculture and Mechanic Arts.

Be it enacted by the Senate and House of Representatives of the United States of America in Congress assembled, That there be granted to the several States, for the purposes hereinafter mentioned, an amount of public land, to be apportioned to each State a quantity equal to thirty thousand acres for each senator and representative in Congress to which the States are respectively entitled by the apportionment under the census of eighteen hundred and sixty: Provided, That no mineral lands shall be selected or purchased under the provisions of this Act. . . .

SEC. 4. And be it further enacted, That all moneys derived from the sale of the lands aforesaid by the States to which the lands are apportioned, and from the sales of land scrip hereinbefore provided for, shall be invested in stocks of the United States, or of the States, or some other safe stocks, yielding not less than five per centum upon the par value of said stocks; and that the moneys so invested shall constitute a perpetual fund, the capital of which shall remain forever undiminished, (except so far as may be provided in section fifth of this act,) and the interest of which shall be inviolably appropriated, by each State which may take and claim the benefit of this act, to the endowment, support, and maintenance of at least one college where the leading object shall be, without excluding other scientific and classical studies, and including military tactics, to teach such branches of learning as are related to agriculture and the mechanic arts, in such manner as the legislatures of the States may respectively prescribe, in order to promote the liberal and practical education of the industrial classes in the several pursuits and professions in life.

Source: An Act Donating Public Lands to the Several States and Territories Which May Provide Colleges for the Benefit of Agriculture and the Mechanic Arts, *U.S. Statutes at Large,* 37th Cong., 2nd sess., July 2, 1862, pp. 503–505.

TESTIMONY OF BENJAMIN SINGLETON, BEFORE THE SENATE SELECT COMMITTEE INVESTIGATING THE "NEGRO EXODUS FROM THE SOUTHERN STATES," WASHINGTON, D.C., APRIL 17, 1880

The Homestead Act opened land-owning opportunities for many people. Freed slaves in several southern states availed themselves of this chance to acquire their own farms free of the brutal persecution they were facing at the time. One of the main promoters of relocating African-Americans to the grasslands was Benjamin "Pap" Singleton, and his motives and aspirations can be gleaned from the following excerpts of his testimony before Congress in 1880.

Q. Yes; What was the cause of your going out, and in the first place how did you happen to go there, or to send these people there?

A. Well, my people, for the want of land—we needed land for our children—and their disadvantages—that caused my heart to grieve and sorrow; pity for my race, sir, that was coming down, instead of going up—that caused me to go to work for them. I sent out there perhaps in '66—perhaps so; or in '65, any way—my memory don't recollect which; and they brought back tolerable favorable reports; then I jacked up three or four hundred, and went into Southern Kansas, and found it was a good country, and I thought Southern Kansas was congenial to our nature, sir; and I formed a colony there, and bought about a thousand acres of ground—the colony did—my people. . . .

Q. They had some means to start with?

A. Yes; I prohibited my people leaving their country and going there without they had money—some money to start with and go on with a while. . . .

Q. Have they any property now?

A. Yes; I have carried some people in there that when they got there they didn't have fifty cents left, and now they have got in my colony—Singleton colony—a house, nice cabins, their milch cows, and pigs, and sheep, perhaps a span of horses, and trees before their yards, and some three or four or ten acres broken up, and all of them has got little houses that I carried there. They didn't go under no relief assistance; they went on their own resources; and when they went in there first the country was not overrun with them; you see they could get good wages; the country was not overstocked with people; they went to work, and I never helped them as soon as I put them on the land. . . .

Q. Well, tell us all about it.

A. These men would tell all their grievances to me in Tennessee—the sorrows of their heart. You know I was an undertaker there in Nashville, and worked in the shop. Well, actually, I would have to go and bury their fathers and mothers. You see we have the same heart and feelings as any other race and nation. (The land is free, and it is nobody's business, if there is land enough, where the people go. I put that in my people's heads.) Well, that man would die, and I would bury him; and the next morning maybe a woman would go to that man (meaning the landlord), and she would have six or seven children, and he would say to her, "Well, your husband owed me before he died" and they would say that to every last one of them, "You owe me." Suppose he would? Then he would say, "You must go to some other place; I cannot take care of you." Now, you see, that is something I would take notice of, that woman had to go out, and these little children was left running through the streets, and the next place you would find them in a disorderly house, and their children in the State's prison. . . .

I then went out to Kansas, and advised them all to go to Kansas; and, sir they are going to leave the Southern country. The Southern country is out of joint. The blood of a white man runs through my veins. That is congenial, you know, to my nature, that is my choice. Right emphatically, I tell you today, I woke up the millions right through me! The great God of glory has worked in me. I have had open air interviews with the living spirit of God for my people; and we are going to leave the South. We are going to leave it if there ain't an alteration and signs of change. . . .

Source: PBS—The West. "Testimony of Benjamin Singleton before the Senate Select Committee Investigating the 'Negro Exodus from the Southern States,' April 17th, 1880," http://www.pbs.org/weta/thewest/resources/archives/seven/w67singl.htm (accessed June 15, 2006).

DAWES ACT, 1887

Many Euro-American reformers sought to transform Indian peoples into Christian, English-speaking, yeoman farmers. This was a part of the larger effort to domesticate the grasslands of the United States. The Dawes Act was an important contribution to this overall endeavor, as it broke up reservations and parceled the land to individuals within the tribes and sold the remaining land on the open market. The key provisions of this act are given below.

An Act to provide for the allotment of lands in severalty to Indians on the various reservations, and to extend the protection of the laws of the United States and the Territories over the Indians, and for other purposes.

Be it enacted by the Senate and House of Representatives of the United States of America in Congress assembled, That in all cases where any tribe or band of Indians has been, or shall hereafter be, located upon any reservation created for their use, either by treaty stipulation or by virtue of an act of Congress or executive order setting apart the same for their use, the President of the United States be, and he hereby is, authorized, whenever in his opinion any reservation or any part thereof of such Indians is advantageous for agricultural and grazing purposes, to cause said reservation, or any part thereof, to be surveyed, or resurveyed if necessary, and to allot the lands in said reservation in severalty to any Indian located thereon in quantities as follows: To each head of a family, one-quarter of a section; To each single person over eighteen years of age, one-eighth of a section; To each orphan child under eighteen years of age, one-eighth of a section; and To each other single person under eighteen years now living, or who may be born prior to the date of the order of the President directing an allotment of the lands embraced in any reservation, one-sixteenth of a section: . . .

SEC. 2. That all allotments set apart under the provisions of this act shall be selected by the Indians, heads of families selecting for their minor children, and the agents shall select for each orphan child, and in such manner as to embrace the improvements of the Indians making the selection. . . .

SEC. 5. That upon the approval of the allotments provided for in this act by the Secretary of the Interior, he shall cause patents to issue therefor in the name of the allottees, which patents shall be of the legal effect, and declare that the United States does and will hold the land thus allotted, for the period of twenty-five years, in trust for the sole use and benefit of the Indian to whom such allotment shall have been made. . . . That all lands adapted to agriculture, with or without irrigation so sold or released to the United States by any Indian tribe shall be held by the United States for the sole purpose of securing homes to actual settlers and shall be disposed of by the United States to actual and bona fide settlers only tracts not exceeding one hundred and sixty acres to any one person, on such terms as Congress shall prescribe, subject to grants which Congress may make in aid of education: . . .

SEC. 7. That in cases where the use of water for irrigation is necessary to render the lands within any Indian reservation available for agricultural purposes, the Secretary of the Interior be, and he is hereby, authorized to prescribe such rules and regulations as he may deem necessary to secure a just and equal distribution thereof among the Indians residing upon any such reservation; and no other appropriation or grant of water by any riparian proprietor shall be permitted to the damage of any other riparian proprietor.

Source: An Act to Provide for the Allotment of Lands in Severalty to Indians on the Various Reservations, and to Extend the Protection of the Laws of the United States and Territories over the Indians, and for Other Purposes. *U.S. Statutes at Large,* 49th Cong., 2nd sess., February 8, 1887, pp. 388–391.

JEROME COMMISSION, 1892

Indian peoples understood that their way of living on the land would be dramatically changed with the implementation of the Dawes Act. They also knew that they possessed few means to realize the well-intentioned goals of Euro-American reformers. Below, the excerpts illustrate the concerns of some, and especially take note of Big Tree's horse analogy for the Earth.

Before the beginning of the Council, Howea, a Comanche Chief, presented to the Commission a printed copy of the treaty made at Medicine Lodge, concluded October 21, 1867; ratification advised July 25, 1868; proclaimed August 25, 1868.

(Mr. Jerome) The Council will now open. . . .

In 1889 the great council of the United States passed a law authorizing the President to appoint three commissioners to come out and trade with the Indians about their lands. Under that law the President appointed Judge Wilson, Judge Sayre and myself to come out here and talk to you for the Government. We come here this morning as strangers to these Indians, but we hope before we get ready to leave, before we get through with these councils, that our conduct will be such that we will go away the friends of the Indians. . . .

After having thought of these matters for a great many years at Washington they have adopted a plan for all the Indians that are in the country and that is what we are here to tell you about. . . .

The plan in a few words is this, that every Indian on the reservation, every man, woman and child, shall have ample land set off for a home that can never be taken away from him that they can live and cultivate forever. In selecting these homes and setting them apart to you, your children and yourselves will always have as much land as they can use, even if they get ten times as many as you have now, and I do not think that time will ever come. These homes are to be selected from the best lands in the reservation to the end that you may make a living on the best of this most fruitful country. You should make your own selections, each for yourself and your wife and your children, take the very best lands you can find.

(Quannah Parker, Com.) I have talked with a great many officers and men of learning and the agent and have taken their advice about taking farms and building houses, and it can readily be seen that my people take after me by looking at their places. My friends here, the Kiowas and Apaches, have talked the matter over with me and I know that some of the people have made up their minds that it would be a good thing to sell our country and are in a hurry because of the money, but I want a thorough understanding and thought it would be better to wait until the expiration of [The Treaty of Medicine Lodge]. I have been doing business for myself and people for quite a length of time and am a leader of my people and they are pressing this matter too close. I do not want them to do anything until they know about it. . . .

(Big Tree, Kiowa) . . .

Yesterday the Commission told us that they were here to trade for our land. I want you to look at the three tribes that are before you. They give a very good picture of what they have been, what they are and what they will be in the future. They do not know how to take care of themselves. Take pity on them because they are ignorant, if I was in your place I would have some good feeling for these Indians. Whenever these Indians travel they generally look for good places to cross, they do not drive over hills and banks but they look for good places, good crossings. They do not care to jump across a big gap. These Indians that are sitting before you have only one horse; it is very large and very fat, it is a working horse, it can plow the ground and bring us some grain. When we want to haul wood we have to use him, when we have to do anything that is heavy we have to use him and now you take this horse away from us, it is very hard for us to give this horse up. We hope that you will not force us to give him up. For some time past we have heard that this Commission has been trading with the different tribes of Indians in the north, trying to get them to dispose of their surplus lands.

A year ago this Commission came to the Cheyenne and Arapahoe Indians. They talked to those Indians very good. These Indians came to the Kiowa and Comanche reservation. We saw tears in their eyes, we saw that they had nothing to their name. They are poor, they will be poor in the future; they had made a mistake in selling their country; that money was given them but it was all gone. You are here today on the Kiowa and Comanche Reservation. Look at these three tribes and think about their clothing. They are not civilized, their hands are not trained to work, they are not taught to take care of themselves and for that reason we desire the Government not to force us to sell the country. In [The Medicine Lodge Treaty] we are told that we are to live upon this reservation thirty years, and during that period we are to be furnished clothing and other things . . . we can not throw it away, we can not burn it, we must look into it a little. If I were to come to your house and your place and attempt to buy something that you prize very highly, you would probably laugh at me and tell me you were not anxious to sell it. So I tell you, I am not anxious to sell this useful horse.

(I-see-o, Kiowa) . . .

You look at these Indians, since you speak to them you frighten them, you made them feel uneasy. You can see that by what they say of each other a few days ago, and this kind of talk is not very helpful to these Indians at this time, instead of doing them good it is doing them harm. While some of the old men say they are ready to take the good road, yet, if the young men talk of selling the country they will be the first to die, they are not taught to work. They have only a few hundred cattle and they will not last many days, they will starve. It is something that they are not ready for now. . . . You probably told the [Cheyennes] what you have told us for the past few days. And the Cheyennes talked amongst themselves and decided to take their allotments. This they did and it is only a few months ago that the Cheyennes came to this military reservation and brought their wagons and fancy shawls, velvet blankets and carriages and told us that the money that the great father had given them was all gone; that the money they got was invested in these things. Now the wagons are old, being used very hard and the velvet shawls will be worn out. What were they here for?

They came down to get some cattle and ponies from the Kiowas. They gave us a big dance, so we gave them some ponies. In a few years those Cheyenne Indians will be the poorest Indians, and they will be coming all the time for ponies. Look at them today surrounded by white men, they will get the Indians drunk and get his money; they will make him sign a contract to get anything that the Cheyenne has got, and the Cheyenne's life in the next three years will be worse than when he was an Indian. So that is why we say wait. . . .

Source: "Minutes of the Jerome Commission Meeting with the Kiowa, Comanche and Apache." September 28–October 17, 1892. RG 75, Kiowa Agency Files. National Archives Southwest Region, Texas, Fort Worth.

KANSAS V. COLORADO (1907)

It was just a matter of time before the hundreds of irrigation systems built throughout the grasslands would cause noticeable depletions and interruptions to interstate river flows. This naturally led to conflicts and litigation. Such was the case for Coloradans and Kansans as they fought each other over the flows in the Arkansas River. Their contest was the first of this nature to appear as a suit of original jurisdiction before the U.S. Supreme Court. Justice David Brewer wrote the opinion in which he simultaneously restricted the power of the federal government to regulate interstate stream flows and in which he established the doctrine of equity. This doctrine is still the fundamental creed for interstate water cases involving the federal courts.

Mr. Justice Brewer . . . delivered the opinion of the court. . . .

Turning now to the controversy as here presented, it is whether Kansas has a right to the continuous flow of the waters of the Arkansas River, as that flow existed before any human interference therewith, or Colorado the right to appropriate the waters of that stream so as to prevent that continuous flow, or that the amount of the flow is subject to the superior authority and supervisory control of the United States. . . .

The primary question is, of course, of national control. For if the Nation has a right to regulate the flow of the waters, we must inquire what it has done in the way of regulation. If it has done nothing the further question will then arise, what are the respective rights of the two States in the absence of national regulation? . . .

This very matter of the reclamation of arid lands illustrates this: At the time of the adoption of the Constitution within the known and conceded limits of the United States there were no large tracts of arid land, and nothing which called for any further action than that which might be taken by the legislature of the State, in which any particular tract of such land was to be found, and the Constitution, therefore, makes no provision for a national control of the arid regions or their reclamation. But, as our national territory has been enlarged, we have within our borders extensive tracts of arid lands which ought to be reclaimed, and it may well be that no power is adequate for their reclamation other than that of the National Government. But if no such power had been granted, none can be exercised.

It does not follow from this that the National Government is entirely powerless in respect to this matter. These arid lands are largely with the Territories, and over them by virtue of [the Constitution] Congress has full power of legislation, subject to no restrictions other than those expressly named in the Constitution, and, therefore, it may legislate in respect to all arid lands within their limits. As to those lands within the limits of the States, at least of the Western States, the National Government is the most considerable owner and has power to dispose of and make all needful rules and regulations respecting its property. . . .

Now the question arises between two States, one recognizing generally the common law rule of riparian rights and the other prescribing the doctrine of the public ownership of flowing water. Neither State can legislate for or impose its own policy upon the other. . . . It does not follow, however, that because Congress cannot determine the rule which shall control between the two States or because neither State can enforce its own policy upon the other, that the controversy ceases to be one of a justiciable nature, or that there is no power which can take cognizance of the controversy and determine the relative rights of the two States. . . .

It cannot be denied in view of all the testimony (for that which we have quoted is but a sample of much more bearing upon the question), that the diminution of the flow of water in the river by the irrigation of Colorado has worked some detriment to the southwestern part of Kansas, and yet when we compare the amount of this detriment with great benefit which has obviously resulted to the counties in Colorado, it would seem that equality of right and equity between the two States forbids any interference with the present withdrawal of water in Colorado for purposes of irrigation. . . .

Summing up our conclusions, we are of the opinion that the contention of Colorado of two streams cannot be sustained; that the appropriation of the waters of the Arkansas by Colorado, for purposes of irrigation, has diminished the flow of water into the State of Kansas; that the result of that appropriation has been the reclamation of large areas in Colorado, transforming thousands of acres into fertile fields and rendering possible their occupation and cultivation when otherwise they would have continued barren and unoccupied; that while the influence of such diminution has been of perceptible injury to portions of the Arkansas Valley in Kansas, particularly those portions closest to the Colorado line, yet to the great body of the valley it has worked little, if any, detriment, and regarding the interests of both States and the right of each to receive benefit through irrigation and in any other manner from the waters of this stream, we are not satisfied that Kansas has made out a case entitling it to a decree. At the same time it is obvious that if the depletion of the waters of the river by Colorado continues to increase there will come a time when Kansas may justly say that there is no longer an equitable division of benefits and may rightfully call for relief against the action of Colorado, its corporations and citizens in appropriating the Waters of the Arkansas for irrigation purposes.

Source: State of Kansas v. State of Colorado, et al., The United States of America, Intervener, 27 Sup. Ct. 653 (1906).

WINTERS V. THE UNITED STATES (1908)

Indian peoples needed to have access to water if they were ever going to have a chance to farm their reservations. The United States Supreme Court awarded the Indians living on the Fort Belknap Reservation "reserved" rights to water for their farming regardless of the claims and legal arguments of upstream Euro-American farmers. This important decision has had far-ranging consequences for the development and distribution of water rights throughout the semi-arid and arid grasslands.

Mr. Justice McKenna . . . delivered the opinion of the court. . . .

The case, as we view it, turns on the agreement of May, 1888, resulting in the creation of Fort Belknap Reservation. In the construction of this agreement there are certain elements to be considered that are prominent and significant.

The reservation was a part of a very much larger tract which the Indians had the right to occupy and use, and which was adequate for the habits and wants of a nomadic and uncivilized people. It was the policy of the government, it was the desire of the Indians, to change those habits and to become a pastoral and civilized people. If they should become such, the original tract was too extensive; but a smaller tract would be inadequate without a change of conditions. The lands were arid, and, without irrigation, were practically valueless. And yet, it is contended, the means of irrigation were deliberately given up by the Indians and deliberately accepted by the government. The lands ceded were, it is true, also arid; and some argument may be urged, and is urged, that with their cession there was the cession of the waters, without which they would be valueless, and "civilized communities could not be established thereon." And this, it is further contended, the Indians knew, and yet made no reservation of the waters. We realize that there is a conflict of implications, but that which makes for the retention of the waters is of greater force than that which makes for their cession. The Indians had command of the lands and the waters—command of all their beneficial use, whether kept for hunting, "and grazing roving herds of stock," or turned to agriculture and the arts of civilization. Did they give up all this? Did they reduce the area of their occupation and give up the waters which made it valuable or adequate? And, even regarding the allegation of the answer as true, that there are springs and streams on the reservation flowing about 2,900 inches of water, the inquiries are pertinent. If it were possible to believe affirmative answers, we might also believe that the Indians were awed by the power of the government or deceived by its negotiators. Neither view is possible. The government is asserting the rights of the Indians. But extremes need not be taken into account. By a rule of interpretation of agreements and treaties with the Indians, ambiguities occurring will be resolved from the standpoint of the Indians. And the rule should certainly be applied to determine between two inferences, one of which would support the purpose of the agreement and the other impair or defeat it. On account of their relations to the government, it cannot be supposed that the Indians were alert to exclude by formal words every inference which might militate against or defeat the declared purpose of themselves and the government, even if it could be supposed that they had the intelligence to foresee the "double sense" which might some time be urged against them.

    . . . That the government did reserve [the water appropriations for the reservation] we have decided, and for a use which would be necessarily continued through years. This was done May 1, 1888, and it would be extreme to believe that within a year Congress destroyed the reservation and took from the

Indians the consideration of their grant, leaving them a barren waste—took from them the means of continuing their old habits, yet did not leave them the power to change to new ones.

Source: *Winters v. United States*, 207 Sup. Ct. 564 (1908).

MAGGIE AND DENNIS DAVIS, LETTERS FROM THEIR MONTANA HOMESTEAD, 1910–1926

Moving onto and farming the open grasslands had great appeal to Euro–Americans. Even in the semi-arid northern grasslands, many thought they could reap fantastic profits by using dryland farming techniques. Maggie and Dennis Davis were such optimistic emigrants to the eastern grasslands of Montana. As the excerpts from their letters below reveal, they, and countless others, failed to find their pot of gold.

[June 1910, Maggie] There are houses all around us. Coming in from the river one night I counted over forty. . . . I think that story we read was a pretty true picture of Montana life. The boys say that all the people they have met who are coming in are nice but the people who already lived here are the ones to look out for. There are several big ranchmen near here. Some are friendly and some of them went to Fort Benton and tried to get the merchants to promise not to sell to the dryland farmers as they call them.

[July 1911, Dennis] There were thousands of acres [of wheat] sown in the vicinity this fall and prospects for a bumper crop are fine so far this winter. I am one of the small farmers, of course, but I have sown 120 acres of winter wheat and expect or at least I am planning to put in about 70 acres this spring.

[February 1913, Maggie] You don't need to think there is any danger of Dennis getting discouraged out here for wild horses couldn't drag him away from here. He says for the first time in his life he has an opportunity to make good. Mr. Wiley [a neighbor and Burnettsville, Indiana, family friend] was offered a loan of $4,500 on his farm and I see no reason why we could not get that much so you see we could get that anytime and pull out, *but we'll never get another farm given to us* so we had better stay with it till land advances to $50 to $60 then we can sell and retire or travel or something.

[July 1915, Maggie] It was dry, not a drop of rain in April, one rain in May but on the first day of June it began to rain and has rained nearly every day since. . . . It isn't supposed to rain in July on account of harvest and never has before. Gardens, corn and oats are not very good. . . .

There have been two ranches sold for $8500 each. Do you think that is a good price or not? One of the men is sorry already that he sold. Dennis has 130 acres of summer fallow ready to sow. But we have to work awful hard and help is so hard to get, that is good help and so expensive. We now have the third man this summer and none very good and the one we have now wants $3 per day through harvest so I think we'll have to let him go. I think some time we are foolish to work so hard. By the time we get enough ahead to buy an automobile we'll be too old to enjoy it.

[June 1919, Maggie] The winter wheat and rye had all been ruined. . . . And of all the poor horses I ever saw ours were the poorest. . . . We were over south of Benton and a little cloud came up. Well it rained quite hard. It was pathetic to see the joy of the people. . . . It rained even harder out here. Dennis had already sowed 48 acres of spring wheat and reseeded most of 56 acres to oats and wheat and we went right to work and sowed 34 acres of flax. That was all the seed we had and we were afraid to risk buying more. There were 14 acres more in the field. There have been several more good rains since. The flax is looking nice and the wheat was until Monday and we had a dreadful sand storm and when it was over the field looked bare again. We have millions of grasshoppers again. They are taking the garden as fast as it comes up. . . . Dennis is trying to get a piece of ground ready for millet. But he is so crippled and the horses are so poor and the thistles grow so fast it is uphill business. . . . Dennis says all we lack of selling is to find a sucker. . . .

They say the dry years are over for Montana now for four or five years but I don't know what to think. We have land that ought to be summer fallowed now. I wanted him to get 45 acres ready but he gets discouraged so easy and give up. We have 6 heifers and cows that will have to go I guess. But we cant get rid of the horses unless we eat them.

[October 1919, Maggie] Not a thing except a three or four year old straw pile occasionally for the stock. Not one family in a hundred has potatoes and not a bite from the garden. . . . If Dennis could have got some one to have kept the cattle he would have gone to work in [Great Falls]. . . . The Unions are howling for more money all the time but from $5 to $7 a day looks pretty good to a poor farmer who hasn't had a crop for 3 years.

[April 1923, Maggie] Our prospects in every way were never poorer. We have had a dry winter. Winter wheat is going or gone. Fields that should be green are brown as can be. Everybody that was left is planning on going now. . . . Even the bankers have nothing encouraging to say anymore. . . . I don't know what will become of us. If we could sell the land I would be for coming but as it is I don't see any hopes of ever coming.

[*Burnettsville* (Indiana) *News*, July 1926] Dennis Davis arrived here yesterday morning from Carter, Montana, with all of his cattle, horses, farming tools and household goods to make his home in Hoosierland again. He left here sixteen years ago and took up a 320 acre claim near Carter and has met the same fate that has befallen all of those who settled in that semi-arid region. The soil is extremely fertile and produces immense crops when there is enough rainfall, otherwise they are a failure. Unfortunately the barren years outnumber the fruitful ones and the farms in the locality are being rapidly deserted and their owners are seeking other localities in which to recoup their ill fortune.

Source: Kohl, Seena B. 2001. "'Well I Have Lived in Montana for Almost a Week and Like It Fine': Letters from the Davis Homestead, 1910–1926." *Montana, Magazine of Western History* 51 (Autumn): 32–45.

WILLIAM ALLEN WHITE, "KANSAS: A PURITAN SURVIVAL," 1922

While domesticating the grasslands seemed a good and noble ideal, some Americans began to wonder if turning all of the prairies into ranches and farms was such a good idea. William Allen White, the "Sage of Emporia," who reveled in the beauty of the grasslands, expressed his own misgivings in the excerpt below.

. . . Our sense of humor saves us, but not entirely whole; we have never laughed ourselves out of our essential Puritanism. . . . Joy is an incident, not the business of life. Justice as it works out under a Christian civilization is the chief end of man in Kansas.

But alas, this is begging the question. For who can say that the establishment of justice is the chief end of a state? Indeed who can say even what justice is? Is it just that every man should earn what he gets and get what he earns? Or is it just that those who see and feel and aspire to do great things—to make life beautiful for themselves and others—should be pared down to the norm in their relations with mankind? Is it justice to establish a state where the weak may thrive easily and the strong shall be fettered irrevocably in their most earnest endeavors? Should a state brag of the fact that it distributes its wealth equitably—almost evenly—when it has produced no great poet, no great painter, no great musician, no great writer or philosopher? Surely the dead level of economic and political democracy is futile if out of it something worthy—something eternally worthy—does not come. The tree shall be known by its fruit.

What is the fruit of Kansas? Is happiness for the many worth striving for? What is the chief end of a civilization? What is the highest justice?

What we lack most keenly is a sense of beauty and the love of it. Nothing is more gorgeous in color and form than a Kansas sunset; yet it is hidden from us. The Kansas prairies are as mysterious and moody as the sea in their loveliness, yet we graze them and plow them and mark them with roads and do not see them. The wind in the cottonwoods lisps songs as full of meaning as those the tides sing, and we are deaf. The meadow lark, the red bird, the quail live with us and pipe to us all through the year, but our musicians have not returned the song. The wide skies of night present the age-old mystery of life, in splendor and baffling magnificence, yet only one Kansas poet . . . has ever worn Arcturus as a bosom pin. . . .

Yet why—why is the golden bowl broken, the pitcher at the fountain broken, and in our art the wheel at the cistern still? This question is not peculiarly a Kansas question. It is tremendously American.

Source: White, William Allen. 1922. "Kansas: A Puritan Survival." *Nation* (April 19): 460–462.

TAYLOR GRAZING ACT, 1934

By the early 1930s, people became more aware of denuded public ranges, starving cattle, and economically depressed ranchers, and they called for some means to regulate the use of public grasslands. The federal government, as part of New Deal conservation policies, stepped in to bring greater ecological regulation and management of the public range. The act below was one of the tools used to achieve this end.

Be it enacted, That in order to promote the highest use of the public lands pending its final disposal, the Secretary of the Interior is authorized, in his discretion, by order to establish grazing districts or additions thereto and/or to modify the boundaries thereof, not exceeding in the aggregate an area of eighty million acres of vacant, unappropriated, and unreserved lands from any part of the public domain of the United States (exclusive of Alaska), which are not in national forests, national parks and monuments, Indian reservations, [etc.] and which in his opinion are chiefly valuable for grazing and raising forage crops: Provided, That no lands withdrawn or reserved for any other purpose shall be included in any such district except with the approval of the head of the department having jurisdiction thereof. . . .

SEC. 2. The Secretary of the Interior shall make provision for the protection, administration, regulation, and improvement of such grazing districts as may be created under the authority of the foregoing section, and he shall make such rules and regulations and establish such service, enter into such cooperative agreements, and do any and all things necessary to accomplish the purposes of this Act and to insure the objects of such grazing districts, namely, to regulate their occupancy and use, to preserve the land and its resources from destruction or unnecessary injury, to provide for the orderly use, improvement, and development of the range; and the Secretary of the Interior is authorized to continue the study of erosion and flood control and to perform such work as may be necessary amply to protect and rehabilitate the areas subject to the provisions of this Act, through such funds as may be made available for that purpose, and any willful violation of the provisions of this Act or of such rules and regulations thereunder after actual notice thereof shall be punishable by a fine of not more than $500.

SEC 3. That the Secretary of the Interior is hereby authorized to issue or cause to be issued permits to graze livestock on such grazing districts to such bona fide settlers, residents, and other stock owners as under his rules and regulations are entitled to participate in the use of the range, upon the payment annually of reasonable fees in each case to be fixed or determined from time to time. . . . Such permits shall be for a period not more than ten years, subject to the preference right of the permittee to renewal in the discretion of the secretary of the interior, who shall specify from time to time numbers of stock and seasons of use. During periods of range depletion due to severe drought, natural or other unnatural causes, or in case of a general epidemic of disease, during the life of the permit, the Secretary of the Interior is hereby authorized, in his discretion to remit, reduce, refund in whole or in part, or authorize postponement of payment of grazing fees for such depletion period so long as the emergency exists. . . .

SEC. 7 That the Secretary is hereby authorized, in his discretion, to examine and classify all lands within such grazing districts which are more valuable and suitable for the production of agricultural crops than native grasses and forage plants, and to open such lands to homestead entry in tracts not exceeding three hundred and twenty acres in area. Such lands shall not be subject to settlement or occupation as homesteads until after same have been classified and opened to entry after notice to the committee by the Secretary of the Interior, and the lands shall remain a part of the grazing district until patents are issued there for the homesteader to be, after his entry is allowed, entitled to the possession and use thereof. . . .

Source: An Act to Stop Injury to the Public Grazing Lands by Preventing Overgrazing and Soil Deterioration, to Provide for their Orderly Use, Improvement, and Development, to Stabilize the Livestock Industry Dependent upon the Public Range, and for other Purposes. *U.S. Statutes at Large*, 73rd Cong., 2nd sess., June 28, 1934, p. 1269.

SOIL CONSERVATION AND DOMESTIC ALLOCATION ACT, 1935

Great, billowing dust storms engulfed large portions of the Central Grasslands during the early to mid-1930s. Hugh Hammond Bennett understood that the farming techniques of the time, when coupled with dry conditions, led to disastrous social, economic, and ecological results. Through his efforts, this landmark legislation empowered the federal government to address these mounting problems.

An Act to provide for the protection of land resources against soil erosion, and for other purposes.

Be it enacted by the Senate and House of Representatives of the United States of America in Congress assembled, That it is hereby recognized that the wastage of soil and moisture resources on farm, grazing, and forest lands of the Nation, resulting from soil erosion, is a menace to the national welfare and that it is hereby declared to be the policy of Congress to provide permanently for the control and prevention of soil erosion and thereby to preserve natural resources, control floods, prevent impairment of reservoirs, and maintain the navigability of rivers and harbors, protect public health, public lands and relieve unemployment, and the Secretary of Agriculture, from now on, shall coordinate and direct all activities with relation to soil erosion and in order to effectuate this policy is hereby authorized, from time to time—

1. To conduct surveys, investigations, and research relating to the character of soil erosion and the preventive measures needed, to publish the results of any such surveys, investigations, or research, to disseminate information concerning such methods, and to conduct demonstrational projects in areas subject to erosion by wind or water;

2. To carry out preventive measures, including, but not limited to, engineering operations, methods of cultivation, the growing of vegetation, and changes in use of land;

3. To cooperate or enter into agreements with, or to furnish financial or other aid to, any agency, governmental or otherwise, or any person, subject to such conditions as he may deem necessary, for the purposes of this Act; and

4. To acquire lands, or rights or interests therein, by purchase, gift, condemnation, or otherwise, whenever necessary for the purposes of this Act. . . .

5. The Secretary of Agriculture shall establish an agency to be known as the "Soil Conservation Service," to exercise the powers conferred on him by this Act.

Source: An Act to Provide for the Protection of Land Resources against Soil Erosion, and for Other Purposes. *U.S. Statutes at Large*, 74th Cong., 1st sess., April 27, 1935, pp. 163–164.

THE FUTURE OF THE GREAT PLAINS, 1936

Dust storms, denuded rangelands, and economically depressed farmers and ranchers led many New Dealers to advocate other ways of domesticating the grasslands. This landmark report set the guidelines for reform that the federal government would pursue well into the twenty-first century.

The Purpose of the Report

In 1934 and again in 1936 drought conditions in the Great Plains area of the United States became so severe that it was necessary for the Federal Government to take emergency steps to rescue dying cattle, relieve destitute families, and safeguard human life. The experience of the two tragic years made it evident that the drought had merely accentuated a situation which had been long developing. The agricultural economy of the Great Plains had a perilously narrow reserve. Its prosperity depended on favorable weather and markets, neither of which could be expected to be continuously present.

Droughts could not be prevented. They were admittedly part of a weather cycle which runs its course beyond the range of human interference. Agriculture must adapt itself to the cycle and make the most of what Nature has to offer. This it had largely failed to do. It became clear that unless there was a permanent change in the agricultural pattern of the Plains, relief always would have to be extended whenever the available rainfall was deficient. Current methods of cultivation were so injuring the land that large areas were decreasingly productive even in good years, while in bad years they tended more and more to lapse into desert. The water supply, which literally meant life or death to human activities in the Region, was being in part permitted to run to waste, in part put to uses which did not extract all its values. . . .

These factors led to the creation of the Great Plains Drought Area Committee, which rendered a preliminary report last August; and to the appointment by the President of the Great Plains Committee, whose further studies are summarized and recommendations submitted in the present report.

The Nature of the Problem

The present situation in the Great Plains area is the result of human modification of natural conditions. Prior to the coming of the white man, and to a large extent prior to about 1866, man did not greatly alter conditions on the Plains. The Indians did two things: they killed buffalo and they sometimes set fire to the grass. They do not seem to have reduced the number of the buffalo seriously, and though their fires may have influenced the nature of the vegetation they did not destroy the primitive grass cover. . . .

Nature has established a balance in the Great Plains by what in human terms would be called the method of trial and error. The white man has disturbed this balance; he must restore it or devise a new one of his own. . . .

The [Great Plains] has certain common characteristics: relatively light rainfall, high summer temperature, high winds, and fine-grained soils which blow and drift when not held together by vegetative cover. There is little natural growth of trees, except in river bottoms, on the eastern slopes of the Rockies, and on occasional outcroppings such as the Black Hills. Fluctuations in rainfall may be not greater than in most other parts of the United States—perhaps not even so great, measured in inches—but they are all-important because the rainfall hovers around and at times falls below the critical point at which it is possible to grow crops without irrigation.

The soils are among the richest on the continent. With water they are highly productive. . . .

All these facts points point to the urgent necessity of more detailed knowledge of the land than we now possess and a more thorough inventory of the water resources which make it valuable. . . .

The Present Situation

As a productive resource, as a place to work and as a place to live, the Great Plains therefore present a disquieting picture. If there were no hope of restoration, with benefit both to the population of the area and to the Nation, the present report would be only a brief final chapter in a record of failure and disaster. No such conclusions need be arrived at, yet certain facts must be faced. There are perhaps 24,000 crop farms, covering a total of 15,000,000 acres, which should no longer be plowed. Of the range lands probably 95 percent have declined in forage value, this decline varying from 25 to 50 percent of its original value in southwestern North Dakota to from 50 to 75 percent in southwestern Nebraska and northwestern Kansas.

These physical changes unavoidably have been accompanied by social and economic changes. There had been a marked decline in the quality of living which could be achieved by a stalwart and energetic population, which in stock and physique is not excelled in the western world. Farmers have met the problem of holdings too small to support a family by renting additional acreage and there had grown up a confusing, intricate, and inefficient pattern of ownerships and tenures. . . .

The Region as a whole has not maintained its economic position; the return for energy expended has been less than for similar expenditures of energy upon the land in most other sections of the country. Despite its energy and self-reliance the population of the Great Plains has found itself in a position in which it was compelled to ask or accept outside assistance out of proportion to its numbers.

That is a matter of direct concern to the Federal Government and to the country as a whole. Between April 1933 and June 1936 the accumulated amounts of Federal aid expended in the area ran as high in some counties as $200 per capita. . . .

The Nation can afford such relief when it is needed. It cannot afford *not* to give it when it is needed. But the integrity of community life on the Plains, the solvency of the Plains economy, and the welfare of the Nation, which suffers indirectly as the population of the Plains suffers directly, demand that here, as elsewhere, a secure and stable substitute for relief be found. . . .

The people of the Plains are finding their way toward an attitude of mind, deep-seated and not often brought out into the open, which will affect both their thinking and their doing. Many of the assumptions which the pioneers had found workable in other regions, under other conditions, have proved unworkable on the Plains. The Plainsman cannot assume that whatever is for his immediate good is also good for everybody—only of his long-run good is this true, and in the short run there must often be sacrifices; he cannot assume the right always to do with his own property as he likes—he may ruin another man's property if he does; he cannot assume that the individual action he can take on his own land will be sufficient, even for the conservation and best use of that land. He must realize that he cannot "conquer Nature"—he must live with her on her own terms, making use of and conserving resources which can no longer be considered inexhaustible.

Steps Toward a Solution

The problem of the Great Plains offers no simple solution. Yet enough is known about conditions and their causes generally throughout the Region, and in detail with respect to certain parts, to permit immediate and vigorous execution of a program of readjustment and development. Further studies of details should proceed simultaneously with the execution of the program, but the beginning of action should not be permitted to await these studies, which should in fact be a part of the program.

I. Lines of Federal Action

1. *Investigations and Surveys.* It is recommended that provision be made promptly for the requisite investigations and surveys to determine, insofar as it has not been done, the best uses of land, waters, and other natural resources throughout the Great Plains. . . .

2. *Federal Acquisition of Land in Range Areas.* It is recommended that the Federal Government continue the policy of purchasing scattered crop farms and other appropriate lands in areas devoted largely to grazing and most suitable for that purpose. . . .

3. *Control and Use of Lands Acquired by the Federal Government in Range Areas.* In conjunction with the policy of land acquisition it is recommended that the control of purchased lands situated within the

limits of Federal grazing districts be retained by a Federal agency free to distribute range rights in accordance with the objective of general rehabilitation as well as of existing priorities. . . .

4. *Measures to Increase the Size of Farms.* It is recommended that assistance in the enlargement of undersized operating units be provided . . .

5. *Development of Water Resources.* The water supply of the Great Plains cannot be increased by any practicable means within human control. The best that can be done is to regulate the varying supply at our disposal, and to adjust the land and water economy to that supply. The Soil Conservation Service has demonstrated that generally water can be stored by suitable farm practices in the soil itself in sufficient quantities to increase growth of grass and farm crops and to resist drought. Every effort should be made to acquaint farmers with the water-conserving measures which have been found effective. . . .

6. *Resettlement.* Excessive droughts in the Great Plains have resulted in the aimless and desperate migration of thousands of families in search of some means of livelihood. . . . The adoption of the recommendations of the Committee would necessarily result in a certain measure of resettlement. . . . Suitable opportunities should be found, if possible, within the Region, but each case should receive individual consideration for its best solution.

7. *Compensation to Local Governments on Account of Federal Land Acquisition.* . . .

8. *Control of Destructive Insect Pests.* The control and possible eradication of insect pests which ravage periodically sections of the Great Plains should be a part of the long-range rehabilitation program. . . .

9. *Development of Other Resources.* . . .

III. Local Action and Its Importance

The success of a long-time plan for essential readjustments in the Great Plains economy in the final analysis will depend on local action even more than on Federal or State action. The Federal agencies may advise, assist, and coordinate. State agencies may administer permissive or mandatory legislation, but in the end local attitudes, policies, and actions are bound to decisive. These can be guided and influenced—indeed, they must be if the downward trend is to be stopped—but they cannot be coerced. . . .

In a sense the Great Plains afford a test of American ways of dealing with matters of urgent common concern. They have not responded favorably to a purely individualistic system of pioneering. The Committee is confident that they will respond to an altered system which will invoke the power of voluntary cooperation without sacrificing any of the virtues of local initiative and self-reliance.

Source: The Great Plains Committee. 1936. *The Future of the Great Plains.* Washington DC: U.S. Government Printing Office.

HUGH HAMMOND BENNETT, "SOIL CONSERVATION AND FLOOD CONTROL," 1936

Bennett was a key reformer in guiding the initial work of the Soil Conservation Service and in establishing the future role of the federal government as outlined in The Future of the Great Plains. *The excerpts from his speech below give insights into Bennett's values and aspirations.*

Let me say a few words at this point in reference to our American habits of waste. The forces which drove our pioneering forbears westward in history's greatest march of agricultural occupation, gave rise to a misconception of the extent and durability of the land and other natural resources of this continent. It would be useless to dwell at length upon these earlier misconceptions with respect to the permanency of our soil, our streams, our forests and wildlife— misconceptions that have cost us much. Probably no nation or race has been so negligent and wasteful of its land. Civilizations have disappeared because the same kind of mistakes were made on the land during thousands of years. But think of the short time it has taken us to ruin 50,000,000 acres, seriously damage another 50,000,000 acres, strip the soil or most of it from 100,000,000 acres more and get the process of wastage under way on still an additional 100,000,000 acres! Think of the result of this wastage—of the tens of thousands of farmers reduced to the lowly level of bankrupt farming on land hopelessly impoverished by erosion! What has happened to Oklahoma is appalling: A new State with 13 million acres of its 16 million in cultivation already suffering seriously from erosion, half of it having reached the stage of gullying.

Think how quickly we slaughtered for their hides the millions of buffalo that formerly roamed the plains, and the very short time it took us to strip off the grass which for countless centuries had supported those roaming herds without serious damage to the land! I am sure we are not likely soon to forget the dust storms that have carried rich soil from the plains country at the heart of the nation to the Atlantic Ocean, then to the Gulf of Mexico, and again to the Pacific Ocean, according to the direction of the wind; nor are we likely to forget that it was our failure to safeguard the land against the winds that gave birth in this country to the same type of dust phenomena common to the regions bordering the Sahara. . . .

In this manner we have exhausted and continue to exhaust irreplaceable resources. The soil is one of these. When it passes out to sea—and more than a half billion tons enter the oceans every year—it is lost forever. Even that which washes no farther than from the upper to the lower side of a field is essentially lost, since under our American system it is not likely to be hauled back. Soil reproduces from its parent materials so slowly—probably not faster than an inch in 400 to 1,000 years—that we may as well accept as a fact that once the surface layer is washed off, land so affected is generally in a condition of permanent impoverishment.

Source: Natural Resources Conservation Service, United States Department of Agriculture, "Address by H. H. Bennett, Chief, Soil Conservation Service, U.S. Department of Agriculture, before the Connecticut Engineering Congress, Bridgeport, Conn., July 25, 1936," http://www.nrcs.usda/gov/about/history/speeches/19360725.html (accessed June 15, 2006).

FLOOD CONTROL ACT [PICK-SLOAN PLAN], 1944

Federal development of water resources throughout the Central Grasslands was stalled as the Army Corps of Engineers and the Bureau of Reclamation employees contended over which agency would have the greatest responsibilities for building throughout the Missouri River watershed. The Flood Control Act of 1944 resolved these long-standing issues and set into motion a massive wave of dam building in the following decades.

An Act authorizing the construction of certain public works on rivers and harbors for flood control, and for other purposes.

Be it enacted by the Senate and House of Representative of the United States of America in Congress assembled. In connection with the exercise of jurisdiction over the rivers of the nation through the construction of works of improvement, for navigation or flood control, as herein authorized, it is hereby declared to be the policy of the Congress to recognize the interests and rights of the States in determining the development of the watersheds within their borders and likewise their interests and rights in water utilization and control, as herein authorized to preserve and protect to the fullest possible extent established and potential uses, for all purposes, of the waters of the Nation's rivers; to facilitate the consideration of projects on a basis of comprehensive and coordinated development; and to limit the authorization and construction of navigation works to those in which a substantial benefit to navigation will be realized therefrom and which can be operated consistently with appropriate and economic use of the waters of such rivers by other users.

In conformity with this policy: (a) Plans, proposals, or reports of the Chief of Engineers, War Department, for any works of improvement for navigation or flood control not heretofore or herein authorized, shall be submitted to the Congress only upon compliance with the provisions of the paragraph (a). . . . If such investigations in whole or part are concerned with use or control of waters arising west of the ninety-seventh meridian, the Chief of Engineers shall give to the Secretary of the Interior, during the course of the investigations, information developed by the investigations and also opportunity for consultation regarding plans and proposals, and to the extent deemed practicable by the Chief of Engineers, opportunity to cooperate in the investigations. . . .

(b) The use for navigation, in connection with the operation and maintenance of such works herein authorized for construction, of waters arising in States lying wholly or partly west of the ninety-eighth meridian shall be only such use as does not conflict with any beneficial consumptive use, present or future, in States lying wholly or partly west of the ninety-eighth meridian, of such waters for domestic, municipal, stock water, irrigation, mining, or industrial purposes.

(c) The Secretary of the Interior, in making investigations of and reports on works for irrigation and purposes incidental thereto shall, in relation to an affected State or States (as defined in paragraph (a) of this section), and to the Secretary of War, be subject to the same provisions regarding investigations, plans, proposals, and reports as prescribed in paragraph (a) of this section for the Chief of Engineers and the Secretary of War. . . .

SEC. 2. That the words "flood control" as used in section 1 of the Act of June 22, 1936, shall be construed to include channel and major drainage improvements, and that hereafter Federal investigations and improvements of rivers and other waterways for flood control and allied purposes shall be under the jurisdiction of and shall be prosecuted by the War Department under the direction of the Secretary of War and supervision of the Chief of Engineers, and Federal investigations of watersheds and measures for run-off and water-flow retardation and soil-erosion prevention on watersheds shall be under the jurisdiction of and shall be prosecuted by the Department of Agriculture under the direction of the Secretary of Agriculture, except as otherwise provided by Act of Congress. . . .

SEC. 4. The Chief of Engineers, under the supervision of the Secretary of War, is authorized to construct, maintain, and operate public park and recreational facilities in reservoir areas under the control of the War Department, and to permit the construction, maintenance, and operation of such facilities. . . .

SEC. 8. Hereafter, whenever the Secretary of War determines, upon recommendation by the Secretary of the Interior that any dam and reservoir project operated under the direction of the Secretary of War may be utilized for irrigation purposes, the Secretary of the Interior is authorized to construct, operate, and maintain under the provisions of the Federal reclamation laws . . . such additional works in connection therewith as he may deem necessary for irrigation purposes. . . .

SEC. 9. (b) The general comprehensive plan for flood control and other purposes in the Missouri River Basin approved by the Act of June 28, 1938, as modified by subsequent Acts, is hereby expanded to include the works referred to in paragraph (a) to be undertaken by the War Department; and said expanded plan shall be prosecuted under the direction of the Secretary of War and supervision of the Chief of Engineers.

(c) Subject to the basin-wide findings and recommendations regarding the benefits, the allocations of costs and the repayments by water users, made in said House and Senate documents, the reclamation and power developments to be undertaken by the Secretary of the Interior under said plans shall be governed by the Federal Reclamation Laws . . . except that irrigation of Indian trust and tribal lands, and repayment therefor, shall be in accordance with the laws relating to Indian lands.

Source: An Act Authorizing the Construction of Certain Public Works on Rivers and Harbors for Flood Control, and for Other Purposes. *U.S. Statutes at Large,* 78th Cong., 2nd sess., December 22, 1944, pp. 887–907.

ALDO LEOPOLD, "THE ECOLOGICAL CONSCIENCE," 1947

Leopold, as were people such as Hugh Hammond Bennett, was uneasy about the manner in which farmers conducted their business and what that meant for the land itself and the human community which rested upon it. Leopold's ideas excerpted below were a brief forerunner to a fuller expression of his famous "land ethic."

Everyone ought to be dissatisfied with the slow spread of conservation to the land. Our "progress" still consists largely of letterhead pieties and convention oratory. The only progress that counts is that on the actual landscape of the back forty, and here we are still slipping two steps backward for each forward stride.

The usual answer to this dilemma is "more conservation education." My answer is yes by all means, but are we sure that only the *volume* of educational effort needs stepping up? Is something lacking in its *content* as well? I think there is, and I here attempt to define it.

The basic defect is this: We have not asked the citizen to assume any real responsibility. We have told him that if he will vote right, obey the law, join some organizations, and practice what conservation is profitable on his own land, that everything will be lovely; the government will do the rest.

This formula is too easy to accomplish anything worthwhile. It calls for no effort or sacrifice; no change in our philosophy of values. It entails little that any decent and intelligent person would not have done, of his own accord, under the late but not lamented Babbitian code.

No important change in human conduct is ever accomplished without an internal change in our intellectual emphases, our loyalties, our affection, and our convictions. The proof that conservation has not yet touched these foundations of conduct lies in the fact that philosophy, ethics, and religion have not yet heard of it.

I need a short name for what is lacking; I call it the ecological conscience. *Ecology is the science of communities, and the ecological conscience is therefore the ethics of community life. . . .*

I hasten to add that no one has ever told farmers that in land-use the good of the community may entail obligations over and above those dictated by self-interest. The existence of such obligations is accepted in bettering rural roads, schools, churches, and baseball teams, but not in bettering the behavior of the water that falls on the land, nor in preserving the beauty or diversity of the farm landscape. Land-use ethics are still governed wholly by economic self-interest, just as social ethics were a century ago. . . .

The practice of conservation must spring from a conviction of what is chiefly and esthetically right, as well as what is economically expedient. A thing is right only when it tends to preserve the integrity, stability, and beauty of the community, and the community includes the soil, waters, fauna, and flora, as well as people.

It cannot be right, in the ecological sense, for a farmer to drain the last marsh, graze the last woods, or slash the last grove in his community, because in doing so he evicts a fauna, a flora, and a landscape whose membership in the community is older than his own, and is equally entitled to respect.

If we grant the premise that an ecological conscience is possible and needed, then its first tenet must be this: economic provocation is no longer a satisfactory excuse for unsocial land-use (or, to use somewhat stronger words, for ecological atrocities). This, however, is a negative statement. *I would rather assert positively that decent land-use should be accorded social rewards proportionate to its social importance.*

I have no illusions about the speed or accuracy with which an ecological conscience can become functional. It has required nineteen centuries to define decent man-to-man conduct and the process is only half done; it may take as long to evolve a code of decency for man-to-land conduct. In such matters we should not worry too much about anything except the direction in which we travel. The direction is clear, and the first step is to throw your weight around on matters of right and wrong in land-use. Cease being intimidated by the argument that a right action is impossible because it does not yield maximum profits, or that a wrong action is to be condoned because it pays. That philosophy is dead in human relations, and its funeral in land-relations is overdue.

Source: Leopold, Aldo. 1948. "The Ecological Conscience." *Journal of Soil and Water Conservation* 3 (July): 109–112.

REPORT AND RECOMMENDATIONS ON ORGANIC FARMING, 1980

Organic farming techniques had little impact on American techniques in the grasslands until well after 1960. This report was the first recognition on the part of the Department of Agriculture of the importance and practicality of organic methods, and suggested an alternative means of farming for those throughout the domesticated grasslands.

In April 1979, Dr. Anson R. Bertrand, Director, Science and Education, U.S. Department of Agriculture, designated a team of scientists to conduct a study of organic farming in the United States and Europe. Accordingly, the team has assessed the nature and activity of organic farming both here and abroad; investigated the motivations of why farmers shift to organic methods; explored the broad sociopolitical character of the organic movement, assessed the nature of organic technology and management systems; evaluated the level of success of organic farmers and the economic impacts, costs, benefits, and limitations to organic farming; identified research and education programs that would benefit organic farmers; and recommended plans of action for implementation. . . .

It has been most apparent in conducting this study that there is increasing concern about the adverse effects of our U.S. agricultural production system, particularly in regard to the intensive and continuous production of cash grains and the extensive and sometimes excessive use of agricultural chemicals. Among the concerns most often expressed are the following:

1. Sharply increasing costs and uncertain availability of energy and chemical fertilizer, and our heavy reliance on these inputs.

2. Steady decline in soil productivity and tilth from excessive soil erosion and loss of soil organic matter.

3. Degradation of the environment from erosion and sedimentation and from pollution of natural waters by agricultural chemicals.

4. Hazards to human and animal health and to food safety from heavy use of pesticides.

5. Demise of the family farm and localized marketing systems.

Consequently, many feel that a shift to some degree from conventional (that is, chemical-intensive) toward organic farming would alleviate some of these adverse effects, and in the long term would ensure a more stable, sustainable, and profitable agricultural system.

While other definitions exist, for the purpose of this report organic farming is defined as follows:

Organic farming is a production system which avoids or largely excludes the use of synthetically compounded fertilizers, pesticides, growth regulators, and livestock feed additives. To the maximum extent feasible, organic farming systems rely upon crop rotations, crop residues, animal manures, legumes, green manures, off-farm organic wastes, mechanical cultivation, mineral-bearing rocks, and aspects of biological pest control to maintain soil productivity and tilth, to supply plant nutrients, and to control insects, weeds, and other pests. . . .

The following is a brief summary of the principal findings of this study:

1. The study team found that the organic movement represents a spectrum of practices, attitudes, and philosophies. On the one hand are those organic practitioners who would not use chemical fertilizers or pesticides under any circumstances. These producers hold rigidly to their purist philosophy. At the other end of the spectrum, organic

farmers espouse a more flexible approach. While striving to avoid the use of chemical fertilizers and pesticides, these practitioners do not rule them out entirely. . . .

2. Organic farming operations are not limited by scale. . . .

3. Motivations for shifting from chemical farming to organic farming include concern for protecting soil, human, and animal health from the potential hazards of pesticides; the desire for lower production inputs; concern for the environment and protection of soil resources.

4. Contrary to popular belief, most organic farmers have not regressed to agriculture as it was practiced in the 1930's. . . .

5. Most organic farmers use crop rotations that include legumes and cover crops to provide an adequate supply of nitrogen for moderate to high yields.

6. Animals comprise an essential part of the operation of many organic farms. In a mixed crop/livestock operation, grains and forages are fed on the farm and the manure is returned to the land. . . .

7. The study team was impressed by the ability of organic farmers to control weeds in crops such as corn, soybeans, and cereals without the use (or with only minimal use) of pesticides. Their success here is attributed to timely tillage and cultivation, delayed planting, and crop rotations. They have also been relatively successful in controlling insect pests.

8. Some organic farmers expressed the feeling that they have been neglected by the U.S. Department of Agriculture and the land-grant universities. They believe that both Extension agents and researchers, for the most part, have little interest in organic methods and that they have no one to turn to for help on technical problems.

9. In some cases where organic farming is being practiced, it is apparent from a study of the nutrient budget that phosphorus (P) and potassium (K) are being "mined" from either soil minerals or residual fertilizers applied when the land was farmed chemically. . . . [I]t is likely that eventually some organic farmers will have to apply supplemental amounts of these two nutrients.

10. The study revealed that organic farms on the average are somewhat more labor intensive but use less energy than conventional farms. . . .

11. This study showed that the economic return above variable costs was greater for conventional farms (corn and soybeans) than for several crop rotations grown on organic farms. This was largely due to the mix of crops required in the organic system and the large portion of the land that was in legume crops at any one time.

12. There are detrimental aspects of conventional production, such as soil erosion and sedimentation, depleted nutrient reserves, water pollution from runoff of fertilizers and pesticides, and possible decline of soil productivity. If costs of these factors are considered, then cost comparisons between conventional (that is, chemical-intensive) crop production and organic systems may be somewhat different in areas where these problems occur.

In conclusion, the study team found that many of the current methods of soil and crop management practiced by organic farmers are also those which have been cited as best management practices (USDA/EPA joint publication on "Control of Water Pollution from Cropland," Volume I, 1975, U.S. Government Printing Office) for controlling soil erosion, minimizing water pollution, and conserving energy. These include sod-based rotations, cover crops, green manure crops, conservation tillage, strip cropping, contouring, and grassed waterways. Moreover, many organic farmers have developed unique and innovative methods of organic recycling and pest control in their crop production sequences. Because of these and other reasons outlined in this report, the team feels strongly that research and education programs should be developed from a holistic research effort to investigate the organic system of farming, its mechanisms, interactions, principles, and potential benefits to agriculture both at home and abroad.

Source: USDA Study Team on Organic Farming. 1980. *Report and Recommendations on Organic Farming.* Washington, DC: U.S. Department of Agriculture.

TALLGRASS PRAIRIE NATIONAL PRESERVE ACT, 1996

While George Catlin might have championed the creation of a national grassland park as early as 1832, it would require over a century and a half of work before its proponents' efforts bore fruit. Below are excerpts from the act creating the first national grassland park in the country.

SEC. 1001. SHORT TITLE.

This subtitle may be cited as the "Tallgrass Prairie National Preserve Act of 1996."

SEC. 1002. FINDINGS AND PURPOSES.

(a) FINDINGS.—Congress finds that—

1. of the 400,000 square miles of tallgrass prairie that once covered the North American Continent, less than 1 percent remains, primarily in the Flint Hills of Kansas;

2. in 1991, the National Park Service conducted a special resource study of the Spring Hill Ranch, located in the Flint Hills of Kansas;

3. the study concludes that the Spring Hill Ranch—

(A) is a nationally significant example of the once vast tallgrass ecosystem, and includes buildings listed on the National Register of Historic Places pursuant to section 101 of the National Historic Preservation Act (16 U.S.C. 470a) that represent outstanding examples of Second Empire and other 19th Century architectural styles; and

(B) is suitable and feasible as a potential addition to the National Park System; and

4. the National Park Trust, which owns the Spring Hill Ranch, has agreed to permit the National Park Service—

(A) to purchase a portion of the ranch, as specified in the subtitle; and

(B) to manage the ranch in order to—

(i) conserve the scenery, natural and historic objects, and wildlife of the ranch; and

(ii) provide for the enjoyment of the ranch in such a manner and by such means as will leave the scenery, natural and historic objects, and wildlife unimpaired for the enjoyment of future generations.

Source: An Act to Provide for the Administration of Certain Presidio Properties at Minimal Cost to the Federal Taxpayer, and for other P.L. 104–333, 104th Cong., 2nd sess.

RON ARNOLD, "OVERCOMING IDEOLOGY," 1996

Many farmers and residents in the grasslands have little use for environmentalists, who are often depicted as insensitive to the plight of farmers and property rights. Arnold, an advocate of the "Wise Use Movement," articulates a position that many people throughout the grasslands have held in the late twentieth and early twenty-first centuries.

It was 1964, the year of the Wilderness Act. Historian Leo Marx began his classic, *The Machine in the Garden*, with the assertion that "The pastoral ideal has been used to define the meaning of America ever since the age of discovery, and it has not yet lost its hold upon the native imagination." . . .

The pastoral ideal is not simply a location, but also a psychic energy condenser: it stores the charge generated between the polarities of civilization and nature. Ortega y Gasset recognized this as long ago as 1930 in *The Revolt of the Masses*: "The world is a civilized one, its inhabitant is not: he does not see the civilization of the world around him, but he uses it as if it were a natural force. The new man wants his motor-car, and enjoys it, but he believes that it is the spontaneous fruit of an Edenic tree."

. . . Since 1964, the rise of environmentalist ideology has pushed the pastoral ideal increasingly toward nature, striving to redefine the meaning of America in fully primitivist terms of the wild. Eco-ideologists have thrust their metaphoric raging Wolf into every rank and row of our civilized Garden to rogue out both the domesticated and the domesticators. The Wolf howls Wild Land, Wild Water, Wild Air. Whether Wild People might have a proper place in Wolf World remains a subject of dispute among eco-ideologists. . . .

Since the 1970s we've heard increasingly about the competing paradigm, wherein:

Growth must be limited.

Science and technology must be restrained.

Nature has finite resources and a delicate balance that humans must observe. . . .

Environmentalism, like pastoral literature, was about those pastoral rural dwellers who produced dinner, dress and domicile for everyone, but was generated by the educated elite, not by those who lived the pastoral ideal. Environmentalism's ideology was promulgated for the ruling elite, not for the farmer or rancher or family forest owner or mineral prospector. . . .

Although it would be rash to propose wise use's articles of faith—it is a diverse movement—some of the following principles would probably find wide agreement among those who provide the material goods to all of humanity:

1. Humans, like all organisms, must use natural resources to survive. This fundamental verity is never addressed by environmental ideology. The simple fact that humans must get their food, clothing and shelter from the environment is either ignored or obliquely deplored in quasi-suicidal plaints such as, "I would rather see a blank space where I am— at least I wouldn't be harming anything."

If environmentalism were to acknowledge our necessary use of the earth, the ideology would lose its meaning. To grant legitimacy to the human use of the environment would be to accept the unavoidable environmental damage that is the price of our survival. Once that price is acceptable, the moral framework of environmental ideology becomes irrelevant and the issues become technical and economic.

2. The earth and its life are tough and resilient, not fragile and delicate. Environmentalists tend to be catastrophists, seeing any human use of the earth as damage and massive human use of the earth as a catastrophe. An environmentalist motto is "We all live downstream," the viewpoint of hapless victims.

Wise users, on the other hand, tend to be cornucopians, seeing themselves as stewarding and nurturing the bountiful earth as it stewards and nurtures them. A wise use motto is "We all live upstream," the viewpoint of responsible individuals.

The difference in sense of life is striking. Environmentalism by its very nature promotes feelings of guilt for existing, which naturally degenerate into pessimism, self-loathing and depression.

Wise use by its very nature promotes feelings of competence to live in the world, generating curiosity, learning, and optimism toward improving the earth for the massive use of future generations.

The glory of the "dominant Western worldview" so scorned by environmental ideologists is its metaphor of progress: the starburst, an insatiable and interminable outreach after a perpetually flying goal. Environmentalists call humanity a cancer on the earth; wise users call us a joy.

If there is a single, tight expression of the wise use sense of life, it has to be the final stanza of Shelley's *Prometheus Unbound*. I think wise users will recognize themselves in these lines:

> *To suffer woes which Hope thinks infinite;*
> *To forgive wrongs darker than death or night;*
> *To defy Power, which seems omnipotent;*
> *To love, and bear; to hope till Hope itself creates*
> *From its own wreck the thing it contemplates;*
> *Neither to change, nor falter, nor repent;*
> *This, like thy glory, Titan! is to be*
> *Good, great and joyous, beautiful and free;*
> *This is alone Life, Joy, Empire, and Victory!*

3. We only learn about the world through trial and error. The universe did not come with a set of instructions, nor did our minds. We cannot see the future. Thus, the only way we humans can learn about our surroundings is through trial and error. Even the most sophisticated science is systematized trial and error. Environmental ideology fetishizes nature to the point that we cannot permit ourselves errors with the environment, ending in no trials and no learning.

There will always be abusers who do not learn. People of good will tend to deal with abuse by education, incentive, clear rules and administering appropriate penalties for incorrigibles.

4. Our limitless imaginations can break through natural limits to make earthly goods and carrying capacity virtually infinite. Just as settled agriculture increased earthly goods and carrying capacity vastly beyond hunting and gathering, so our imaginations can find ways to increase total productivity by superseding one level of technology after another. Taught by the lessons learned from systematic trial and error, we can close the loops in our productive systems and find innumerable ways to do more with less.

5. Humanity's reworking of the earth is revolutionary, problematic and ultimately benevolent. Of the tenets of wise use, this is the most oracular. Humanity is itself revolutionary and problematic. Danger is our symbiote. Yet even the timid are part of the human adventure, which has barely begun.

Humanity may ultimately prove to be a force of nature forwarding some cosmic teleology of which we are yet unaware. Or not. Humanity may be the universe awakening and becoming conscious of itself. Or not. Our reworking of the earth may be of the utmost evolutionary benevolence and importance. Or not. We don't know. The only way to see the future is to be there.

As the environmental debate advances to maturity, the environmental movement must accept and incorporate many of these wise use precepts if it is to survive as a social and political force.

Source: Arnold, Ron. 1996. "Overcoming Ideology." In *A Wolf in the Garden: The Land Rights Movement and the New Environmental Debate,* edited by Philip D. Brick and R. McGreggor Cawley. Lanham, MD: Rowman and Littlefield, pp. 15–26.

INTRODUCTION TO, AND MISSION STATEMENT OF, THE LAND INSTITUTE, UPDATED AUGUST 24, 2004

Wes Jackson, the founder of the Land Institute near Salina, Kansas, believes that people have a lot to learn from the functioning of wild grassland ecosystems. He believes that if wild prairie plants could be hybridized for grain harvesting, farming throughout the grasslands would be transformed into a more ecologically benign enterprise, one that would reduce soil erosion, lessen the need for chemical inputs, exist in ecological harmony with its climate and geography, and contribute to the well-being of the human community.

The Land Institute has worked for over 20 years on the problem of agriculture. Our purpose is to develop an agricultural system with the ecological stability of the prairie and a grain yield comparable to that from annual crops. We have researched, published in refereed scientific journals, given hundreds of public presentations here and abroad, and hosted countless intellectuals and scientists. Our work is frequently cited, most recently in *Science* and *Nature*, the most prestigious scientific journals. We are now assembling a team of advisors which includes members of the National Academy of Sciences. These scientists understand our work and stand ready to endorse the feasibility of what we have come to call Natural Systems Agriculture.

Our strategy now is to collaborate with public institutions in order to direct more research in the direction of Natural Systems Agriculture. We are seeking funds to construct and operate a research center devoted to Natural Systems Agriculture and to underwrite scientists elsewhere who will engage with us in such research. We estimate the research cost to be $5 million a year for 25 years, which is a small fraction of one percent of the nation's annual agricultural research investment.

Important questions have been answered and crucial principles explored to the point that we feel comfortable in saying that we have demonstrated the scientific feasibility of our proposal for a Natural Systems Agriculture. Because this work deals with basic biological questions and principles, the implications are applicable worldwide. If Natural Systems Agriculture were fully adopted, we could one day see the end of agricultural scientists from industrialized societies delivering agronomic methods and technologies from their fossil fuel–intensive infrastructures into developing countries and thereby saddling them with brittle economies.

Mission Statement

When people, land, and community are as one, all three members prosper; but as competing interests, by consulting Nature as the source and measure of that membership, The Land Institute seeks to develop an agriculture that will save soil from being lost or poisoned while promoting a community life at once prosperous and enduring.

Source: The Land Institute, "Introduction and Mission," http://www.landinstitute.org/vnews/display.v/ART/2000/08/10/37a747b43 (accessed June 15, 2006).

IMPORTANT PEOPLE, EVENTS, AND CONCEPTS

ALTITHERMAL A prolonged period of climatic drying and warming occurred in North America between 7,500 and 3,000 years ago. While not uniform in its effects, it radically altered both the natural and cultural landscape of the Great Plains. Climatologists refer to this episode variously as the Middle Holocene, the Atlantic climate phase, the hypsithermal, or the altithermal. First described by Ernst Antevs in 1948, the altithermal climate shifted the locations of ecosystems throughout the region. The grassland/woodland frontier shifted east of its present margins, as did the tallgrass and shortgrass zones. Surface and ground-water sources became dried or brackish, and unfit for herbivores. Bison, pronghorn deer, and their predators moved to more flourishing locales to the east or the foothills to the west. The hunters who depended on these same animals also followed their trails. The diets of those early archaic peoples also indicated an inability to capture large game animals. Their diets showed a greater use of small animals and plants than appeared in the diets of earlier peoples.

Similar to modern droughts, the effects of this long-term dry period were not uniform in terms of duration or location. For example, the southern plains of contemporary Texas and New Mexico appear to be suffering the greatest effects of current climatic patterns while the northern reaches of the grasslands show comparatively wetter and cooler conditions even though these conditions are generally drier and warmer than in previous decades. During the altithermal, people in the southern climes dug wells in former watercourses to reach a declining water table, and they remained for a considerable period of time before harsh conditions and depleted resources forced them to relocate. Human migrations in the north show evidence of movement into the river valleys and other so-called refugia.

Scientists have debated the nature, extent, and even the existence of the altithermal since Antevs first described it. While they generally agree that the climate significantly warmed between 7500 and 3000 before present, scholars still know that they have much to learn about the human and ecological communities that once inhabited the grasslands during those centuries.

See also Dust Bowl; Malin, James; Sand Hills, Scarp Forests.

Sources

Kay, Marvin, 1998. "The Great Plains Setting." In *Archaeology on the Great Plains*, edited by W. Raymond Wood. Lawrence: University Press of Kansas.

Pielou, E. C. 1991. *After the Ice Age: The Return of Life to Glaciated North America.* Chicago: University of Chicago Press.

AMERICAN FUR COMPANY   One company, more than nearly any other, exemplifies the fur trade throughout the grasslands and its social and ecological consequences. John Jacob Astor created this company and incorporated it in April 1808. Located in New York, he began a serious competition with the British companies operating out of Canada for control of the fur trade throughout the prairies of the Old Northwest. Later, he represented the British Northwest Company and its interests in 1816 after the United States government forbade foreign-owned fur companies from operating within its boundaries. At the same time, Astor turned his attention more to the west, especially after Britain closed fur trading in Canada to Americans, and after Congress abolished its U.S. government–owned trading posts called factories.

St. Louis had become the locus of fur trading in the United States. Small companies were attempting to trade with Indian nations all throughout the greater Missouri River Valley. This was dangerous yet at times highly profitable work. Astor saw his opportunity in consolidating with one particular operation largely directed by the powerful Chouteau family, and this new arrangement became known as the Western Department of the American Fur Company (AFC). Under the direction of Kenneth McKenzie, the AFC built its major trading post, Fort Union, near the confluence of the Yellowstone and Missouri rivers in 1828. The contracts with the Chouteaus continued until 1834 when Astor retired from the western fur trade and sold the Western Department to the Chouteau-dominated company of Pratte, Chouteau and Company. During this time the Western Department began using steamboats to ply the Missouri River and supply all the posts up to Fort Union. In all of this trading, liquor flowed more regularly through the company than did the water in the tributaries of the Missouri River even though the federal government had outlawed its trade in Indian Territory in 1832.

Pierre Chouteau, Jr., more than any one person in St. Louis at the time, directed the operations of the American Fur Company, as most people called the Western Division. Born in January 1789, he had acquired extensive experience in the fur trade with his father, a founder of St. Louis. In 1813 he had formed a partnership with Bartholomew Berthold; and their company not only expanded and largely controlled the bison robe trade, but it also promoted scientific expe-

ditions throughout the grasslands. This trade brought Indian nations throughout the grasslands into the currents of the American economic system, transformed and improved much of their material cultures, altered their diplomatic
ties, and revolutionized their hunting practices. These factors had devastating
ecological consequences with the expansion of horse and robe trade–centered
lifestyles. Wild fur-bearing animal populations began to decline; riparian ecosystems became denuded of trees; pandemics of smallpox, measles, and cholera
carried off huge numbers of people in several nations; and debilitating occurrences of sexually transmitted diseases and alcoholism disrupted kinship ties.

After the Mexican-American War, the profitability of the American Fur
Company, never great, declined along with increasing warfare among Indian
peoples and with Euro-Americans, mounting ecological degradation, and a
growing Euro-American colonization of the grasslands. Pierre Chouteau, Jr., understood these changes were not boding well for the future of the company, and
he began diversifying many of his investments before the fur trade fell into
complete disarray by the beginning of the American Civil War. His son,
Charles, took the helm of the company, but realizing it had no future, sold the
company in 1865, just shortly before the death of his father.

See also Bison; Yellow Wolf.

Sources

Chittenden, Hiram Martin. 1902. *The American Fur Trade of the Far West: A His-
tory of the Pioneer Trading Posts and Early Fur Companies of the Missouri
Valley and the Rocky Mountains and of the Overland Commerce with Santa
Fe.* 3 vols. New Work: Francis P. Harper.

Christian, Shirley. 2004. *Before Lewis and Clark: The Story of the Chouteaus, the
French Dynasty That Ruled America's Frontier.* New York: Farrar, Straus and
Giroux.

Wishart, David J. 1979. *The Fur Trade of the American West, 1807–1840.* Lincoln:
University of Nebraska Press.

BENNETT, HUGH HAMMOND (1881–1960) Driving through the grasslands
one encounters roadside signs that read, for example, "Chase County Conservation District." These districts are governed by an elected, unpaid board that
hires a professional staff to administer its soil and water conservation policies
and programs. The officials of these county districts do this in compliance with
state and national natural resource law and in partnership with the Natural Resources Conservation Service—before 1994, referred to as the Soil Conservation
Service. One person more than anyone else had the vision to create this system
of conservation districts spanning the grasslands today, and that was Hugh
Hammond Bennett.

When growing up in rural North Carolina, Bennett could observe firsthand the debilitating effects of soil erosion in Anson County. A bright and studious young man, he attended the University of North Carolina where he earned a degree in geology and chemistry in 1903. Bennett soon landed a position in the Department of Agriculture as a soil surveyor. In this capacity he became keenly aware of the ecological, social, and economic consequences of poor soil conservation techniques. His work abroad only strengthened his conviction that soil erosion, if not checked, would destroy the ecological base of a people or nation. He, along with his co-author W. R. Chapline, made this message especially clear in a bulletin titled, *Soil Erosion, a National Menace* (1928). This publication caught the attention of some in Congress, and with this recognition came additional funding for the Department of Agriculture to address soil erosion in experiment stations. At the same time, Bennett was working with the Department of Interior as it created the Soil Erosion Service, and in 1933 Bennett assumed its helm.

The massive dust storms arising in the heart of the grasslands gave Congress an additional sense of urgency to deal with the problem of soil erosion. Bennett's crusade to enlist more direct federal aid in addressing the problem, along with detailed local and national press coverage of those storms, propelled Congress to tackle soil erosion then and there. Bennett was exceptionally effective before Congress in once timing one of his testimonies to coincide with a storm of wind-blown dust from the Great Plains as it blanketed the skies over Washington, D.C., in deep hues of red. Congress needed little further prodding that something had to be done to address soil erosion, and in April 1935 it created the Soil Conservation Service within the Department of Interior with Hugh Bennett at its head. When he retired in 1951, Bennett left a powerful legacy, one rooted in work that strove to preserve the agricultural and ecological health and wealth of the grasslands. Numerous national societies and organizations, such as the National Audubon Society, the American Geographical Society, and the United States Department of Agriculture, honored his work. Even in retirement, Bennett continued to be the evangelist of soil conservation. In a speech shortly before his death, he warned that the public and individual good "are tied together in such a completely complementary way, there is no point in pursuing the subject beyond indicating that no man should have the right legally or otherwise, to recklessly or willfully destroy or necessarily waste any resources on which the public welfare is dependent." As far as Bennett was concerned, the one resource, more than other, upon which people depended was soil.

See also Dust Bowl; Malin, James.

Sources

Brink, Wellington. 1951. *Big Hugh: The Father of Soil Conservation.* New York:

Macmillan.

Helms, Douglas. 1999. "Hugh Hammond Bennett." *American National Biography*, vol. 2. New York: Oxford University Press.

BISON   Perhaps more than 600,000 years ago a species of bison, *Bison priscus*, evolved on the steppes of Eurasia. From there this species would eventually migrate to and occupy major ranges in the North American continent between 300,000 to 200,000 years ago. Out of this line evolved *Bison latifrons*, or the giant horned bison, and it flourished south of the Illinoian ice sheet while separaged from *B. priscus* in Eurasia. During the interglacial period between the Illinoian and Wisconsin glaciation, *B. latifrons* and *B. priscus* may have contended for the same North American ranges with *B. latifrons* losing the struggle and passing into extinction, while *B. priscus* adapted to its new environment and evolved into *B. antiquus*, or *B. latifrons* itself evolved into the smaller, shorter horned *B. antiquus*. Either way, around 100,000 years ago, *B. latifrons* passed into extinction, and *B. antiquus* thrived in the grassland parks south of the Wisconsin ice sheet, and became a hunted animal by the humans who were making their appearance in North America at the same time. During the Wisconsin glaciation, in Eurasia, *B. priscus* was evolving into the smaller, fleeter modern bison, *Bison bison*, and at the end of the Wisconsin, *B. bison* migrated into the North American continent while at the same time *B. antiquus* passed into extinction. Some scholars believe *B. priscus* evolved into *B. bison* to confront human hunting tactics and wolf predation through flight rather than fight. The smallest of any bison to populate the grasslands, its physical characteristics endowed it with speed, agility, and a keen sense of smell, and it soon displaced *B. antiquus* throughout the grasslands. Some scholars argue that *B. bison* evolved into two distinct species—one, *Bison bison bison*, occupying the grasslands, and the other, *Bison bison athabascae*, the "wood bison," occupying the eastern and western forests. Other scholars contend that there was too little significant differentiation between the two groups to warrant their being considered two separate species.

Bison may have been the keystone species to the grassland ecosystem that emerged after the hypsithermal. One definition of a keystone species is "one whose impact on its community or ecosystem is large, and disproportionately large relative to its abundance." Given this definition, bison certainly shaped the tallgrass ecosystem by increasing the heterogeneity of cool- and warm-season grasses, and by increasing the number and diversity of forbs. The maintenance of the grasslands, given current research, is enhanced with the periodic grass fires in conjunction with grazing. The question arises then as to whether humans, rather than bison, were the keystone species if indeed humans were the primary cause of grass fires. Regardless, bison were an important component in shaping and keeping the grasslands.

Bison bison, unlike its predecessors, escaped extinction, but only narrowly. Perhaps bison in the grasslands numbered between 30 to 60 million animals. Unquestionably, hunting pressures had nearly annihilated them by 1880. Only a few thousand animals at most had survived, and these would have completely collapsed had it not been for the efforts of a few men and women. While Charles and Mary Goodnight, along with William Hornaday, shared most of the spotlight for preserving bison, far too little acclaim has been given to the labors of such people as Samuel Walking Coyote, a Pend d'Oreille; Charles Conrad from Kalispell, Montana; Charles "Buffalo" Jones from Garden City, Kansas; Frederick Dupree and his Lakota wife Mary; Austin Corbin from Newport, New Hampshire; or James "Scotty" Philip and his Cheyenne wife Sarah Larrabee who lived in South Dakota. Without the endeavors of these people, and many others like them, *Bison bison* would have met the same dismal fate as all of its ancestors.

Some ecologists believe that more than 200,000 acres at minimum is required to re-create a viable grassland ecosystem. Within such a range wild bison could flourish once again. The question here is one of wildness, as nearly 90 percent of all bison in the United States are now raised in domestic settings. This has led Dale F. Lott, a professor emeritus of biology at the University of California, Davis, to label such animals "buffattle." Given the ease of cross-breeding, even cattle DNA has been worked into the genetic composition of bison. The preservation of "wild" bison apart from domestic ones is a growing concern among many scholars and wildlife advocates. In short, wild bison "are not out of the woods yet."

See also American Fur Company; Endangered Species; Goodnight, Charles and Mary Ann; Yellow Wolf.

Sources

Guthrie, R. D. 1970. "Bison Evolution and Zoogeography in North America during the Pleistocene." *Quarterly Review of Biology* 45 (March): 1–15.

Lott, Dale F. 2002. *American Bison: A Natural History.* Berkeley: University of California Press.

Knapp, Alan, et al. 1999. "The Keystone Role of Bison in North American Tallgrass Prairie." *Bioscience* 49 (January): 39–50.

BONANZA FARMS   While the 160-acre homestead may be the mythic archetype of an American farm, those men and women who championed "bonanza farms" blazed the path of modern agriculture. They intended to apply corporate business techniques to farming and to manage farming in a manner similar to that used in industrial factories. Clearly, by the early 1870s large-scale farm operations were showing great economic promise. The pioneering winter wheat growing of Theodore C. Henry along the Kansas Pacific Railroad line exhibited great success. In the mid-1870s, Henry's wheat production on over 10,000 acres had made him the richest man in Kansas.

Others knew that farming on a massive scale employing corporate techniques could result in riches. As an attorney, Oliver Dalrymple had made considerable money handling Euro-American's claims against the Sioux as a result of the Minnesota Uprising of 1862. Dalrymple took his earnings and invested in wheat farming to the south of St. Paul, Minnesota, and through a lucrative law practice and sensible farming techniques, he rapidly became known as the "wheat king" of Minnesota until he wiped out his fortune playing the futures market.

At the same time, the managers of the Northern Pacific Railway Company were managing an enterprise poorly administered and deeply in debt. James B. Power, the land agent for the company, devised a plan for exchanging the bonds of the company for its land with the prediction that the eventual inflation of land values would compensate the investor for losses tied to the horribly depreciated worth of the bonds. Consequently, a few people acquired huge tracts of land for pennies on the dollar throughout the fertile, 300-mile-long Red River Valley. Knowing Dalrymple's farming reputation, Power and others called on him to manage their extensive holdings and to invest in railroad lands. Dalrymple eagerly accepted their offer.

From the beginning, these holdings were operated as "wheat factories." Dalrymple began work in 1875, and in a couple of years his undertakings showed great promise. Managing his own land as well as the land of others, Dalrymple employed professional management, mechanization, and specialized production techniques to work thousands of acres. In 1877, to place into wheat production some 4,000 acres required 26 breaking plows, 21 seeders, 60 harrows, 30 self-binding harvesters, 5 steam threshers, 80 horses, a work crew of 50 men during planting, and another 80 to 100 men at harvest time. Dalrymple divided this farm into four divisions, each overseen by a foreman who had a house provided for him and his family. Quarters were built to house the laborers, stables for the draft animals, machine sheds for the equipment, a blacksmith shop for repairs, and elevators to store wheat. In time, Dalrymple managed around 100,000 acres of wheat production in a region marked by scores of massive operations similar in scope and reach.

However, the nature of bonanza farms differed considerably from corporate industries, and as a result, these agricultural ventures could not control their operations in the same manner that industries did. Dalrymple could not set the price of goods bought or sold; moreover, the heavily capitalized nature of his enterprise left it vulnerable to anything that affected crop prices—fluctuations in international markets or crop failures as a result of disease or drought. While many of the bonanza farms encountered irreversible losses in the late 1880s and 1890s, Dalrymple and his family managed to survive until the onset of the farm depression after the end of World War I. Interestingly, the family reacquired nearly 75 percent of its former holdings by the mid-1920s and took up farming again.

Bonanza farming did not end with its collapse in the Red River Valley. During World War I, Thomas Donald Campbell, who was raised on a farm in the Red River Valley and who understood bonanza farming techniques, leased huge tracts of land in the Big Horn River on the Crow Indian Reservation in Montana and began the "manufacture of wheat." By 1928 he was harvesting over 95,000 acres of wheat and was featured on the cover of *Time* magazine as the biggest farmer in the United States. The success of his corporate farming techniques, "farming a la Detroit," caught the attention of Soviet planners, and he advised them on refining their own collective farming methods.

The efficiencies of scale in this type of farming captured the attention of New Dealers, had enthusiastic federal support, and nearly resulted in ending farming as a "way of life." The accomplishments of farmers such as Henry, Dalrymple, and Campbell foretold the coming of contemporary corporate farm practices.

See also Bennett, Hugh Hammond; Carleton, Mark; *Future of the Great Plains* (1936); Hightower, Jim; Jackson, Wes.

Sources

Drache, Hiram. 1967–1968. "Bonanza Farming in the Red River Valley." *Transaction of the Historical and Scientific Society of Manitoba* (24): 53–63.

Edwards, Douglas M. 2001. "'The Greatest Hazard of All Is the Human Element': Manning the Machines of the World's Greatest Wheat Farm." *Montana: The Magazine of Western History* 51 (Winter): 26–37.

Fite, Gilbert C. 1966. *The Farmers' Frontier, 1865–1900*. Norman: University of Oklahoma Press.

CARLETON, MARK (1866–1925)   Mark Carleton, as a ten-year-old boy, arrived in Kansas with his father and mother, who began farming in Cloud County in 1876. This boy would later contribute more to devising economically viable crops for the grasslands than nearly anyone else. His most notable achievements came in his introduction of new varieties of hard red wheats and durum wheats to the farmlands throughout the grasslands. He received his B.S. degree from Kansas State Agricultural College in 1887. Later, after some short-lived teaching positions and itinerant work, he returned to Manhattan, Kansas, and completed his master's degree at the college.

While at the college, he began working with A. S. Hitchcock, the botanist of the experiment station, on treating rust disease in grain crops. His work landed him a position as assistant pathologist in the Division of Vegetable Physiology and Pathology of the United States Department of Agriculture. From this office he traveled throughout the grasslands examining and studying rust in domestic grasses and wheat fields. His work led him to search for new varieties that had a natural resistance to rust and more drought hardiness. He avidly examined the practices of the Mennonites in central Kansas, who had planted Russian varieties of hard winter wheat. He saw a connection between the climates of the Russian steppes and the grasslands of North America. Consequently, he began a search for small grain species in Russia suitable for raising in the grasslands of his native home. He began his travels in 1898 and returned to the United States in 1899. His research resulted in the publication of *Russian Cereals Adapted for Cultivation in the United States* (1898). This, and other reports, led to an appointment as the head cerealist for the Bureau of Plant Industry in the Department of Agriculture in Washington, D.C. From there he promoted varieties of durum and hard winter wheats for American farmers.

His fame propelled him to become the first president of the American Society of Agronomy in 1908. While reaching the pinnacle of his career, his personal life had descended into tragedy and financial ruin. By 1918, he was sacked by the secretary of agriculture, David Houston, for unethical behavior and conflict of interest. In the following years, he had a series of positions with various grain and fruit companies. While in Peru, he succumbed to complications associated with a bout of malaria. In spite of a life marked by personal failures and tragedies, his professional accomplishments yielded some of the most valuable and reliable cash crops grown throughout the grasslands.

See also Bonanza Farms; Henry, T. C.; Jackson, Wes; Malin, James; Popper, Deborah and Frank; Winter Wheat.

Sources

Isern, Thomas. 2000. "Wheat Explorer the World Over: Mark Carleton of Kansas." *Kansas History* 23 (Spring–Summer): 12–25.

CHASE, AGNES (1869–1963) Mary Agnes (Meara) Chase pioneered both systematic botany and scientific education. While working with Albert S. Hitchcock of the U.S. Department of Agriculture, she collected, identified, and cataloged a comprehensive collection of grasses of the Americas for the U.S. National Herbarium. In the process Agnes Chase became an acknowledged world expert on American grasses. She also mentored and taught botanists from across the Americas about agrostology (grasses) and plant classification. Active in social causes, Chase worked to further the careers of African-American and women scientists.

Agnes Chase, unlike many other Victorian woman botanists, did not come from a well-off family. Her father died while she was a young child, leaving her mother with five children to support. Once Chase finished grammar school she worked as a clerk and newspaper proofreader and typesetter, and these skills would serve her well in her later career. Chase took classes at the University of Chicago and the Lewis Institute but never received a degree. As with her botanical studies, she was largely self-taught. In 1888 Agnes married William Ingraham Chase, who died only a year later, leaving her to manage his heavily indebted business.

Despite financial obstacles, Chase developed an interest in botany during the early 1890s. Plant collection and preservation was considered a suitable pursuit for women of all social classes and provided them access to the sciences that other fields often denied women. Chase made acquaintances with other local botanists and gained a mentor in the Reverend Ellsworth Jerome Hill. Hill hired Chase to illustrate his publications and taught her how to use microscopes, among other skills. Hill also introduced Chase's work to Charles F. Millspaugh of the Field Museum, who further encouraged Chase's career. Millspaugh assisted her with finding work in the U.S. Department of Agriculture (USDA), first as a meat inspector in Chicago in 1901, and later as a scientific illustrator for the Bureau of Plant Industry in Washington, D.C.

Agnes Chase's career as a botanist flourished when she began working for Albert Spear (A. S.) Hitchcock as an illustrator in 1903. At Millspaugh's insistence, Hitchcock encouraged Chase to study the collections of the National Herbarium after working hours. Chase soon became an expert in the taxonomy of all forms of grasses, from bamboo to lawn grasses, and she published her first scientific paper on grasses in 1906. In 1910 Chase and Hitchcock published the magisterial *North American Species of Panicum*, a complete collection of all varieties of North American millet grasses. Hitchcock pushed for Chase's promotion, and she eventually became senior botanist on Hitchcock's death in 1936. Among her academic publications are the *First Book of Grasses: The Structure of Grasses Explained for Beginners*, published in English in 1922 and later translated into Spanish and Portuguese; "Botanical Expedition to Brazil" in 1926; and "Eastern Brazil through an Agrostologist's Spectacles" in 1927. Because government officials were reluctant to fund a woman, Chase used her own funds or worked as a paid plant collector for various museums to conduct field research in Puerto Rico and South America.

Agnes Chase's interests ranged far beyond purely scientific pursuits. She belonged to the National Association for the Advancement of Colored People, the Women's Christian Temperance Union, and several socialist and pacifist groups. Chase worked for the suffragist cause and was arrested for her actions on two separate occasions. In January 1915 she and others protested Woodrow Wilson's lack of action on women's suffrage by burning copies of any of his speeches that had references to "liberty" or "freedom" in a public bonfire. Chase was arrested again in August 1918 while picketing the White House, and with other suffragists went on a hunger strike while in jail. A. S. Hitchcock kept the USDA from firing Chase after the second arrest, in part by arguing that he could not complete his work without her skills. During Chase's later career she encouraged, mentored, and sponsored minority agrostological students like Maria Bandiera, Clarissa Rolfs, George Black (an African-American working in Brazil), and Zoraida Luces de Febres.

Chase continued to botanize, consult, and work after officially retiring from the USDA in 1939. Until her death at age ninety-four, Agnes Chase served as a volunteer custodian of grasses and as a very active honorary curator of grasses at the National Herbarium. The Venezuelan Ministry of Agriculture invited her to assist in the development of plans for systemic agrostology, including range management programs, in 1940 when Chase was seventy-one. Chase spent six weeks in Venezuela, observing, consulting, and botanizing and would later continued to correspond with the director of the Venezuelan Botanical Service, among other Latin American agrostologists and botanists.

Agnes Chase made major contributions to the classification and identification of grasses of the Americas and to agrostological research across the Western Hemisphere. Author of a still-vital introduction to grasses and specimen collection, Chase encouraged other botanists to collect, identify, and publish the grasses of their own regions. Agnes Chase worked for social justice, women's suffrage, and equal rights for minorities while opening doors for younger women. When institutional prejudice prevented her from receiving funds for field research, Chase used her own money and personal contacts to collect grasses in South America, at the same time developing a network of collaborators and students that ably assisted her work in classifying the grasses of the Americas. Agnes Chase died in 1963, leaving a legacy of scientific knowledge, social progress, and active students and peers.

See also C_3 and C_4 Grasses; Flint Hills; Hitchcock, A. S.; Johnson, Claudia Alta Taylor "Lady Bird"; Malin, James; National Tallgrass Prairie Park; Ordway, Katherine; Popper, Deborah and Frank; Wildflowers.

Sources

Pamela M. Henson. 1981. "Agnes Chase." In *Dictionary of American Biography:*

Supplement 7, edited by John A. Garrety. New York: Charles Scribner's Sons, 119–120.

Pamela M. Henson. 2003. "'What Holds the Earth Together': Agnes Chase and American Agrostology." *Journal of the History of Biology* 36 (Winter): 439–455.

Michael T. Stieber. 1980. "Agnes Chase." In *Notable American Women: The Modern Period*, edited by Barbara Sicherman and Carol Hurd Green. Cambridge, MA: Harvard University Press, 146–148.

CHEYENNE BOTTOMS   Cheyenne Bottoms is one of the few remaining grassland wetlands in the central Great Plains. This unique topographic basin in what is now central Kansas provides shelter and food for most of the migratory waterfowl in central North America during their spring and fall migrations along the North American Central Flyway. For thousands of years, humans and other animals have made constant use of it. An international conference on wetlands, the Ramsar (Iran) Convention, listed wetlands of global importance. Cheyenne Bottoms made the list because it serves as an example of the possibilities of reserving water rights for wildlife and habitat preservation, and of the economic benefits of wildlife-based tourism.

This wetland most likely formed between 80 million and 3 million years ago. Cretaceous bedrock ridges 100 feet (30 m) high form the basin's north, west, and south boundaries, while dunes of windblown silt and sand block the eastern edge of the Bottoms. Although located only a few miles north of the Arkansas River Valley, both paleo-pollen studies and modern species distributions suggest that the basin never contributed to the Arkansas River watershed. Geologic research suggests that the Smoky Hill River, which currently flows north of the Bottoms, flowed through the basin prior to the Kansan glacial stage, roughly a million years ago. Since then, only Blood and Deception Creeks and precipitation contribute water to the wetlands, which have a total drainage area of 104 mi^2 (272 km^2).

Throughout time, Cheyenne Bottoms has been an ephemeral wetland, often drying out for long periods of time. Pollen studies performed on soil cores taken in the Bottoms show the 166-km^2 basin to have been drier 25,000 years ago, and possibly totally dry between 22,000 and 24,000 years ago. Around 11,000 years ago water returned in depth to the Bottoms, but since then water levels have fluctuated frequently, and grasslands replaced the forests that had previously surrounded the basin. Cheyenne Bottoms has gone at least partly dry two out of every three years for the last several thousand years. At other times—for example, when in 1927 fourteen inches (35 cm) of rain fell in the vicinity—the Bottoms becomes a lake covering over 64,000 acres (40.4 hectares). Groundwater from subsurface aquifers contributed to the Bottoms, too, by augmenting the flows of Blood Creek and Deception Creek. Even with this additional moisture, water levels in the basin have varied seasonally, peaking during the spring storm season and declining over summer and fall when the waters evaporate. This expansion and contraction has contributed to the variety of plants and animals that inhabit or make use of Cheyenne Bottoms.

The varieties of plants found around the Bottoms generally fall into four communities: open water–mudflat, cattail, wheatgrass-saltgrass, and spikesedge communities. Cattails are the dominant plants within the central marsh of Cheyenne Bottoms. They inhabit the deepest pools and provide food for muskrats besides sheltering nesting herons, bitterns, and other birds. Although algae are common, few submerged plants grow in the Bottoms because of the cyclical drying of the marsh and the turbidity (muddiness) of the water, which limits the amount of sunlight below the water's surface. Also, bulrushes, arrowhead, smartweed, and spikesedge border the pools, and these plants provide food and cover for migrating and nesting birds, and for resident small and larger mammals. When evaporation exposes the mudflats, annuals like goosefoot, aster, pigweed, and barnyard grass take root. Cottonwood and black willow trees tap the shallow groundwater and edge Blood Creek and Deception Creek, while tall- and midgrass species like big and little bluestem (*Andropogon gerardi* and *A. scoparius*), wild plum, and prairie cordgrass (*Spartina pectinata*) grow on the surrounding uplands.

These plants in turn support a large variety of birds, reptiles, amphibians, and mammals. Of the vertebrate animals known to live in Kansas, excluding fish, Cheyenne Bottoms is home to 371 of 610 total vertebrate varieties, or 61 percent. Bloodworms, midge larvae, and rotifers form the bottom of the animal food chain, feeding on decaying plant matter and animal waste and remains. Migrating shorebirds, fish, and resident waterfowl feed on these superabundant invertebrates, while frogs and other reptiles feed on larvae and on the adult insects, too. Herons, raccoons, mink, snapping turtles, and kingfishers make use of the fish that swim in from the creeks, while muskrats control the cattail population and white-tailed deer browse the marsh edges and take cover in the dense vegetation around the marshes within Cheyenne Bottoms. Numerous mice and voles feed on seeds and berries, and in turn feed harrier hawks, mink, and other predators.

Humans have made use of Cheyenne Bottoms since the first Paleoindians entered the area. Peoples like the Wichitas and Cheyennes, for whom the Bottoms were named, regularly visited the basin. In addition to providing excellent hunting opportunities, the Bottoms' rich plant communities offered a variety of edible, medicinal, and useful plants. These plants included arrowhead, prized for its starchy tubers, along with goosefoot seeds, wild plums, and probably cattails as well. The first Europeans to visit the Bottoms were part of the *entrada* led by Francisco Vásquez de Coronado in 1541. Juan Jaramillo commented favorably on both the Arkansas River Valley and the Bottoms, noting their lushness. Later Anglo-American explorers and settlers found Cheyenne Bottoms to be less promising. Lieutenant Zebulon Pike visited the Bottoms in October 1806, and Stephen Long studied the area during late summer in 1820. Neither man found much to recommend the area, aside from locally lush forage for their animals. Although homesteaders and others settled and farmed the land around Cheyenne Bottoms, no one took up land within the basin. The soil was too boggy, alternately flooding or baking, and the soil probably would not produce enough crops to justify the cost and effort of trying to drain the area. Professional and recreational hunters, however, took advantage of the wetlands and used Cheyenne Bottoms as a source of waterfowl for feathers and meat.

Only in the twentieth century did serious attempts to drain Cheyenne Bottoms begin, while at the same time sportsmen and conservationists began working to preserve it for its wildlife. A serious effort to drain the Bottoms began in the late 1920s, when local farmers formed a drainage association and floated bonds to raise funds for ditch construction. Opposition to this plan came from amateur and professional hunters, conservationists, local businessmen, and the Chambers of Commerce from nearby Great Bend and Hoisington. The farmers persisted and in April 1928 filed association articles for Cheyenne Bottoms Drainage District #1. Opponents protested the action, but a local court found that the Drainage District was doing nothing illegal and allowed the operation to continue.

As the Drainage District formed and began work, other groups tried to preserve Cheyenne Bottoms as a wetland. Amateur ornithologist Frank Robl had begun banding birds in Cheyenne Bottoms in 1923 and soon discovered that the Bottoms formed a vital part of the North American migration corridor for many waterfowl. His work, along with other factors, persuaded the head of the Kansas Forestry, Fish and Game Agency (KFFGA) to begin work to protect the wetlands. While representatives of conservation groups like Ducks Unlimited and the KFFGA worked in the courts to prevent ditching, local and national organizations like the Izaac Walton League, readers of *Forest and Stream* magazine, and the national Audubon Societies joined KFFGA to lobby Congress for funds to buy the Bottoms and designate it a National Wildlife Refuge. The Kansas attorney general filed suit in April 1930 to block the pending Drainage District contracts, arguing among other things that draining Cheyenne Bottoms would hurt the local environment and cause flooding problems downstream; that the soil was unsuitable for agriculture because of its high concentration of salt; and that the action was probably in violation of the state constitution. This time

Judge Roy Beal agreed and blocked the Drainage District's actions. Soon thereafter, the House and Senate passed legislation to fund Cheyenne Bottoms land purchases, and although major land deals were not completed until 1942, the physical integrity of the wetlands was preserved.

Before the formation of Cheyenne Bottoms Wildlife Area, several people made efforts to create a constant flow of water into the wetland to form a permanent lake. The Grand Lake Reservoir Company dammed and diverted water from the Arkansas River between 1899 and 1902 via the Koen canal but failed to meet the goal of a lake large enough for a steamboat. The State of Kansas constructed dikes, canals, and other water control structures along Walnut Creek and the Arkansas River beginning in 1949, and added a diversion dam on the Arkansas River in 1957 to increase the flow of water into the Bottoms. These measures failed because the flow of water upstream had been greatly reduced by irrigation diversions and by pumping the groundwater from the Ogallala Aquifer. This pumping lowered the water table and prevented groundwater from flowing into the river and creeks. This effectively dried up the streams, leaving only precipitation to refill the wetlands.

Cheyenne Bottoms Wildlife Area is unusual in that senior water rights on Walnut Creek of 20,000 acre feet of water are attached to it. Kansas is one of the few states that allow water rights to be claimed for the benefit of recreation. This became crucial during a drought in the early 1990s. In January 1992 the chief engineer for the Kansas State Board of Agricultural Water Division set up an Intensive Groundwater Use Control Area (IGUCA), limiting how much junior water users could pump for irrigation so as to protect the senior user's water supply. The junior right holders, farmers who depended on irrigation for protection from the drought, protested that the state was favoring ducks over people, and that since the Bottoms had not perfected its senior right by claiming all 20,000 acre feet every year, that right should be reduced or even denied. However, under the doctrine of prior appropriation, the senior right holder has priority over the juniors whether that holder is a wildlife area, individual, or corporation. As senior, Cheyenne Bottoms Wildlife Area (or the State of Kansas acting on behalf of the Bottoms) had the legal right to request its water. The 1992 decision to allow the IGUCA to stand did not set a precedent of placing the welfare of wildlife over that of humans; it merely reaffirmed the right of senior water right holders to limit the amount junior users can withdraw. That the right holder was a wildlife area did not change the law.

Although still at risk of further future water depletion, Cheyenne Bottoms remains a crucial oasis of wetland habitat in the central Great Plains. In 1990 Cheyenne Bottoms was designated a Wetland of International Importance under criteria set up by the Ramsar Convention, and some consider this geologic basin to be the most important single wetland in the Western Hemisphere for migratory waterfowl. It is certainly the most crucial remaining wetland along the Central Flyway. The bird-watchers and hunters who visit the Bottoms contribute over $2 million per year to the local economy, helping ensure regional support for continuing preservation and expansion of the protected area in and around Cheyenne Bottoms.

Cheyenne Bottoms is a unique geologic basin containing some of the last remnants of the many grassland wetlands once found on the Great Plains of North America. Like the salt marshes of nearby Quivara National Wildlife Refuge, Cheyenne Bottoms provides food and shelter to millions of migrating waterfowl each year, as well as rich year-round habitat for birds, mammals, reptiles, and amphibians. Although threatened by drainage in the past and groundwater mining today, Cheyenne Bottoms serves as an example of both the positive economic effects of habitat preservation and the ecological value of wetlands to both wildlife and humans.

See also Endangered Species; Ogallala Aquifer; Playa Lakes.

Sources

Duram, Leslie Aileen. 1995. "Water Regulation Decisions in Central Kansas Affecting Cheyenne Bottoms Wetland and Neighboring Farms." *Great Plains Research* 5 (February): 1052–1065.

Harvey, Douglas S. 2001. "Creating a 'Sea of Galilee': The Rescue of Cheyenne Bottoms Wildlife Area, 1927–1930." *Kansas History* 24 (Spring): 2–17.

Zimmerman, John. 1990. *Cheyenne Bottoms: Wetland in Jeopardy*. Lawrence: University Press of Kansas.

CLEMENTS, FREDERICK E. (1874–1945) AND EDITH SCHWARTZ (1874–1971)
Raised in Lincoln, Nebraska, Frederick Clements was stirred at the age of sixteen to seek an interview with Professor Charles Bessey. Bessey, trained by Asa Gray, a nationally renowned Harvard botanist and early advocate of Darwin's theory of evolution, had been at the university for six years and also served as the state botanist. Like his compatriot A. S. Hitchcock to the south at Kansas State Agricultural College, Gray was intent on documenting the indigenous grass species of his state before they were lost to the plow. He already had a young enthusiast, Roscoe Pound, out tramping the state taking grass samples, and recognizing something in this young Clements boy, he had him join Pound in the field. Pound and Clements later published their findings in *The Phytogeography of Nebraska* (1898). Apparently Pound soon had enough of ecology and headed to the East Coast to begin a prestigious career in law. Clements, on the other hand, became enthralled with ecology, and after receiving his Ph.D. degree remained as a professor at Nebraska until 1907 when he took a new position at

the University of Minnesota. There he remained for another ten years before accepting a position at the Carnegie Institution of Washington where he worked until retiring in 1941.

Edith Schwartz, on the other hand, showed little early promise of becoming a botanist. Raised in Omaha, Nebraska, she left after graduating from high school to attend the University of Minnesota, but returned to finish her undergraduate degree in Germanic languages from the University of Nebraska. She then taught German at the university, and a year later her relationship with Frederick Clements blossomed into marriage. Frederick whetted her appetite for studying botany, and Edith began working toward a Ph.D. while pursuing a parallel study of geology and continuing a lifelong study of language. She conducted her field research at the Nebraska Alpine Laboratory in Minnehaha, Colorado. She became the first woman to receive a Ph.D. with the completion and defense of her dissertation, "Relation of Leaf Structure to Physical Factors." Obviously a gifted scientist, Edith subsumed her professional career by becoming Frederick's assistant, albeit an assistant who worked more as an equal intellectual partner. They shared a tight and thoroughly compatible relationship, one marked by mutual respect, intellectual reciprocity, and constant companionship. As one of their friends described them: "Both vigorous workers, they had the advantage of constant counsel and advice, each of the other—a potent scientific partnership."

While at the University of Nebraska, Clements's publications gained him national attention and respect. His *Research Methods in Ecology* (1905) was born out of the fieldwork that he and Pound had conducted years early. This quickly became a standard methodological text for student ecologists. By 1916, *Plant Succession: An Analysis of the Development of Vegetation*, written while he was at the University of Minnesota, became his most notable achievement. In this work Clements fully developed his thinking about succession. Any biome, or life community, went through several stages of plant and animal mixes before reaching a stable, self-sustaining equilibrium; this equilibrium then became marked by a well-defined assemblage of species that formed a complex set of reciprocal, nurturing relationships. Of course this "climax ecosystem" could be disturbed or wrecked by climatic change or human interference, but remove the disturbance and the former balance of plants and animals would restore itself.

This thinking had particular application to events such as the Dust Bowl, which seemed to demonstrate the destructive capacity of humans to alter, if not destroy, the grasslands climax system. The grassland climax ecosystem offered, so Clements thought, the keys for unlocking the secrets to successful farming or ranching in the grasslands. Consequently, for Clements it was of the utmost importance to preserve this climax system so as to understand the relationships creating its balance, and how the biome responded and adjusted itself to climatic fluctuation. What this did, in essence, was to overlook the role that humans, especially Indian peoples, played in shaping, managing, and maintaining a grasslands that Clements believed came into existence without the aid or conscious activities of humans. His thinking also tended to give what he perceived as nonhuman-shaped climax systems a moral worth greater than that accorded to farm fields and ranch pastures. The backlash against Clements's thinking led to the initiation of some of the earliest environmental history such as that written by James Malin, who emphasized the ecologically adaptive creativity of people through science and technology. Clements's thinking, however, has formed a fundamental building block of contemporary ecosystem theory.

See also Chase, Agnes; C_3 and C_4 Grasses; Dust Bowl; Hitchcock, A. S.; Jackson, Wes; Malin, James; Popper, Deborah and Frank; Webb, Walter Prescott.

Sources

Allred, B. W. 1953. "Edith Schwartz Clements." *Journal of Range Management* 6 (November): 379–381.

Bonta, Marcia Myers, ed. 1995. "Edith Clements: Ecology and World War I." In *American Women Afield: Writings by Pioneering Women Naturalists.* College Station: Texas A&M University Press, 161–170.

Gollely, Frank Benjamin. 1993. *A History of the Ecosystem Concept in Ecology: More Than the Sum of Its Parts.* New Haven: Yale University Press.

Toby, Ronald C. 1981. *Saving the Prairies: The Life Cycle of the Founding School of American Plant Ecology, 1895–1955.* Berkeley: University of California Press.

Worster, Donald. 1977. *Nature's Economy: A History of Ecological Ideas.* Cambridge: Cambridge University Press.

C_3 AND C_4 GRASSES An interesting word, "grass." It is a derivative of an Old Aryan word, *ghra-*, which means to grow, and much later some English words like *grain, green,* and, of course, *grow* would trace their ancient origins to *ghra-*. The Romans probably used it to form their word *gramen,* or grass. Within this word lies the health, and perhaps sum, of human life. John James Ingalls, a senator from Kansas, observed this connection in an article that he wrote and published in the *Kansas Magazine* in 1872. "Grass feeds the ox: the ox nourishes man: man dies and goes to grass again; and so the tide of life, with everlasting repetition, in continuous circles, moves endlessly on and upward, and in more senses than one, all flesh is grass." While Senator Ingalls understood, even if in a somewhat overstated manner, the importance of the grasslands to humanity, much about this life community, or biome, remained a mystery to him and the people of his times, and to those who had passed before him.

When Euro-Americans first described grasses throughout North America they usually grouped them into two categories: "warm-season" and "cool-season" grasses. Not surprisingly, warm-season grasses were observed growing late in the spring, thriving throughout most of a dry summer, maturing in the fall, and going dormant throughout the winter. "Cool-season" grasses on the other hand flourished well in wetter, cooler environments. Such places might encompass the northern prairies early in the spring or later in the fall, or the more southeasterly grasslands throughout the winter.

In 1968 two biologists, W. J. Downton and E. Tregunna, figured out the different ways in which "warm-season" and "cool-season" grasses combined visible light and carbon dioxide to form carbohydrates, or the basic fuels of all biological life. A simple, but crucial, problem confronts grasses: they must take in carbon dioxide while at the same time retaining the water necessary for their lives. Plant leaves regulate the absorption of carbon dioxide, which is indispensable for photosynthesis, through microscopic holes in their leaves called stomata. When a climate is wet there is little need for a leaf to close its stomata to prevent water loss, so it can easily capture as much carbon dioxide as its stomata can inhale. But if the climate is hot and dry, then keeping the stomata open can prove exceptionally costly for a plant. Water loss through open stomata will result in dehydration, and plants easily die as a result. Shutting off the stomata for a long period, however, causes another problem: the plant has no means to absorb carbon dioxide for photosynthesis.

Downton and Tregunna observed how warm-season and cool-season grasses differed in the ways they delivered carbon dioxide to the site where photosynthesis occurs. Cool-season grasses bind carbon dioxide to an enzyme to produce a molecule with three carbons; this is distributed throughout the plant and powers photosynthesis. Most ecologists now label these plants "C_3 grasses." The warm-season grasses attach carbon dioxide to an acid, which results in a four-carbon molecule while in addition still employing the C_3 pathway of the cool-season grasses. Consequently, these grasses are labeled "C_4 grasses." This extra step in forming an additional carbon-based molecule makes these plants very efficient at grabbing carbon dioxide from the atmosphere when they have their stomata open. This capacity allows these plants to shut their stomata more often and for longer periods of time than C_3 plants, and allows C_4 grasses to thrive in much drier climes. There is a cost involved to C_4 grasses, however. The extra pathway for internal plant distribution of the molecules responsible for photosynthesis makes them less efficient at this activity than the cool-season grasses when water or temperature are not controlling factors.

C_4 plants once thoroughly dominated the landscape of the central grasslands. Throughout the eastern extent, or the tallgrass prairies, Big Bluestem

(*Andropogon gerardi*) flourished; in the mixed-grass prairies to the west Little Bluestem (*Andropogon scoparius*) was the main species; and everyone knew the shortgrass plains by the carpet-like appearance formed by the prevalence of Buffalo Grass (*Buchloe dactyloides*). Of course, other plants lived alongside these mainstays, but any half-observant traveler knew when he or she had entered one plant association or the other by simply detecting which of these three grasses was most abundant.

See also Flint Hills; Hitchcock, A. S.; Jackson, Wes; Johnson, Claudia Alta Taylor "Lady Bird"; Malin, James; National Tallgrass Prairie Park; Ordway, Katherine; Popper, Deborah and Frank; Wildflowers.

Sources

Redmann, R. E., and E. G. Reekie. 1982. "Carbon Balance in Grasses." In *Grasses and Grassland: Systematics and Ecology*, edited by James R. Estes, Ronald J. Tyrl, and Jere N. Brunken. Norman: University of Oklahoma Press.

Reichman, O. J. 1987. *Konza Prairie: A Tallgrass Natural History.* Lawrence: University Press of Kansas.

DUST BOWL   Rain follows the plow, or at least many colonizers on the Great Plains believed it so before the beginning of World War I. Dryland farming, they thought, held the key. This was the practice of farming without the aid of irrigation in the semi-arid regions of the grasslands. By 1900, many dryland tillage techniques and machinery had been developed that allowed farmers to raise crops on very large acreages of land. Even though the record of dryland farming was spotty at best, enough success in the eastern reaches of Montana, Wyoming, Colorado, and New Mexico, and the western portions of Kansas, Oklahoma, and Texas encouraged the continuation of it after the war. Yet by the mid-1930s, drought, high winds, and the pulverization of topsoil into brown, swirling clouds of dust led many to question the practice.

Who or what created the Dust Bowl in the grasslands, and can it happen again? Scholars, farmers, and bureaucrats have debated the answer to these questions long and hard, and without resolution to date. The Depression Era artist, Alexander Hogue, had firsthand experience with blowing dust while living in Denton and Dallas, Texas. In a stark painting he certainly made clear to all who saw it who he blamed for causing the "dirty thirties." In his *Mother Earth Laid Bare* (1936), he revealed a highly eroded landscape, devoid of any life, with a delapidated farm house and outbuildings in the background. Upon a closer look, the coarsely cut, windblown sandy soils in the middle take the form of an exposed woman, Mother Earth Laid Bare. Mankind, Hogue depicted, had with his technology and economic greed transformed the "garden into the desert that the early explorers had mistakenly thought it to be." This was one view.

Others had a more sanguine view of the Dust Bowl. As the Kansas scholar James C. Malin wrote, the grasslands had been subject to severe blowing soils long before farmers ever sank their plows into its sod. He knew soil scientists had discovered deep deposits of loess, fine windblown soil particles, underlying several distinct soil profiles. Winds had deposited these layers thousands of years ago during the hypsithermal and demonstrated to Malin that plowing was not necessary for winds to lift soils. Malin held that so long as a range of climate existed, people had a great amount of freedom within it to devise their means of living. He firmly believed that farmers had not exceeded this range during the 1930s; rather, the climate had temporarily transcended the farm technology capable of existing within it. The rains would return, and people would devise new means, new systems, new methods for farming in the grasslands. In essence, Malin held capitalism and the mechanistic approach to agriculture free from any culpability for creating the Dust Bowl.

Any of the methods or policies for ameliorating the Dust Bowl depended entirely on the manner in which anyone understood its causation. New Dealers became convinced that poor care of the land resulted in blowing fields and human impoverishment on the Great Plains. This was certainly illustrated in four major ways: the production of the film *The Plow That Broke the Plains*; the creation of the Soil Conservation Service with Hugh Bennett at its helm; the passage of the Taylor Grazing Act; and the report titled *The Future of the Great Plains*. All of these approaches or acts pinpointed human mismanagement of the grasslands as the cause of the Dust Bowl and called for the enactment of centralized, scientifically driven federal planning and conservation measures.

Not until the 1950s, with an emerging backlash against federal planning, would a call for free market and technological solutions for soil conservation and dryland farming be heard. Still, the Soil Conservation Service, called the Natural Resources Conservation Service today, continues to play an important role in seeking ways to farm the Great Plains. And yet, as droughts in the 1990s and early 2000s still illustrated, dryland farming in the Great Plains is a precarious occupation.

See also Clements, Frederick E., and Edith Schwartz; C$_3$ and C$_4$ Grasses; *Future of the Great Plains* (1936); Jackson, Wes; Malin, James; Popper, Deborah and Frank; Webb, Walter Prescott; Zybach, Frank.

Sources

Bonnifield, Paul. 1979. *The Dust Bowl: Men, Dirt, and Depression.* Albuquerque: University of New Mexico Press.

Hargreaves, Mary W. M. 1993. *Dry Farming in the Northern Great Plains: Years of*

Readjustment, 1920–1990. Lawrence: University Press of Kansas.

Hurt, R. Douglas. 1981. *The Dust Bowl: An Agricultural and Social History.* Chicago: Nelson Hall.

Svobida, Lawrence. 1986. *Farming the Dust Bowl: A First-Hand Account from Kansas,* foreword by R. Douglas Hurt. Lawrence: University Press of Kansas.

Worster, Donald. 1979. *Dust Bowl: The Southern Plains in the 1930s.* New York: Oxford University Press.

ENDANGERED SPECIES President Richard Nixon called for a much stronger response toward protecting endangered species. In 1972, he recounted how the first act passed in 1966, and later revised in 1969, simply failed to provide the type of protection these animals and plants required in order to survive. The 1966 legislation simply created a list of endangered species and left any action to save them to the initiatives of local, state, and federal agencies. While some have questioned the president's motives, his call for action led Congressman John Dingell to introduce a bill in the House of Representatives that passed by a 390 to 12 vote margin; Senator John Turney's companion bill in the Senate passed unanimously, 90 to 0. President Nixon signed the bill into law with provisions significantly strengthening the 1966 and 1969 legislation. One provision protected every plant and animal on the list. The act also provided protection to subspecies and isolated populations, and species considered threatened, as opposed to endangered, also received protection. This legislation greatly enlarged the number of animals and plants eligible for protection—from 114 in 1973 to over 1,100 by 2000. Success rates as administered and charted by the U.S. Fish and Wildlife Service (FWS) have revealed that the 1973 act has at best only been partially successful when it reported that only one in ten species were improving by 1995. In the grasslands, this is particularly true.

Each state creates a list of endangered species, and these are telling indicators of the ecological health of their biosystems, or as the United States Geological Survey terms them, they are "environmental monitors." For example, the six fish species on the list in North Dakota are telling evidence about stream quality changes. Such changes can, and often do, have direct consequences for the humans who use and consume the same water. The seven bird species on the same list may be a gauge of plant and soil toxicity arising from chemical applications in farming. Water pollution has been a great concern throughout the grasslands, and most states' endangered species list certainly confirms this growing problem. In New Mexico, for example, of the 118 species and subspecies on its endangered or threatened list in 2002, 56 were exclusively aquatic animal species and this does not include birds. Kansas has a similar story. In

2005, over half of the mussel species in the state were in serious trouble, and mussels, as they remain in place and filter water for their food, are excellent indicators of water quality. In fact, the majority of endangered or threatened species in Kansas are either amphibians, fish, or invertebrates.

Two grassland species in particular have caused a ruckus as to whether they should be considered endangered. One is on the endangered species list in several states, and the other is not at all. There is public support in states such as Idaho to have gray wolves (*Canus lupus*), currently listed as a "threatened" species on the federal endangered species list, regulated solely by the state rather than the federal Fish and Wildlife Service. Ranchers throughout the northern grasslands are concerned about their property rights in cattle and want the liberty to hunt wolves. Whether wolves pose a significant threat to cattle ranching remains an open question, and in some areas a hotly debated one.

The other species, the black-tailed prairie dog (*Cynomys ludovicianus*), once occupied around 100 million acres of grasslands that had been reduced to roughly 360,000 acres by 1960. Today, the prairie dog has made something of a comeback and now inhabits an estimated 1.8 million acres. Consequently, the federal Fish and Wildlife Service removed the prairie dog as a "candidate" for listing as either threatened or endangered. With an estimated population of ten animals per acre, the FWS believes that in 2004 there were over 18 million prairie dogs in the United States. Still, some ecologists contend that the black-tailed prairie dog is the keystone species of the grasslands and deserves better protection from states and the federal government than it has received. In fact, their research suggests that a healthy grassland ecosystem simply cannot exist without the presence of prairie dogs, and that several other species, specifically grassland birds, are at risk without widespread populations of prairie dogs. Critics question the ecological value of prairie dogs and regard them as little more than pests. As of 2005, the black-tailed prairie dog had received no protection under the federal Endangered Species Act.

See also Bison; Cheyenne Bottoms; Goodnight, Charles and Mary Ann; Johnson, Claudia Alta Taylor "Lady Bird"; National Tallgrass Prairie Park; Playa Lakes; Wildflowers.

Sources

Bean, Michael. 1999. "Endangered Species, Endangered Act?" *Environment* 41 (February): 12–18, 34–38.

Bean, Michael, and Melanie J. Rowland. 1997. *The Evolution of Wildlife Law*, 3rd ed. Westport, CT: Praeger.

Kotlair, Natasha B., et al. 1999. "A Critical Review of Assumptions about the Prairie Dog as a Keystone Species." *Environmental Management* 24 (August): 177–192.

FLINT HILLS   What a fantastic site Zebulon Pike beheld. His expeditionary command, along with its Osage guides, had entered a region of tallgrasses, their roots meshing together in a thin layer of topsoil covering deep, sedimentary limestone deposits. These limestone layers had been over 500 million years in the making, the result of massive accumulations of sediments mixed with the skeletal remains of the marine animals that once flourished in the inland seas. The last of those seas formed a thick, limestone deposit over 60 million years ago. A passage of nearly 40 million years would occur before grasses started taking their place atop these thick limestone layers as the uplift of the Rocky Mountains began creating a rainfall shadow to the east, which produced a drier climate more conducive to grasses than to trees. More recently, some 2 million years ago, the ice ages began, and these intense periods of cold certainly left their marks on the hills traversed by Pike. One glacier advance, the Nebraskan, arrived to the north of these limestone-clad hills, which oak woodlands covered.

During the next three glacial episodes, spruce and oak competed to occupy the hills, and by end of the last glaciation, the Wisconsinian, spruce forests formed a protective mantle over the hills. However, these forests were short-lived once the ice blanket had retreated, and within a few hundred years the grasses Pike observed had come to spread over the hills. Also, the megafauna of the ice ages had given way to an assemblage of animals that Pike held in view, such as bison, elk, deer, brown bears, cougars, gray wolves, coyotes, bobcats, turkeys, and pronghorn deer. Water erosion had sharply cut limestone deposits and had created the impression of hills, all of which Pike labeled the "very ruff flint hills." From September 1806 on, the name Flint Hills has stuck as the identifier of this region, and remarkably, much of the ecosystem that Pike witnessed has been able to reproduce itself into the twenty-first century.

Of course, depending upon how one draws the map, the Flint Hills encompass a broad band of tallgrass prairie of around 5 million acres in an area about 50 miles wide and around 300 miles long located in 13 east-central counties of Kansas and in Osage County in north-central Oklahoma. Humans have continuously resided in these hills for at least 11,000 years, and they have played a key role in how the ecosystems of the place have evolved over that time. From the Clovis culture to modern ranchers, fire has been a key tool used by humans to shape the landscape. While in the past some scholars have argued that the thin soils are not suitable for agriculture, thereby preserving the tallgrass prairies of these hills largely intact, farming, whether by Indian peoples or Americans, has always existed as an important land use. Still, the pastoral uses, with people either managing the hunt of wild ungulates or raising domestic ones, have always required a vast sweep of these bluestem prairies. In a few counties, such as Chase County in Kansas, agriculture has had little, if any, effects on these grasslands—in short, a few areas have retained their ecological integrity for centuries if not thousands of years.

Many ecologists, environmentalists, outdoor recreationists, and artists have seen priceless, enduring beauty and worth in the Flint Hills. Others, too, such as urban developers, ranchers, farmers, and miners—rock quarries, sand dredging, oil drilling—have given the region a distinctive economic worth, one that normally has transcended any aesthetic or scholarly values recognized by these same people. In more recent years, the social and political tensions have grown between many diverse groups and have led to bitter contests over the creation of national, state, or local parks, the right to float non-navigable streams, or the building of wind power–generation farms. Even people who on the surface appear to share common goals, such as the Sierra Club and an Audubon affiliate, have divided sharply over whether to locate wind-powered generators in the Flint Hills. Given the imperiled nature of grassland ecosystems throughout North America, the future preservation and maintenance of the Flint Hills, "last stand of the Tallgrass Prairies," will require careful study, planning, and cooperation among a myriad of people with different interests. In many respects, the fate of the Flint Hills epitomizes that of grasslands throughout North America.

See also Bison; Chase, Agnes; Jackson, Wes; Malin, James; National Tallgrass Prairie Park; Ordway, Katherine; Wildflowers.

Sources

Heat-Moon, William Least. 1991. *PrairyErth (A Deep Map)*. Boston: Houghton Mifflin Company.

Kollmorgen, Walter M., and David S. Simonett. 1965. "Grazing Operations in the Flint Hills–Bluestem Pastures of Chase County, Kansas." *Annals of the Association of American Geographers* 55 (June): 260–290.

Larabee, Aimee, and John Altman. 2001. *Last Stand of the Tallgrass Prairie*. New York: Freidman/Fairfax.

THE FUTURE OF THE GREAT PLAINS (1936)   At the height of the Dust Bowl years during the Great Depression, the future of American life in the Great Plains looked bright to a great many New Deal reformers. Here, in wind- and heat-ravaged lands, was to be the Great Plains of the future where careful planting would restore forests in the foothills; soil conservation and sensible grazing practices would restore soil fertility and grass productivity; and scientific agriculture would yield fewer, more prosperous, and comfortable farmers tilling larger, more efficiently managed acreage.

This was the collective foresight of five prominent New Dealers, who in 1936 published *The Future of the Great Plains*, a guidebook for achieving sustainable farming in the grasslands. Secretary of Agriculture Henry A. Wallace, director of the Works Progress Administration Harry Hopkins, director of the Soil Conservation Service Hugh Bennett, administrator of the Resettlement Administration Rexford G. Tugwell, and director of the Rural Electrification Administration Morris Cooke, who also served as the committee chairman, were the signatories of the report. These men served in critical governmental capacities, and the reverberations of their and their staffs' thinking still shape the general contours of agricultural policy throughout the grasslands today. The manner in which they understood the social, economic, and ecological problems of the Great Plains and the way they attempted to alleviate these difficulties lend insight into the predicaments faced by people living in the grasslands now.

The ecological, economic, and social problems relating to the Great Plains prompted President Franklin D. Roosevelt to request this study in the first place. He wanted concrete recommendations for future legislation to ameliorate the crisis engulfing the region. The president saw the problem rather simply. In short, traditional farming practices in the eastern portion of the United States bore poor fruits when applied to the grasslands. The president, in fact, called such practices "unsuitable." For him, dust storms, the economic collapse of small family farmers, the growing rate of farm tenancy, and the influx of corporate agriculture all attested to alarming distress.

Unwittingly, the writers of the report viewed Indian peoples as "savages," and consequently the authors failed to appreciate or understand that Indian peoples' management practices were critical to the creation and maintenance of the grasslands in the first place. They concluded that Indian peoples had always lived in ignorant but "productive harmony" with their environments without polluting rivers or depleting wildlife or other natural resources. For them, it was Euro-American farmers' inabilities to work in harmony with nature that led to the problems in the semi-arid grasslands.

The committee had several specific notions for a curative. It advocated more scientific studies of the region in order to understand its ecology. It recommended a greater role for the federal government in the acquisition and management of rangelands. It wanted federally supported soil conservation practices and water management techniques implemented. It wished to remove small, failing farmers to other occupations and locations. The committee championed a full-scale "attack" on if not outright eradication of "insect pests." For those farmers who would remain, the authors expected to see outright "readjustments" in farm organization and practices through the application of greater scientific agriculture in order to achieve greater economies of scale. In short, the committee summarized its position as substituting an "intelligent adjustment to nature for futile attempts to conquer her."

Over the past six decades, these recommendations have largely guided federal policies despite continued declining soil fertility, increasing water pollution, spreading stream depletions, and mounting simplifications of plant and animal species. Moreover, recommendations of this report have failed to stem rural depopulation, as family farms have continued to disappear, replaced by corporate entities with an attendant influx of immigrant labor and the social disruptions stemming from it.

See also Clements, Frederick E. and Edith Schwartz; Dust Bowl; Malin, James; Popper, Deborah and Frank; Webb, Walter Prescott.

Sources

Great Plains Committee. 1936. *The Future of the Great Plains.* Washington, DC: Government Printing Office.

White, Gilbert F. 1986. "The Future of the Great Plains Re-Visited." *Great Plains Quarterly* 6 (Spring): 84–93.

GOODNIGHT, CHARLES (1836–1929) AND MARY ANN (1839–1926)   Charles and Mary, or Mollie as she was known to her friends, worked to preserve the southern Great Plains herd of bison from extinction. Their efforts combined a sympathy for those animals along with visions of commercial gain in raising them. They nourished a herd that in time formed and added to the foundation of several other herds such as the one now found in Yellowstone Park. Along with the likes of Charles J. "Buffalo" Jones and Scotty Phillips, they drew national attention to the needs of bison and assisted in rescuing them from oblivion.

Moving with his family to Texas in 1845, the nine-year-old Charles saw his first bison along the Trinity River. The grasslands were a great opportunity for ranchers, and Charles began his career west of present-day Fort Worth, Texas, in 1857. At the time, this area formed the frontier with Comanches, who frequently raided Anglo outposts in the vicinity. During the Civil War, Charles joined the Texas Frontier Regiment (the Rangers) and traveled the edges of the Llano Estacado in the pursuit of Comanches and their allies. While in the region he took careful note of the grasslands of northwest Texas, bison grazing habits, and the potential for cattle ranching.

After the Civil War, Goodnight joined Oliver Loving and together they drove a herd of cattle from east Texas south and around the Comanche-controlled Llano Estacado to the Pecos River, and then north toward Denver and the goldfields of the Front Range. At Fort Sumner, the partners decided to sell their herd to the U.S. government for supplying concentrated Navajos and Apaches living there. So good was the trade that Goodnight returned to his ranch for more cattle, and so began a lucrative trade at Fort Sumner. Even though Goodnight lost his good friend Loving, he became enamored of the land and started the Apishapa Ranch some forty miles northeast of Trinidad, Colorado.

Mollie was always an important partner to Charles. She had had a rough upbringing in Texas, a life that prepared her well for the rigors of ranching. As a young woman, she raised five of her brothers after both of her parents died shortly after migrating to Texas from Tennessee. She supplemented the farm income by teaching during the Civil War. Through her brother, who worked on the Goodnight ranch just to the east of Pueblo, Colorado, she met Charles years later. In 1870, Charles sold his Trinidad ranch and together the newly married couple moved to the Pueblo ranch house.

The spread of farming in the region made ranching more difficult for Charles in the early 1870s. Consequently, he and Mollie decided to move their operations eastward, and eventually to the south, where they located on what would be called the JA Ranch. This enterprise was on the northern edge of Palo Duro Canyon in the Panhandle of Texas. It was there that Mollie viewed the last remnants of the great southern bison herd as Charles's cowboys drove those animals out of the canyon to make room for cattle.

Both Charles and Mollie lamented the destruction of this vast herd, and wondered if some future for bison could be had other than extinction. New Mexicans, they realized, had attempted to domesticate bison for nearly three decades without success. Even Charles had once sought to tame bison calves, but one of his partners at the time had sold the animals before Charles had an opportunity to judge the outcome of his experiment. Charles knew the land well, and a budding fear for the future of bison took root in his mind. By 1878, the army and the Bureau of Indian Affairs had restricted Comanches to a reservation. Euro-American hunters had nearly exterminated all the remaining bison, and now those fellows had vacated the region for other haunts. Still, Charles knew that enough of the grasslands remained intact for supporting a small herd if enough animals could be found.

Mollie pushed Charles and his brothers to take an active role in preserving bison rather than just simply thinking about doing it. Charles rode out onto the plains and captured a few bison calves and returned to headquarters with them. They were turned into pens with cows that nursed them, and when that failed to work, Mollie would bottle-feed them. Local ranchers began taking "orphaned" calves to the Goodnight ranch, and by 1910 Goodnight's Tule Canyon/Quitaque Ranch supported a bison herd of 125 animals. From this stock, the Goodnights sold bison to the National Park Service and Buffalo Bill's Wild West Show, among other buyers.

Mollie lived the rest of her life near this herd, and Charles continued to maintain the herd after she died in 1926. When Charles died in 1929, The Texas legislature authorized its Fish and Game Commission to buy the animals but without providing a place to sustain the herd. A syndicate of private investors with some land intervened, bought the herd, and gave it a new place to live. Descendants of those animals now live in parks, ranches, zoos, and the Caprock Canyon State Park nearby the Quitaque Ranch.

The Goodnights' work to preserve the bison, and to encourage others to do the same, was an important contribution toward saving this animal from extinction. Motivated by more than an urgent desire to preserve an endangered species, Mollie and Charles proved to be good conservationists. Through their work, the Goodnights preserved bison as a viable link to grasslands of the past and as a guide to its future.

See also Bison; Endangered Species.

Sources

Snapp, Harry F. 1996. "West Texas Women: A Diverse Heritage and a Succession of Frontiers." *West Texas Historical Association Yearbook* 72:21–38.

Wood, Judith Hebbring. 2000. "The Origin of Public Bison Herds in the United States." *Wicazo Sa Review* 15 (Spring): 157–182.

Zontek, Ken. 1995. "Hunt, Capture, Raise, Increase: The People Who Saved the Bison." *Great Plains Quarterly* 15 (Spring): 133–149.

HENRY, T. C. (1841–1914)   Once called the "Father of Irrigation" in Colorado, the "Wheat King" of Kansas, and the richest man in Kansas, T. C. Henry died penniless in a Denver hotel in 1914. While it is questionable just how much Henry personally contributed to growing winter wheat in Kansas and to the development of irrigated agriculture throughout Colorado, no doubt he received many public accolades for his successes besides condemnation and financial ruin for his failures.

Raised in upstate New York, Henry's early childhood was shaped by onerous work on his family farm. Certainly this formed his notion that it was better to finance and promote agriculture than it was to work in the field. Soon after the Civil War, Henry put this belief into practice by taking up cotton growing in Alabama. This venture, like many of his to come, quickly failed. Perhaps being regarded as a carpetbagger by the locals also had something to do with his lack of success. Nonetheless, Henry possessed an undauntable entrepreneurial spirit that led him to Abilene, Kansas, in 1867.

Remarkable changes were coming to the Sunflower State in 1867. Workers labored to build the Kansas Pacific Railroad, and by the spring they had reached nearly the center of the state at the small, nondescript town of Salina. At the same time, W. W. Suggs was managing a group of Texas cowboys driving a large herd of cattle toward the one-hotel town of Abilene, Kansas, some twenty miles east of Salina. There, with Joseph McCoy, Henry helped to launch the Texas cattle trade. In quick order, Henry decided that the cattle trade was not for him, and he again took up farming, but this time wheat. While many farmers in the state were still attempting to raise either corn or spring wheat, Henry turned to experimenting with varieties of winter wheat. He found winter wheat well adapted to growing conditions in the Sunflower State, and by 1878 he had over 15,000 acres in crops, most of it in winter wheat. He had reached the pinnacle of his success in wheat farming when a dry 1879 destroyed his crops and fortunes.

Perhaps because of his experience with drought, Henry turned his attention to irrigated farming and began promoting it throughout the state of Colorado in 1883. He believed in "cooperative" enterprises, or as he phrased it, the transference of the ownership of "the ditches to the farmers under them." Henry thought that irrigators in Colorado could "more safely rely upon the production of ten acres of irrigated land in Colorado for a livelihood than upon one hundred and sixty acres of non-irrigated land . . . in Kansas." But building large irrigation works required huge capital expenditures, and few regional capitalists could afford such investments. This led Henry to tap Eastern and European sources for the wherewithal to build his dreams. All together Henry was responsible for initiating twelve major irrigation projects in the Colorado, Uncompahgre, North and South Platte, Rio Grande, and Arkansas River valleys. His projects always became terribly underfunded and mired in debt, which often resulted in local merchants holding thousands of dollars in worthless credit from unpaid contractors and laborers. Tellingly, the residents in the town of Henry, Colorado, angrily changed the name to Monte Vista.

Predictably, Henry soon found himself in numerous court suits with eastern investors such as the Travelers Insurance Company of Boston. In other state supreme court cases the farmers of the ditch systems themselves wrested control away from Henry, as did the Fort Lyon Canal Company in the Arkansas River Valley. Such suits gave federal district judges, such as David Brewer, extensive experience with the problems of western irrigation. This familiarity with water issues played a decided role in the United States Supreme Court when then justice David Brewer wrote the decision for *Kansas v. Colorado* (1907). Despite Henry's many setbacks, he retained enough respect that in 1889, Governor Job Cooper appointed him as part of a committee to revise and codify the irrigation laws of Colorado. This commission traveled to southern Europe to study the uses of irrigation there. Their recommendations, which expanded the concept of beneficial uses beyond domestic irrigation and industrial uses, fell largely on deaf ears.

Reduced to a Denver hotel room, Henry died completely broke. But he also died with a larger-than-life reputation, as indicated in the obituary printed in the *Denver Post* (3 February 1914): he "taught Coloradans to make Gardens of deserts." In the same piece, Colorado Supreme Court justice Tully Scott emoted, "A great man this day has fallen in Israel . . . who transformed "barren wastes into blooming fields." He was eulogized as the "Father of Irrigation" in Colorado and the "Wheat King" of Kansas. Henry's actual accomplishments measured less than these appellations; however, his legacy has amply filled them.

See also Bonanza Farms; Carleton, Mark; Jackson, Wes; Malin, James; Winter Wheat.

Sources

Isern, Thomas D. 2005. "Theodore C. Henry: Frontier Booster and Nostalgic Old Settler." In *John Brown to Bob Dole: Movers and Shakers in Kansas History*, edited by Virgil Dean. Lawrence: University Press of Kansas, 81–90.

Malin, James C. 1944, reprinted 1973. *Winter Wheat in the Golden Belt of Kansas: A Study in Adaption to Subhumid Geographical Environment.* New York: Octagon Books.

Sherow, James E. 1992. "Marketplace Agricultural Reform: T. C. Henry and the Irrigation Crusade in Colorado, 1883–1914." *Journal of the West* 31 (October): 51–58.

HIGHTOWER, JIM (1943–) A stinging rebuke to the social, ecological, and economic harvests of land-grant universities came in 1972 with the publication of *Hard Tomatoes, Hard Times.* Jim Hightower railed against what he saw: "Had the land-grant community chosen to put its time, money, its expertise and its technology into the family farm, rather than into corporate pockets, then rural America today would be a place where millions could live and work in dignity" (p. 139). Hightower was convinced that land-grant institutions had become handmaidens of a corporate elite that had little, if any, shared interests with the common farm folks of America, who were supposedly the intended beneficiaries of the 1862 Morrill Act. This trend had devastating consequences for both the ecological and economic health of small family-owned farms across the nation, but particularly so in the grasslands where Hightower had been raised and still lives today.

Hightower spent his youth in Sherman, Texas. In this community he acquired an enduring empathy for working-class folks and small family farmers. A popular kid in high school, his classmates elected him as their senior class president. He deeply felt a reformist zeal influenced by the youthful presidency of John F. Kennedy. At the University of North Texas, he achieved prominence as a student politician and was elected student body president in his senior year. It was then that Hightower became more socially aware as he confronted racial segregation.

After graduation, Hightower tried law school, but after one week he found it completely unsatisfying. Politics, especially the politics of agriculture, was much more appealing to his nature. In the late 1960s he served as an aide to Texas senator Ralph Yarborough, and in this setting he was introduced to national power politics. In 1972, he spearheaded the formation of the Agribusiness Accountability Project (AAP). It was styled on the consumer investigatory model pioneered by Ralph Nader. In fact, one of Nader's assistants, Susan De-Marco, worked with Hightower in launching the project, and out of this developed a romantic and professional relationship still flourishing to this day. It was through AAP that Hightower and DeMarco researched and wrote *Hard Tomatoes, High Times.* Of course, established farm interests such as the Farm Bureau, Eli Lilly, and the Case Tractor Company took a dim view of Hightower's pronouncements and research. By the 1978 printing of *Hard Tomatoes, High Times,* Hightower had become well known for opposing the chemical, mechanized approach to agriculture, and a firm social proponent of agricultural laborers and small family farm operations.

Hightower dabbled for a while in muckraking as publisher and editor of *The Texas Observer.* But being in the thick of politics flows in his blood, and in 1982 he successfully ran for election as the Texas Agricultural Commissioner. He won re-election four years later as a down-home populist who championed the small against the rich and mighty. In what he called "percolate up growth," he oversaw greater regulation of petrochemical use in farming; worked to create and enhance farmers' markets statewide; encouraged and promoted organic farming; and stressed self-help and mutual cooperation. He created programs such as "Taste Texas" to popularize nationally the agricultural production of his state. In 1990 he lost his bid for a third term to Rich Perry, a client of political adviser Karl Rove.

Since then Hightower and DeMarco have waged a campaign designed to challenge the political elite within both the Democratic and Republican Parties in an effort to create a broadly based popular politics. By airing his views on television, radio shows, and the Web, and touring the country in his "Rolling Thunder," Hightower continues to agitate by enlisting people "at the grassroots level around issues, bring people in across lines, like privacy, like NAFTA, and the WTO." Hightower's voice is one continuing to yell into the prevailing winds of farm practices and politics for a change in course.

See also Jackson, Wes; Popper, Deborah and Frank.

Sources

Buttel, Frederick H. 2005. "Ever since Hightower: The Politics of Agricultural Research Activism in the Molecular Age." *Agriculture and Human Values* 22 (September): 275–283.

Smith, Evan. 1995. "Interview with Jim Hightower." *Mother Jones* 20 (November/December): 58–60.

HITCHCOCK, A. S. (1865–1935)   A. S. Hitchcock was one of the leading botanists in the United States at the turn of the twentieth century. An avid plant collector, scholar, and educator, Hitchcock contributed greatly to the development of botany in the Americas. He taught collectors and taxonomists like Agnes Chase and others, both as a professor in Iowa, Missouri, and Kansas, and through his work for the United States Department of Agriculture (USDA) and the Smithsonian at the National Herbarium.

Albert S. Hitchcock was born in Owasso, Michigan, on September 4, 1865. He received a bachelor of science degree from Iowa State Agricultural College at Ames, Iowa, in 1884, and a master of arts in botany from the same college in 1886. He taught chemistry at the University of Iowa from 1886 to 1889, then moved to the Missouri School of Botany in St. Louis. On March 16, 1890, Hitchcock married Rania Bell March, and later in 1891 he was hired as the chair of the botany department at Kansas State Agricultural College where he worked for ten years. While at Kansas State, Hitchcock designed textbooks and pamphlets for students as well as giving students access to his personal botanical library. He was a popular teacher, and several student groups issued "resolutions of regret" when he left the college. In 1901 he departed academics and became an agrostologist (grass classification specialist) for the USDA at the National Herbarium in Washington, D.C., rising to become senior botanist in 1924 and head of agrostology, a position he held until his death in 1935.

Hitchcock was an avid plant collector as well as an academic botanist. During one nine-week trip across western Kansas in 1895, he and an assistant collected over 11,000 specimens for the Kansas State Herbarium. Sites of his other collecting trips included the Bahamas in 1893, Mexico and Central America 1910–1911, British Guiana in 1919, the highlands of South America in 1923, and an expedition to Newfoundland in 1928, as well as numerous trips through the United States. He sent out calls through publications like Kansas State Agricultural College's campus newspaper, *The Industrialist*, requesting samples and information about plant ranges.

Hitchcock published widely, in both scholarly and semi-popular journals. He contributed stories about botanizing in Jamaica and Florida to *The Industrialist*, along with articles aimed at farmers and agricultural agents that discussed plant physiology. Hitchcock's books include *The Textbook of Grasses*, published in 1914, *A Manual of Farm Grasses* in 1921, and *The Manual of Grasses of the United States*, published shortly before his death in 1935. He also had over 150 articles to his credit, several of which have been translated into other languages.

In addition to botanizing and publishing, A. S. Hitchcock participated in the larger national and international scientific community. He served as the president of the Botanical Society of America in 1914 and president of the Biological Society of America in 1923. In addition he was a fellow of the American Association for the Advancement of Science and a member of the National Research Council, American Academy of Arts and Sciences, and several international botanical associations. He and his wife were returning from the International Botanical Congress in Amsterdam aboard the S.S. *City of Norfolk* when he died suddenly on December 16, 1935.

> *See also* Chase, Agnes; Clements, Frederick E. and Edith Schwartz; C$_3$ and C$_4$ Grasses; Jackson, Wes; Malin, James; Popper, Deborah and Frank.
>
> *Sources*
>
> Barnand, Iralee. 2003. "The 137-Year History of the Kansas State University Herbarium." *Transactions of the Kansas Academy of Science* 106 (Spring): 81–91.
>
> Henson, Pamela M. 2003. "'What Holds the Earth Together': Agnes Chase and American Agrostology." *Journal of the History of Biology* (Winter): 439–455.
>
> *The* (Kansas State Agricultural College) *Industrialist*, October 24, 1891; October 26, 1891; January 14, 1893; January 1898, November 1898, October 24, 1899; February 6, 1900; January 22, 1901; March 7, 1901; and March 25, 1935.

INTERSTATE WATER COMPACTS   During the first original suit of jurisdiction before the United States Supreme Court involving which state had rights to the flows of an interstate stream, one of those who testified was Frederick Newell, first commissioner of the federal Reclamation Service. Newell was an engineer who had considerable experience with irrigation in the American West. He, like many others, was appalled at the extraordinary costs borne by the states of Colorado and Kansas in waging this type of interstate litigation. At the time, *Kansas v. Colorado* was the largest and most expensive case to have ever come before the Court. The justices were wading through thousands of pages of testimony by hundreds of witnesses, besides hundreds of documents, many of them highly technical in terms of hydrologic engineering. Newell believed he knew a more efficient way to settle interstate water controversies. He recommended that officials and experts from each state come together to resolve their differences, and then enter into a legal, binding agreement regulating the stream flows. Such an accord would be treated much like an international treaty and would require the ratification of the United States Congress.

At the time, Delph Carpenter was a young lawyer, state legislator, and irrigator with extensive ties to the ditch companies around Greeley, Colorado. He saw great potential in Newell's idea for solving mounting interstate stream flow controversies. He further elaborated Newell's ideas and was soon appointed by the governor to negotiate interstate water compacts for Colorado. Carpenter first tried to resolve some outstanding issues with Kansas by negotiating with its designated representative, George Knapp. Those negotiations bore little fruit in the legislatures of either state. Still, Carpenter believed the concept was sound, and so did engineers and politicians throughout the West. In 1922, a commission chaired by Secretary of Commerce Herbert Hoover, along with representatives from seven western states, signed the most monumental interstate water compact ever negotiated in the United States. The Colorado River Compact was then ratified by each state, approved by Congress, and signed into law by President Coolidge.

With this apparent successful resolution of interstate differences outside of litigation, interstate compact negotiations became the *modus operandi* for settling interstate disputes. Such agreements have bound states throughout the grasslands with compacts regulating the interstate flows of such rivers as the Canadian, Rio Grande, Republican, South Platte, and Arkansas, to name a few.

Still, water compacts have not completely eliminated interstate litigation. For example, Texas has filed suit in the United States Supreme Court claiming that New Mexico had failed to comply with the terms of the compact regulating the flows of the Canadian River. In the late 1980s, Kansas initiated similar suits against Colorado and Nebraska over the flows of the Arkansas and Republican rivers, respectively. Even the Colorado River Compact, Carpenter's crowning achievement, has come under considerable strain with the phenomenal urban growth of cities throughout the desert grasslands such as Phoenix, Arizona, and Las Vegas, Nevada. In short, Carpenter's hopes for water compacts have proven nearly as illusionary as a mirage.

See also Kansas v. Colorado (1907); Water Use Doctrines.

Sources

Hundley, Norris, Jr. 1975. *Water and the West: The Colorado River Compact and the Politics of Water in the American West.* Berkeley: University of California Press.

Tyler, Daniel. 2003. *Silver Fox of the Rockies: Delphus E. Carpenter and Western Water Compacts.* Norman: University of Oklahoma Press.

JACKSON, WES (1936–) The recipient of a MacArthur Genius Fellowship, the Pew Conservation Scholar Award, a Right Livelihood Award (the "alternative Nobel Peace Price"), and listed by *Life* magazine as one of the 100 Americans who are likely to shape the twenty-first century, Wes Jackson has achieved considerable fame and recognition for his work in creating "sustainable" farming and communities. A fourth-generation native of the grasslands and raised on a highly diversified truck farm near Topeka, Kansas, Jackson acquired a true passion for and love of the grasslands.

Jackson spent his early years mastering football and working on his family farm. When he entered a small Methodist college to play football, he showed little promise for becoming a future leader in the American environmental movement and alternative agriculture. Even though he declared biology his major, not until his junior year did he become serious about his academic work. After graduation with a master's degree from the University of Kansas, he spent time as a public school teacher and coach, and later as an instructor in biology at Wesleyan College, his former undergraduate school.

Aspiring now to become a full-fledged university professor, he began his Ph.D. studies at North Carolina State University in the mid-1960s. It was there that his interests were piqued by the idea of maintaining wild grassland systems as a gauge against which domestic agricultural practices could be measured and judged. When he returned to Wesleyan in 1967 with his degree in hand, he entered a college environment shaped by concerns with civil rights, the Vietnam War, and ever more apparent environmental problems. He worked to integrate these issues into his biology classes, and from this experience he identified three interrelated problems: (1) population/food, (2) resource depletion, and (3) environmental pollution. These themes became the core of his first book, *Man and the Environment*, (1971). The publication of this book launched his career as an agricultural reformer and environmentalist.

The fine public reception of this book led to a job offer at California State University at Sacramento. There Jackson worked in organizing the Center for Environmental Studies and began climbing the ladder of success in academia. But he felt a strong tug on his heartstrings to return to the grasslands where he and his family were more comfortable. In 1974, he turned his back on California and returned to his grassland state of Kansas where he launched the Land Institute, an organization dedicated to the perfection of perennial polyculture agriculture.

American farmers by and large have always centered their domesticated grass production on monocultures such as wheat, oats, rye, and sorghums. Grown separately in fields and harvested, these plants could not reseed themselves, so farmers have had to replant them the following season. Prairie grasses and forbs, in contrast, will reproduce themselves year after year in close proximity. Some plants, such as eastern grama grass (*Tripsacum dactyloides*), show yet unrealized harvest potential. The promise lies in the better nutritional value of its seeds over that found in traditional cereal crops; the liability is that it produces far less poundage per acre than either corn or wheat. Nonetheless, if hybrids of prairie plants could be propagated, Jackson believes, then a variety of mutually beneficial species could be sown together in one field, harvested, and left to proliferate for the following season. In the wake of such production techniques would come a substantial reduction of machinery, petrochemicals, soil erosion and depletion, water pollution, and crop-destroying insects. Farms would tend toward smaller units, which would mean a repopulation of the American grasslands and a revitalization of small, rural community life.

Jackson's views have certainly engendered controversy and put him at odds with modern thought. The course of the Land Institute, in Jackson's view, "challenges the fundamental philosophical cornerstone of modern science . . . that to achieve the benefits of science we must 'bend nature' to our will. Our work, therefore, is not just the restructuring of agriculture, but a restructuring of the scientific paradigm itself." In short, to continue industrial, capitalistic agriculture, an outgrowth of the modern scientific philosophical tradition, results in calamity for both land and the human community. Not only is agricultural production weakened by this system, but so is agricultural education, as enrollments in land-grant agricultural colleges have plummeted over the last couple of decades.

To date, promising, tantalizing results have pointed the way at the Land Institute without reaching the hoped-for ultimate results. Consequently, to many critics, Jackson appears the prophet rather than the accomplished researcher. Still, the Land Institute has seen some tangible results, as it has produced a steady stream of published, scientific research. Jackson's work continues, as "the agriculture I promote depends on the information of a community which has evolved an information system that allows the possessors of the information to occupy a middle place between, for all practical purposes, two infinite resources—the sun and the rocks." In this pursuit Jackson hopes to re-create a self-renewing agriculture that engenders and nurtures the human community.

See also Hightower, Jim; Popper, Deborah and Frank; Winter Wheat.

Sources

Jackson, Wes. 1985. *New Roots for Agriculture,* new edition with foreword by Wendell Berry. Lincoln: University of Nebraska Press.

Sherow, James E. 2005. "Wes Jackson: Kansas Ecostar." In *Kansans Who Made a Difference,* edited by Virgil Dean. Lawrence: University Press of Kansas.

JOHNSON, CLAUDIA ALTA TAYLOR "LADY BIRD" (1912–)   Claudia Alta Johnson, better known as Lady Bird Johnson, used her position as First Lady to bring attention to environmental issues. She also played an often underestimated role in making the general public aware of the importance of maintaining healthy and attractive ecosystems. In several ways, she bridged the gap between the traditional upper-middle-class interest in civic beautification and the new middle-class interest in environmentalism.

Claudia's Tennessee mother raised her in the tradition of southern womanhood. During World War I and in the years afterward, she absorbed the business values of her father, who worked as a cotton merchant and store owner. Her privileged background allowed her to attend the University of Texas where she earned her B.A. degree, cum laude, in history. She continued her education by earning an additional degree in journalism in 1934 and acquiring her teaching certificate in 1935.

During this time she met a young, aspiring politician by the name of Lyndon Baines Johnson (LBJ). Johnson was working as an assistant to then congressman Richard M. Kleberg. In 1934 Lyndon and Claudia married, and soon afterward, Lyndon was elected to Congress where he became a staunch New Dealer. At the same time, Claudia became interested in improving the biotic beauty of public spaces with native plants.

She was already aware of how in her home state some steps had been taken to improve the beauty of public highways. At the suggestion of Judge W. R. Ely of the Texas Highway Commission, the department had been planting indigenous plants along the roadside since 1929. In the early 1930s, the department began using wildflower straw as a mulch and to assist in propagating native flowers in the right-of-ways. In an effort to improve weedy and worn soil, Johnson employed these same methods on the ranch, The Martin Place, that Lyndon and she bought.

When Lady Bird traveled the countryside campaigning for her husband, she took note of the sad state of the land. In a commencement address at Radcliffe College in 1964, she expressed the following point of view: "You couldn't keep from thinking that God had done His best by this country, but Man had certainly done his worst, and now its up to Man to repair the damage." The problems that most caught her eye were the strip mines of West Virginia and the proliferation of billboards along highways. She had important company in her camp, too. Stewart Alsop, in a 1962 article in the *Saturday Evening Post* titled "America the Ugly," described the countryside as a "billboard-ridden, slummish, littered, tasteless . . . messy place." After the election of Johnson in 1964, Lady Bird began using the "white-glove pulpit" of the First Lady to achieve urban cleanup (beginning with Washington, D.C.) and national beautification.

In December 1964, Lady Bird began to focus her lobbying efforts on reducing billboards and junkyards along public highways. At President Johnson's request, a citizen forum, the White House Conference on National Beauty, was held in Washington in early 1965. Problems with this campaign, however, appeared quickly. The "beautification" itself posed some serious public relations difficulties. As Lady Bird realized, "[The term] sounds cosmetic and trivial and it's prissy." What she aspired to achieve was a true improvement in the quality of life for Americans and the maintenance of healthy ecosystems.

As a result of the conference, the attendees framed a bill to address some of the central issues raised. Congress, however, was not totally impressed, and the proposed legislation, titled "The National Highway Beautification Act," faced an uphill struggle for enactment. The bill was designed to give states funding for regulating billboards, and not surprisingly, it faced serious opposition from the Outdoor Advertising Association of America, as well as state's rights Democrats who saw too much federal intrusion and some environmentalists who believed the legislation was too anemic to be effective. Moreover, LBJ's administration had consulted only advertisers about the bill, and this move cost the White House support from its potential allies among environmentalists and civic beautification groups.

Realizing the difficulties in seeing this legislation enacted, Lady Bird broke with tradition, and as First Lady she actively lobbied for the bill. She met with members of Congress, arranged social functions for key supporters, and even participated in strategy sessions. Still, opposition grew in strength. Some critics, such as Senator Everett M. Dirksen, were incensed with Lady Bird's public activity and called the bill "a whim of Mrs. Johnson." Further problems arose because of her interests in broadcast stations, which were held in a blind trust at the time, as opponents claimed that she stood to gain personally from advertising that would shift from billboards to radio and TV.

As Congress considered the highway act, Lady Bird turned her attention to civic beautification and lobbied to win over support from both the tourist industry and inner-city residents. Ecological issues, she believed, were also quality of life issues, and improving one improved the other. By working both with traditional groups such as city garden clubs and with environmentalists, Lady Bird bridged the gap separating traditional middle-class social clubs with environmental activism.

Critics loudly objected to Lady Bird's departure from what they considered proper decorum for a First Lady. Even old-style conservationists disliked her objections to the Grand Canyon dam project. At the same time she served as a role model for others in the general public who became active in environmentalism. Her personal influence led prominent people such as Lawrence Rockefeller to become active environmentalists. In short, many found her an inspiration.

After leaving the White House, Lady Bird remained an active environmentalist, but quietly so. She helped to launch the Texas Highway Department's awards program for roadway beautification. Such initiatives led to James R. Even's creation of an "Adopt-a-Highway" program in Tyler, Texas, and later state highway departments across the nation began similar programs. In 1982, Lady Bird used her own funds to underwrite the creation of the National Wildflower Research Center, the goal of which was to encourage and develop the use of native plants. Now states routinely sow indigenous plants along highways. Besides beauty, the plants have an additional benefit of needing less maintenance than exotic plants. This has resulted in cost savings in mowing and weed spraying expenses.

Lady Bird Johnson connected traditional ideas about civic beautification with environmentalism, which strengthened the public's interest in ecological preservation. Her belief that people shared a reciprocal relationship with their environments is a commonplace idea today. Moreover, she made a lasting contribution to the way individuals and governments worked to preserve indigenous plants throughout the grasslands.

See also Endangered Species; Ordway, Katherine; Wildflowers.

Sources

Gould, Lewis L. 1999. *Lady Bird Johnson: Our Environmental First Lady.* Lawrence: University Press of Kansas.

Koman, Rita G. 2001. "'To Leave This Splendor for Our Grandchildren': Lady Bird Johnson Environmentalist Extraordinaire." *Magazine of History* 15 (Spring): 30–34.

KANSAS V. COLORADO (1907)   In May 1907 the justices of the United States Supreme Court announced their decision on the first suit of original jurisdiction involving an interstate conflict over river flow. The decision established the doctrine of equity in an attempt to resolve the contest and to establish a legal precedent for settling all future interstate stream conflicts.

The suit originated in the 1890s as Kansans around Garden City nervously watched the yearly diminution of the Arkansas River flow, and they blamed Colorado irrigators to the west. At the same time, Marshal Murdock, a powerful Wichita, Kansas, newspaper editor, also feared the effects of irrigation in both western Kansas and eastern Colorado on the flows of the Arkansas River through his city. Murdock envisioned Wichita as an inland port if given some hefty help from the United States Army Corps of Engineers, but he saw his dream withering as the river flows through the city dwindled yearly.

The suit unfolded as follows. Kansas filed its opening briefs in May 1901, and Colorado attorney general Charles Post quickly refuted any right of Kansas to bring suit. The Supreme Court justices announced their right to hear the case in April 1902, and in March 1904 they allowed the U.S. attorney general to intervene in the case to represent the interests of the newly created Reclamation Service. In May 1904 the justices appointed a commissioner to oversee and take testimony, and by the end of 1905, his stenographer had recorded more than 120 court exhibits and 8,500 pages of testimony from more than 300 witnesses. In October 1906, the attorneys made their respective positions clear. Kansas believed the riparian doctrine, which protected the quality and quantity of river flow, giving it the right to an undiminished river flow as a result of upstream Colorado uses; Colorado, on the other hand, claimed that the prior appropriation doctrine, which protected beneficial uses of water in a "first in time, first in right" manner, allowed its state citizens to use all of the water within its boundaries regardless of the effect to downstream flows into Kansas; and the United States attorney general argued for congressional control over all interstate, non-navigable rivers.

In May 1907, the court announced its decision through an opinion written by Justice David Brewer, a former Kansan. He refused to favor either state's water code and instead formulated the doctrine of equity. Brewer had devised an accounting procedure to assess the relative gains each state had made in using the Arkansas River flow. In that manner, each state could maintain its own water doctrine, and the federal government could be prevented from asserting congressional control over interstate rivers. Brewer found that the citizens from both states had made economic gains throughout the valley, and consequently as a result of Kansas prosperity he ruled that it had not made a case for restraining Coloradans' water uses.

Ever after the court has used the doctrine of equity as the governing principle for allocating stream flows in interstate rivers. Cases such as *Nebraska v. Wyoming* (1945) and *Colorado v. New Mexico* (1982 and 1984) reveal just how important this doctrine is in settling feuds between states over limited, interstate stream flows in the grasslands.

See also Interstate Water Compacts, Pick-Sloan Plan; State Engineers; Water Use Doctrines.

Sources

Pisani, Donald J. 2002. *Water and American Government: The Reclamation Bureau, National Water Policy, and the West, 1902–1935.* Berkeley: University of California Press.

Sherow, James E. 1990. "Contest for the 'Nile of America': *Kansas v. Colorado* (1907)." *Great Plains Quarterly* 10 (Winter): 48–61.

LITTLE ICE AGE The period of global cooling and precipitation shifts known as the Neo-Boreal or "Little Ice Age" affected North America from roughly 1350 to 1850 CE. This period was marked by generally cooler summers and autumns and stormier, wetter summers on the central and northern plains, harsher winters on the southern plains, and a somewhat wetter period in the Great Basin as compared to today. While better known for its effects on northern Europe, the Little Ice Age caused human population shifts throughout the grasslands.

Tree-ring studies suggest that from 1400 to the 1930s, the areas surrounding the Great Plains averaged between 1.5 and 0.5 degrees Celsius cooler than the 1860–1980 average and fell to a temperature minimum during the late 1500s. Within this 500-year period, however, temperature and moisture variations were common. The Rocky Mountain glaciers seem to have reached their greatest Little Ice Age extent around 1860, while those in the Cascade Mountains seem to have peaked in the late 1700s–early 1800s. Journal entries and weather observations from trading posts and military expeditions dating from the late 1700s also suggest a cooler, moister regime across the central and northern Great Plains than is found there in the early twenty-first century.

These climatic and weather changes probably contributed to changes in human occupation of the grasslands. Archaeological research suggests that the colder and often drier periods in the central and southern Great Plains led to the disruption of the agricultural Late Village Farmer culture that abandoned the Republican, Smoky Hill, and Niobrara River valleys around 1400 CE, with these inhabitants possibly turning to a more nomadic form of subsistence. The southward movement of the Athapaskan peoples along the eastern slope of the Rocky Mountains toward what is now Texas and Oklahoma may have been accelerated by harsher weather conditions. The population of the Great Basin grasslands seems to have increased as the long droughts of the late Medieval Warm Period ended.

Due to sketchy data, climatologists remain uncertain about the extent of climatic changes in North America during the Little Ice Age. They agree that droughts remained common in the western half of the continent, spanning longer periods of time than those of the twentieth century but encompassing smaller areas. When and where these occurred and what effects they had on the peoples living on the grasslands are questions that still require further research.

See also Altithermal; Scarp Forests.

Sources

Bamforth, Douglas B. 1990. "An Empirical Perspective on the Little Ice Age Climatic Change on the Great Plains." *Plains Anthropologist* 35 (November): 359–366.

Bowden, Martin J. 1977. "Desertification of the Great Plains: Will It Happen?" *Eco-*

nomic Geography 53 (October): 397–406.

Bozell, John R. 1994. "Late Precontact Village Farmers." *Nebraska History* 75 (Spring): 121–131.

Grove, Jean M. 1988. *The Little Ice Age.* London: Routledge, 1988.

LONE WOLF V. HITCHCOCK (1903)   In 1901, Kiowa tribal leader Lone Wolf sued the secretary of the interior of the United States in what would become a landmark decision that redefined the relationship of Indian nations to Congress. Unfortunately, the decision against Lone Wolf in the January 5, 1903, ruling was not based on law or legal precedent; rather it reflected the desire of whites to assert control over more of the grasslands.

When Lone Wolf filed his initial injunction against the Dawes Commission in the Reno County courthouse, the Kiowas, Comanches, and Apaches were in the midst of a sea change in relationship to the lands they had long occupied. Placed on reserves set aside for them in the Oklahoma Territory after the Civil War, the tribes had worked hard to recreate a life for themselves based on stock raising and group farming as well as trade and the occasional raid into Mexico. They were making steady progress in becoming self-supporting when they were visited by the congressional Cherokee commission led by David Jerome. This commission, known as the Jerome Commission, was charged with allotting every Indian 160 acres of land, to be held in trust by the federal government, as a way of encouraging them to farm in the same way as white homesteaders. The "surplus" land within the original reservation that was not allotted could then be sold or "opened" for white settlement. As Lone Wolf and other tribal leaders quickly discerned, the transfer of lands from Indians to whites was the commission's primary objective.

Lone Wolf v. Hitchcock was not simply a matter of congressional authority; it was a clash of visions over what the southern plains grasslands should provide. In the transcription of the negotiations between the Kiowa, Comanche, and Apache leaders and the Jerome Commission, tribal leaders expressly stated that allotment was ultimately a redistribution of resources from Indians to non-Indians. They also understood that unlike other treaty meetings, the Dawes Commission was not there to negotiate so much as to dictate the terms, and the tribes resented this new paternalistic stance. Most insulting to tribal leaders was the commission's gross underestimate of the economic and cultural value of the mountains, creeks, valleys, and especially grasslands that would be lost through allotment. Speaking on behalf of himself and the assembly of Indians from three tribes, Quanah Parker asked the commission point-blank what the price per acre would be for the so-called worthless lands and what might happen if gold or silver was found in them. The commission declined to answer.

Quanah Parker had good reason to resist allotment; he controlled over 40,000 acres of tribal land that he leased to Texas ranchers, but tribal resistance to allotment went far deeper than greed. Access to shared watercourses, timberlands for heating and cooking, wild animal habitat, open grassland pastures for horses, and sacred places for worship and ceremony were foremost on the minds of the Indians but completely ignored by the commissioners. Even Lone Wolf understood that his suit against the Interior Department would not end white encroachment on Indian lands but only delay it until the expiration of the Medicine Lodge Treaty. His was a gamble that the government would change its mind or focus in the intervening years and spare his tribe from allotment and the loss of control over the grasslands.

See also Yellow Wolf.

Sources

Clark, Carter Blue. 1994. *Lone Wolf v. Hitchcock: Treaty Rights and Indian Law at the End of the Nineteenth Century.* Lincoln: University of Nebraska Press.

Hagan, William T. 1993. *Quanah Parker, Comanche Chief.* Norman: University of Oklahoma Press.

Lynn-Sherow, Bonnie. 2004. *Red Earth: Race and Agriculture in Oklahoma Territory.* Lawrence: University Press of Kansas.

MALIN, JAMES (1893–1979)   Raised on the Great Plains of western Kansas, the iconoclastic, innovative, free-ranging James C. Malin proffered a new understanding of history. As an undergraduate college student, Malin acquired three majors: history, biology, and psychology-philosophy. He honed these skills as he pursued his graduate work and later professional career as a historian. In many respects, he developed the methodology of environmental history decades before it became a more recognized field of historical understanding. His history, however, was more than people's ideas about their surroundings, or legislative enactments designed to regulate or conserve natural resources. His was an *ecological* history, one that stressed the interrelationships of people with other actors normally excluded from traditional studies. In other words, rocks, plants, wind, rain, fire, microorganisms, and wild and domestic animals all took the historical stage alongside humans. Malin understood clearly that the study of history was its ecology, and that ecology was a historical science. As he put it: ecology "deals with groups or assemblages of living organisms in all their relations, living together, the difference between plant, animals and human ecology or history being primarily a matter of emphasis." In short, it is just as feasible to write an environmental history of tall bluestem prairie as it is one of ranchers in the same area. It is not simply the story of people, but a story that depicts the fluid relationships binding biotic entities (of which humans are merely one), human cultures, and the abiotic physical forces that constantly rework and re-shape the systems of Earth through time.

In doing this sort of history, Malin incorporated the most recent findings in ecology, meteorology, economics, statistics, and demographics. For Malin, all history was local, bottom-up study. He had an undying faith in the objectivity of history based on a social science approach to research. He rejected any label-ing or classification of his own work even when it was described as an "ecologi-cal" interpretation of history. For Malin, the only thing limiting human beings was their own creative ability. Individual freedom was paramount for him.

Malin's theoretical approach to history rejected Turner's frontier thesis as a closed system idea. Malin's study of the grasslands convinced him that the re-gion was an open system, one in constant change and flux given the historical forces shaping it. It was a region that required continuing, imaginative adapta-tion by the people populating it. In contrast to Webb who believed that the en-vironment shaped human responses, Malin postulated that the grasslands had a range within which human beings could create many ways of living. In essence, there was no end of the frontier as Turner believed; rather, there was ever-changing adaptation and technological exploitation of resources within a region such as the grasslands. Counter to Frederick Clements, Malin also held that the grasslands, or for that matter any ecosystem, never obtained anything resem-bling a "climax." All life groupings were fixed in a time-space continuum, and as the Earth's climate and biotic species are ever in flux, so are their historical relationships. For example, this had implications for whom, or what, created the Dust Bowl. Such storms, as Malin understood it, had been commonplace throughout the grasslands long before the arrival of Euro-American farmers. If there was a human-related problem in the creation of the Dust Bowl, it was in the lack of human creativity in suitably adapting their farming technology to the region.

Malin served as a professor of history at the University of Kansas from 1921 until 1963. Some of his most influential works are *The Contriving Brain and the Skillful Hand in the United States* (1955); *Winter Wheat in the Golden Belt of Kansas* (1944); and *The Grassland of North America* (1948). All together he wrote eighteen books and scores of articles. He had ninety-seven students finish their master's degrees under his direction, and he directed seven students to-ward the completion of their Ph.D. degrees.

See also Clements, Frederick E. and Edith Schwartz; Dust Bowl; *The Future of the Great Plains* (1936); Jackson, Wes; Popper, Deborah and Frank; Webb, Walter Prescott.

Sources

Johannsen, Robert W. 1972. "James C. Malin: An Appreciation." *Kansas Historical Quarterly* 38 (Winter): 457–466.

Malin, James C. 1984. *History and Ecology: Studies of the Grassland*, edited by Robert P. Swierenga. Lincoln: University of Nebraska Press.

Williams, Burton L. 1979. "James C. Malin—In Memoriam." *Kansas History* 2 (Spring): 65–67.

NATIONAL TALLGRASS PRAIRIE PARK   Of all the biomes in North America, the tallgrass prairie is one of the last to be represented by a national park. Perhaps surprisingly, the tallgrass prairie biome was one of the earliest landscapes for which pleas were made for its preservation. One of the first, if not the first, came from George Catlin, a painter who traveled with the army on marches across the grasslands in the 1830s. In 1832, while reflecting on the future of Indian peoples living throughout the prairies, he feared that the rapid advance of American culture threatened the future existence of both peoples and their lands. He called for a "nation's park," one that would preserve the "wild and freshness" of Nature.

Unfortunately, Catlin's voice was lost in the bustle of converting the grasslands into ranches, farms, and cities. Not until 1930 would another person make a concerted pitch for the preservation of the prairies. Dr. V. E. Shelford, an ecologist at the University of Illinois, with the endorsement of the Ecological Society of America sent a proposal for preserving the tallgrass prairie to the National Park Service. Shelford noted two prime areas where this biome still existed: one was throughout the Flint Hills of Kansas, and the other was in portions of the Osage Hills in northeastern Oklahoma. By 1960, one site in particular drew the attention of the National Park Service: nearly 60,000 acres located in Pottawatomie County just to the east of Manhattan, Kansas, and the Blue River.

During the Eisenhower administration, the Pottawatomie site had considerable support from the Manhattan area. Secretary of the Interior Frederick Seaton endorsed a prairie park in Pottawatomie County, and so did the editor of the *Manhattan Mercury*, Bill Colvin. The state legislature even appropriated $100,000 earmarked for purchasing the land if Congress would pay the remaining purchase cost. Public hearings were held, and the residents and political leadership of Manhattan stood solidly behind creating the park. However, the 1961 Pottawatomie park bill never made it out of the congressional subcommittee considering it.

One event more than any other heralded the demise of this proposal. In December 1961, Secretary of the Interior Stewart Udall and National Park Service Director Conrad Wirth were observing the proposed land for the prairie park from their helicopters. Their pilots landed on Twin Mound, and there to greet them was a shotgun-wielding Carl Bellinger. Bellinger, a rancher, leased the pastures where the helicopters landed, and after a brief, less than cordial conversation with the secretary of the interior, the rancher ordered Udall to get off "his" land. In fact, Udall not only left the site, but he and Park Director Wirth never returned.

Two other factors underlay the demise of the Pottawatomie Prairie Park idea. First, the building of Tuttle Creek Dam and Reservoir had generated a great amount of local distrust in the federal government. The farm families who once lived throughout the Blue River Valley actively opposed the building of the dam by the Army Corps of Engineers. They lost the political battle and their farms after nearly a decade of bitter protest during both the Truman and Eisenhower administrations. Second and more important, ranchers and their organizations actively opposed the park. This is one of the reasons that Professor Lloyd Hulbert emphasized over and over again that the creation of what is today the Konza Prairie Botanical Station was a scientific field laboratory and not a prairie park. Hulbert knew that a scientific station had economic overtones in that its research findings could have value for ranchers and farmers. A park, as far as many ranchers were concerned, took productive agricultural lands out of production.

The position of ranchers toward the park was clearly understood by Hulbert, who, incidentally, did campaign for the creation of a tallgrass prairie park in addition to Konza. Hulbert had been trying to enlist the support of state politicians for a park, and this was the response he received from state senator Don Christy from Scott City, Kansas, in 1975. "If you wish to see all facets of the prairies as they once were except for the few wild animals which can be seen in the zoo, all you have to do is drive through the tall grass system now. . . . Consequently, I see no justification for removing that kind of acreage from the food-producing capabilities for this nation. Thanking you for writing, but do not count on my support for this piece of federal or state legislation." Over twenty more years would pass before President Clinton signed the bill creating the Tallgrass Prairie Preserve, a dream realized by the tireless campaign waged by dedicated supporters such as former Kansas senator Nancy Landon Kassebaum.

See also Bison; Cheyenne Bottoms; Flint Hills; Ordway, Katherine; Wildflowers.

Sources

Conard, Rebecca, and Susan Hess. 1998. "Tallgrass Prairie National Preserve Legislative History, 1920–1996." Iowa City, IA: Tallgrass Historians L.C.

Heat-Moon, William Least. 1991. *PrairyErth (A Deep Map).* Boston: Houghton Mifflin.

Shelford, Victor E. 1933. "Preservation of Natural Biotic Communities." *Ecology* 14 (April): 240–245.

OGALLALA AQUIFER   Rhinoceros roamed throughout the area, and easterly windblown volcanic ash often blanketed the land. At the same time, thick layers of deposited sands and gravels took shape across the surface. Eventually, other deposits of eroded materials during the Ice Ages covered these unconsolidated strata overlying impermeable layers of older limestone and shale formations left from a time when shallow inland seas once covered much of the contemporary Great Plains. The unconsolidated sands and gravels sandwiched between the ancient sea floor and the glacial deposits collected and stored glacial melts. Today, this water-bearing stratum, one that covers nearly 174,000 square miles under the Great Plains, is the Ogallala Aquifer, the lifeblood of contemporary farming in the region.

In places this aquifer is recharged by rainfall and snowmelts, but mostly the water in this vast underground storehouse is the ancient remnants of the Ice Ages, and the precipitation of today contributes only minutely to its store. Scientists with the United States Geological Survey (USGS) have estimated that this aquifer contains around 3.25 billion acre-feet of water that current pump technology can access. An acre-foot of water is the amount of water that it takes to cover an acre of land one foot deep in water, or 325,851 gallons. Most of this water is about 100 to 400 feet below the surface, and it remained largely untapped until after World War II.

With Frank Zybach's invention of the center-pivot irrigation system in the late 1940s, pumping the aquifer for agricultural production quickly became economically profitable with 3.5 million acres under irrigation by 1950. By 1990, this aquifer nourished around 20 percent of all the irrigated farm acreage in the United States and represented approximately 30 percent of the pumped water in the country. In 1980, 170,000 wells pumped 18 million acre-feet of water onto over 14 million acres of farmland, which rose to around 16 million acres by 2000. One state alone, Nebraska, has nearly two-thirds of the volume of water under its surface and in places has the capacity to sustain current pumping rates at 2000 fuel costs for well into the twenty-second century. The application of center-pivot-irrigation has turned Great Plains farmers into the nation's biggest producers of wheat, alfalfa, grain sorghum, and corn. Much of this production has fed cattle in some of the largest feedlot facilities in the world and sustains huge meatpacking factories in western Kansas and Nebraska, and eastern Colorado. More stunning is the recent advent of cotton growing in the southern portions of Kansas.

This revolution in farming has not been without its detrimental consequences. The increase in farm acreage has also been attended by increases in chemical fertilizers and pesticides, besides contributing significantly to surface stream non-point pollution, and some of these chemicals have percolated into the groundwater itself. In other places, withdrawals have lowered the ground-water levels close to the substrata deposited when the shallow seas covered the region. These rocks contain high levels of fluorides, chlorides, and sulfates that are beginning to appear in what heretofore had been clear water pumped from the aquifer. Where the Platte, Republican, Niobrara, Smoky Hill, Arkansas, and Canadian rivers cut into the aquifer, traditional recharges to these rivers have been significantly lowered as the aquifer becomes depleted. More alarming, wells have gone dry in several places where the groundwater saturation levels were shallow, and as pumping continues largely unabated, some farmers and hydrologists see similar consequences looming for a vast number of operations throughout the grasslands. Also on the horizon are the effects of ever-increasing fuel costs, especially those attached to natural gas, which powers a majority of the pump engines. Conceivably, even if a farmer's land overlies a deep satura-tion, the fuel cost of pumping from ever-greater depths will at some point make crop production unprofitable.

The legislatures and water bureaus in many states have taken steps to con-serve and protect the Ogallala Aquifer. In Kansas, for example, in 1974 the state passed legislation allowing for the creation of groundwater management dis-tricts. The members of these self-governing entities can devise policies for regu-lating themselves. For example, farmers in the northwestern portion of the state created the Groundwater Management District in 1977, and under the leadership of Wayne Bossart implemented a "zero depletion" policy to forestall the further depletion of groundwater. Achieving this goal has been difficult for those farmers and often consensus has been hard to reach; nonetheless, the dis-trict managers continue to revise this policy to make it ever more effective. Other states have allowed the formation of groundwater management districts, too, such as Texas, and many states, such as New Mexico, have attached the prior appropriation doctrine to groundwater withdrawals. All farmers, scien-tists, environmentalists, and legislators realize one crucial fact: to continue pumping rates at ten times the recharge rate of the aquifer will ultimately re-quire a different way of living on the land.

See also Cheyenne Bottoms; Playa Lakes; Sand Hills; Water Use Doctrines; Zybach, Frank.

Sources

Bowden, Charles. 1977. *Killing the Hidden Waters: The Slow Destruction of Water Resources in the American Southwest.* Austin: University of Texas Press.

Green, Donald E. 1973. *Land of the Underground Rain: Irrigation on the Texas High Plains, 1910–1970.* Austin: University of Texas Press.

Kromm, David E., and Stephen E. White, eds. 1992. *Groundwater Exploitation in the High Plains.* Lawrence: University Press of Kansas.

Opie, John. 1993. *Ogallala: Water for a Dry Land.* Lincoln: University of Nebraska Press.

ORDWAY, KATHERINE (1899–1979)   Oddly enough, the preservation of the tallgrass prairies might have remained little more than a dream had it not been for the efforts of a frail, petite woman. From her outward appearance, Katherine Ordway seemed an unlikely champion of grassland preservation, yet she did more to achieve this than any other individual. The daughter of Lucius and Jessie Ordway, she was raised in a rarified social environment of wealth and privilege in St. Paul, Minnesota. Her father had achieved considerable economic success through his Crane Company of Minnesota and later in creating the Minnesota Mining and Manufacturing Company, the 3-M Company, in 1902. On his death in 1948 he left a trust fund of over $350 million to his five children, and this ultimately served as the source of Katherine's philanthropy. Katherine acquired degrees in botany and art from the University of Minnesota and, as a result, had a lifetime interest in collecting fine art. Not until much later in her life did Ordway's interest in saving grasslands become important.

Katherine's interest in wildland preservation came when in her fifties she sought to preserve the woodland landscape around her home in New York. Her cousin, Samuel Hanson Ordway, Jr., also had an abiding interest in conservation and was keenly involved with the Conservation Foundation (1933–1978). He had written and published *A Conservation Handbook* (1950) and *Resources for the American Dream* (1953), and with Katherine, he established the Goodhill Foundation to fund the work of the foundation. Katherine's conservation efforts in New York also put her in contact with Richard Pough, who along with Richard H. Goodwin, co-founded the Nature Conservancy in 1950. By the late 1950s, Pough served as president of the Goodhill Foundation, which had the mission of population control, conservation, art, and scholarship for original scientific research, and by 1962 the mission of the foundation had shifted somewhat to the preservation of open land with an emphasis on keeping it "wild." Pough always believed that the least protected ecosystems in need of preservation were the grasslands. Through his influence, Katherine became keenly interested in protecting wild prairies, and she played an active role in conserving over 31,000 acres of grasslands in five states—Ordway Prairie Preserve in Minnesota, Konza Prairie in Kansas, Samuel H. Ordway, Jr. Memorial Prairie in South Dakota, Cross Ranch in North Dakota, and Niobrara Prairie in Nebraska.

Perhaps her most notable achievement was the creation of Konza Prairie. Lloyd Hulbert, an ecologist at Kansas State University, was working to create a biological experiment station to measure the role of fire burning and grazing on the tallgrass prairie ecosystem. In the mid-1970s, through the combined efforts of several professors and administrators, Hulbert had managed to acquire from Theo Cobb Landon, Alf Landon's wife, 916 acres to initiate the creation of a tall-grass prairie preserve. By 1977, an adjacent parcel of 7,220 acres of ranch land that had remained largely free of any agricultural development became available. At this moment the Nature Conservancy, through the funds provided by Katherine Ordway, joined with Hulbert and the university to acquire the land, and thereby created Hulbert's dream, a tallgrass biological research station. Yet even this removal of land from economic production raised a protest from some around Manhattan, Kansas. The university hosted a banquet in honor of Katherine Ordway, but she declined to attend because of death threats published in the local newspaper. Sadly, even though the Nature Conservancy awarded Ordway its first Land Guardian Award in 1978, Katherine died before she saw the tallgrass prairie in the heart of the Flint Hills that she had helped to preserve.

See also Flint Hills; Johnson, Claudia Alta Taylor "Lady Bird"; National Tallgrass Prairie Park; Wildflowers.

Sources

Blair, William D. 1989. *Katharine Ordway: The Lady Who Saved the Prairies.* Washington, DC: The Nature Conservancy.

Reichman, O. J. 1987. *Konza Prairie: A Tallgrass Natural History.* Lawrence: University Press of Kansas.

OXNARD SUGAR COMPANY Sugar beet growing and refining had come to the grasslands by 1900. For some time Europeans knew that certain beets could be refined for their sugar. German professor Andreas Sigismund Marggraf devised refining techniques for sugar beets around 1847 at a time when sugarcane was the prevailing source. It would take another four decades before Ebenezer Herrick Dyer and Claus Spreckels perfected the method of growing and refining beets in California. Soon the Oxnard brothers would learn from Spreckels and expand their operations into the grasslands. A family-owned operation, it became the American Beet Sugar Company in 1899, and later the American Crystal Sugar Company in 1934.

The beginning of this operation starts with Thomas Oxnard, who had built a lucrative business in the cane sugar industry of Louisiana. His sons, Robert, Benjamin, Henry, and James, would expand the business until it became one of the dominant sugar companies in the United States. It was Henry who became enthusiastic about the potential of sugar beet refining after a stay in France. The Oxnard brothers' first facility in the grasslands was built in Grand Island, Nebraska, in 1890. Soon afterward they began selecting other sites throughout the region with factories in Colorado, Utah, New Mexico, and Iowa. Henry located one of the more successful ones in Rocky Ford, Colorado, in 1900. This occurred at the urging of George Washington Swink, who had a successful and profitable record in developing irrigated truck farming. Swink took some beets that he had grown and showed them to Henry, who, impressed by the results, decided to locate nearby. Other factories, however, failed to produce the same results. One located at Carlsbad, New Mexico, failed miserably after a few years of horrible beet production in the desert grasslands of the Pecos River Valley.

The profitable results of the Oxnard ventures throughout the grasslands stimulated the expansion of the industry by smaller companies. For example, a group of Colorado Springs businessmen incorporated the United States Sugar and Land Company at Garden City, Kansas, and that factory became the mainstay of the local economy. Sugar beet production, while potentially a money-making undertaking for farmers and capitalists alike, demanded a rigid mode of production. The plants required exact timing in watering, thinning, and harvesting. Reliable irrigation systems became indispensable as did a reliable workforce willing to do arduous, seasonal work. Around Garden City, for example, immigrant Mexican laborers became the ideal, as they were primarily single men who returned to Mexico when the field work was done. Still, race and labor relations with the Garden City company and at the Oxnard operations could, and did at times, turn sour and violent. Also, anything affecting water flows into the irrigation systems often provoked extensive litigation, both intra- and interstate.

By the 1950s, for a number of reasons, the beet industry had weakened to the point that one by one the major plants were shuttered. In 1964, the American Crystal Sugar Company closed its first plant, the Grand Island, Nebraska, factory. Today, only the vacant shells of the former plants remain to remind anyone of an industry that had once promised sustained economic vitality.

See also Kansas v. Colorado (1907); Water Use Doctrines.

Sources

Gutleben, Dan. 1960. "The Sugar Tramp: The Last Chapter" (unpublished manuscript). Denver: Colorado Historical Society.

Markoff, Dena S. 1979. "A Bittersweet Saga: The Arkansas Valley Beet Sugar Industry, 1900–1979." *Colorado Magazine* 56:161–178.

Osborne, Thomas J. 1972. "Claus Spreckels and the Oxnard Brothers: Pioneer Developers of California's Beet Sugar Industry, 1890–1900." *Southern California Quarterly* 54 (2): 117–125.

PICK-SLOAN PLAN Two federal agencies had primary responsibility for building dams on interstate streams throughout the nation. By 1940 the Bureau of Reclamation, established within the Department of the Interior, constructed dams for irrigation and hydroelectric power generation. The Army Corps of Engineers, administered through the War Department, erected dams for flood control and to improve navigation on interstate rivers and intercoastal waterways. The corps had the responsibility for flood control and navigation along the Mississippi River, and of course this generated a lot of concern for regulating the flows of the tributaries to "Old Man River." Prompted by the devastating Mississippi River flood of 1927, Congress authorized surveys that included the entire Missouri River Basin. By 1934 a corps report identified possible projects on the main stem of the Missouri beginning with the Fort Peck Dam in Montana.

In 1943 another large flood occurred along the Missouri River causing extensive damage to Omaha, Nebraska, along with demands from congressmen from the flooded stretches calling for further flood control studies by the corps. Within a few months Colonel Lewis A. Pick had completed a plan recommending levees along the lower reach of the Missouri River, dams on the tributaries to the Missouri, and five more major dams on the "Muddy Mo" itself. The colonel's report raised concerns within the Bureau of Reclamation that its "toes were being stepped on" and that Pick's recommendations would interfere with W. Glenn Sloan's plans for the development of beneficial use projects throughout the upper reaches of the basin. Undoubtedly, flood control projects posed detrimental consequences for any bureau projects focused on beneficial uses such as irrigation or hydroelectric production. Some congressmen wanted the creation of a multiple-purpose authority modeled on the Tennessee Valley Authority to oversee water development throughout the entire basin, and this event raised fears throughout state governments of too much centralized federal control over water.

In a conference held in Omaha, Nebraska, in 1944, the main players reached a compromise solution that reconciled the objectives of the corps and bureau while crushing any plans for the creation of a Missouri River Valley Authority. The corps would continue to plan flood control and navigation projects on the main stem while the bureau would devise the plans for beneficial uses and hydroelectric production. Moreover, states would have the power to review federal planning, beneficial uses would remain paramount over navigation in the rivers west of the 97th meridian, the secretary of war would be allowed to market surplus water, and the secretary of the interior could market hydroelectric power. All of these became components of the Flood Control Act of 1944, especially in Section 9 that established the locations of future dams throughout the basin. This act, more than any other, reworked the grassland hydrology throughout the entire Missouri River Basin.

The work of both the bureau and corps has received an ambivalent reception by the folks most affected by it. Farm families, Indian peoples, and environmentalists have often objected to the building of dams, while authors have referred to many of the projects as "big dam foolishness." Many scholars have questioned the economics of the projects, whether in terms of benefits to navigation or flood control. The 1993 floods showed the limitations of bureau and corps projects to regulate and contain the surging waters, and later some studies concluded that the property damage of over $20 billion and the death toll (at least forty-eight deaths) were worse than had the dams not been built at all. On the other hand, the projects have created jobs, recreational opportunities, and electricity and supplemental water for bourgeoning cities throughout the grasslands. Regardless, there have been, and continue to be, significant ecological and social costs tied to the Pick-Sloan Plan.

See also Interstate Water Compacts; State Engineers; Water Use Doctrines.

Sources

Ferrell, John R. 1993. *Big Dam Era: A Legislative and Institutional History of the Pick-Sloan Missouri Basin Program.* Omaha, NE: Missouri River Division, U.S. Army Corps of Engineers.

Lawson, Michael L. 1982. *Dammed Indians: The Pick-Sloan Plan and the Missouri River Sioux, 1944–1980.* Norman: University of Oklahoma Press.

Schneiders, Robert K. *2003. Big Sky Rivers: The Yellowstone and Upper Missouri.* Lawrence: University Press of Kansas.

PLAYA LAKES "Playas," the pluvial lakes and wetlands found in the southwestern Great Plains of Texas, Oklahoma, and Kansas, serve important hydrologic and biologic functions. For centuries, they have provided water for migrating birds and animals, augmented botanical variety, and served as the primary source of recharge for that section of the Ogallala Aquifer. Although reduced in number by plowing, erosion, and development, playas remain important to life on the High Plains. These scattered basins, whether reflecting blue sky from their shallow waters or holding the last bit of green grass during a droughty October, are one of the critical components of a viable, southern grasslands.

The basins called playas or playa lakes dot the southwestern Great Plains, especially on the Llano Estacado plateau of Texas and New Mexico. Some scholars believe that very name Llano Estacado, or Staked Plain, is a corruption of Llano Estancado, or Plain of Many Ponds. Whatever the derivation, the topographic features called playas share several basic characteristics. They are shallow, rounded depressions that fill with rainwater or snowmelt. Playas release their water into the underlying groundwater or through evaporation, and during dry seasons (October–March) all but the largest playas go dry, occasionally revealing the porous, Randall clay that forms only in playa beds. Playas support a wide variety of plant and animal life even after their standing water has soaked into the ground; they can be identified by the plant communities that grow in them compared to those found on the surrounding uplands. Moreover, playas should not be confused with the pothole lakes of the northern plains, as they are a completely different hydrologic feature.

The exact cause of playa development remains unknown. It is likely that playas were formed by a combination of factors, such as carbonic acid dissolution while there was water in a depression, followed by aeolian erosion during dry episodes and some use by bison. However they were formed, they contribute to the great variety of plant life on the southwestern Great Plains. Depending on the amount of water and overall size of the individual playa, a basin can contain several ring-like zones of plant life. After especially wet winters or if the playas have been modified to recycle irrigation water ("tailwater"), their open water during the April–July rainy season supports muskgrass and widgeon grass, if the water remains long enough. Cattails and bulrushes grow in the marshy area surrounding the open water. Species turnover within the playa basin can be rapid as grasses are flooded out in early spring and replaced with arrowhead (*Sattiga longiloba*), blue mud-plantain, and spikerush once water inundates the inner playa basin. As the playa dries, saltmarsh aster and devilweed aster, barnyard grass, and western wheatgrass and other species replace the plants preferring saturated soil. Buffalo grass and blue grama grass, snow-on-the-mountain, and vine mesquite also begin growing in the playa's outer fringes as the soil dries, eventually blending into the upland plant communities surrounding the basin.

This variety of flora supports an equally large variety of fauna. Some deeper playas have been stocked with fish for recreation or mosquito control, but fish are not native to playas. True toads (*Bufonidae* sp) and spadefoot toads are the most widespread amphibians found in and around playas, although a few true frogs and tiger salamanders live there as well. Reptiles like turtles, horned lizards, and several species of snakes also share the playas, feeding on insects, plants, small mammals, and each other. Mice and pocket gophers abound near playas, as do desert cottontail rabbits. Coyotes, swift and red foxes, and in earlier times wolves, bear, and other predators ate the abundant rodents and rabbits while bison, pronghorn, and elk grazed and watered at the lakes and wetlands. Unlike the northern potholes, playas do not host muskrats because of the lack of permanent water.

Migratory and resident waterfowl are some of the most visible playa users today. The increase of irrigated agriculture over the twentieth century provided more winter food for ducks and geese, while tailwater ponds within the playa basins provided open water surrounded by cattails and other cover plants. The southern plains playas are now an important wintering area for cranes, ducks, geese, and other birds. Canada geese and several kinds of ducks live year-round near urban playas with permanent water, a development greeted with mixed emotions by neighboring humans. The birds are most common in fall and winter, often arriving in large numbers ahead of the winter windstorms locally termed "blue northers."

Besides their biological roles, playas provide an important hydrologic function as the main source of recharge water for the southern section of the Ogallala Aquifer. Playas lack the impermeable substratum of caliche (carbonate) found across the Llano plateau, so precipitation will soak through their bottom Randall clay formation into the groundwater. Total yearly recharge from playas varies but has been estimated at one-half inch (13 mm) to over three inches (82 mm), as compared with almost no groundwater infiltration from the surrounding uplands. Since precipitation is the only source of recharge for the rapidly depleting Ogallala Aquifer, playas remain essential for maintaining numerous life forms in the southwestern Great Plains.

Humans too have used playas since the first migrants crossed the plains thousands of years ago. Many Paleoindian sites on the High Plains have been found near playas, including some that suggest the hunters made use of the boggy basins to trap mammoth and ancient bison. Later historic peoples like the Comanche depended on the playas for water as they traveled across the area hunting and trading. Francisco Vásquez de Coronado's *entrada* over the High Plains in 1541 followed lines of playas and springs, although it was later Spanish speakers who gave the pluvial lakes and wetlands their name. Traders from New Mexico followed known routes from playa to playa as they met with the Comanches to exchange maize and manufactured goods for bison meat and hides. The Anglo-American ranchers who moved into the area in the 1870s discovered that playas provided water and good grazing for their livestock, although springs and later windmill-driven pumps were needed to augment the playas' seasonal water supply.

Despite their importance to the ecosystems and economics of the High Plains, playas face many threats to their existence and functions, including destruction through sedimentation and filling, conversion to tailwater pits, drainage for flood and mosquito control, and chemical contamination. Some farmers deliberately filled in playas to increase their crop acreage, a practice now blocked by the 1985 federal Food Security Act's "swampbuster" provision. Deepening playas by digging pits and channels into them for tailwater collection and return did improve habitat for waterfowl, but it also concentrated agricultural chemicals because of recycling the water. Other pollutants, especially those from feedlots, municipal runoffs, and oil pumping brine disposal, destroy plant growth and kill wildlife. Some playas have been lost to urban development, especially during dry periods when the existence of a playa is harder to detect. Those moving into buildings constructed in former playas can be unpleasantly surprised once precipitation returns—Western Plaza Mall in Amarillo flooded on a regular basis until a deep water catchment pit was dug at the city's expense.

Efforts at playa preservation and restoration have come from several directions. Hunting organizations like Ducks Unlimited, Quail Unlimited, and the Texas Waterfowlers Association have leased playas and funded research aimed at management of moist soil areas. Federal legislation like the "Swampbuster Act" aims at reducing destruction of all kinds of wetland, including playas. National Wildlife Refuges and National Grasslands encompass playas, and the Nature Conservancy owns playa land as part of its conservation goals. All of these interests come under the umbrella institution called the Playa Lakes Joint Venture (PLJV), an organization founded in 1990 to coordinate efforts at playa preservation, to fund education and research, and to encourage protection of playas. Corporations like Phillips Petroleum joined with state fish and game departments from Colorado, Kansas, Texas, Oklahoma, and New Mexico, Ducks Unlimited, the United States Fish and Wildlife Service, the Nature Conservancy, and the Department of Range and Wildlife Management at Texas Tech University to form the PLJV. Efforts at convincing private landowners to see playas as an asset by encouraging hunting leases for both waterfowl and upland birds, especially pheasant, have seen some success.

The playa lakes, although reduced in number and size, remain important to life on the High Plains. These basins provided water, food, and shelter for many of the animals living on the plains and housed much of the botanical diversity once found in the region. Like other wetlands, playas played, and still do play, an important role in filtering and absorbing precipitation and runoff, directing the water down into the Ogallala Aquifer. Although many playas have been damaged or destroyed by farming and urban development, efforts are under way to preserve and protect these High Plains wetlands.

See also Cheyenne Bottoms; Endangered Species; Water Use Doctrines; Zybach, Frank.

Sources

Carlson, Paul. 2005. *Deep Time and the Texas High Plains: History and Geology.* Lubbock: Texas Tech University Press.

Smith, Loren M. 2003. *Playas of the Great Plains.* Austin: University of Texas Press.

Steiert, Jim. 1995. *Playas: Jewels of the Plains.* Lubbock: Texas Tech University Press.

PLEISTOCENE EXTINCTIONS   Ask a random group of paleontologists to give a definitive answer for the extinction of most species of North American megafauna at the end of the Wisconsonian Ice Age, roughly 12,000–10,000 years ago, if you really want to see a good argument without resolution. Although most scientists now agree that there was a major extinction event during this time period, debate has centered on the relative roles of humans and climatic change.

The megafauna (animals larger than 44 kg) that had been relatively common across North America died out rapidly during the end of the Wisconsonian period. This die-off included mammoth (*Mammuthus*), mastadont (*Mammut*), camels (*Camelops*), giant bison (*Bison latifrons*), horses, and giant ground sloths as well as their predators, including cave lions (*Panthera leo*) and saber-toothed cats (*Smilodon*). Survivors included grizzly bear (*U. horribiles*), musk ox, moose, and bison. At the same time, very rapid changes in climate over the North American continent led to habitat reduction as forests changed composition or retreated before advancing grasses. Some scholars suggest that the weather on the Great Plains grew both more seasonal (hotter summers, colder winters) and stormier for a time. At the same time, as indicated in the archaeological record, human hunting of megafauna occurred.

Scholarly debate over the cause of the late Pleistocene magafaunal extinction falls into two major lines of thought. One, first expressed by Paul Martin, focuses on anthropogenic causes: humans hunted North America's big game to extinction. Evidence for this lies in the timing of human-manufactured stone tools and the megafauna's demise as well as archaeological evidence for human hunting of mammoth, mastadont, and other species. The second argument centers on climatic change as the terminal factor, and these proponents argue that while human action probably played a role in the disappearance of the megafauna, habitat loss and fluctuating weather were more important causes.

See also Bison; Endangered Species; Scarp Forests.

Sources

Beck, Michael W. 1996. "On Discerning the Cause of Late Pleistocene Megafaunal Extinctions." *Paleobiology* 22 (Winter): 91–103.

Bryson, Reid A., David A. Baerries, and Wayne M. Wendland. 1970. "The Character of Late-Glacial and Post-Glacial Climatic Changes." In *Pleistocene and Recent Environments of the Central Great Plains*, edited by Wakefield Dort, Jr., and J. Knox Jones, Jr. Lawrence: University Press of Kansas.

Martin, Paul S. 2005. *Twilight of the Mammoths: Ice Age Extinctions and the Rewilding of America*. Berkeley: University of California Press.

POPPER, DEBORAH (1947–) AND FRANK (1944–)   Two New Jersey sociologists have raised a firestorm of debate over the current social and ecological status of the grasslands and what the future holds for the same region. Frank and Deborah Popper seem unlikely scholars to raise the amount of contentiousness that they have in the last two decades. Professor Frank Popper had focused most of his earlier work on questions around urban planning and zoning, and had begun to research the prospects of regional zoning. He turned his attention to the Great Plains and discovered by analyzing the 1980 census that nearly 150 counties had a population of fewer than two people per square mile, and that nearly 400 counties had fewer than six people per square mile. In many respects, it seemed to Popper that large portions of the Great Plains were reverting to a "frontier" condition if frontier meant fewer than seven people per square mile.

After a trip through the Great Plains with his family, Popper co-authored an article with his wife Deborah that caught the attention of people throughout the grasslands. The Poppers called for a reversion of large portions of the grasslands to federal ownership, and then to be managed as a "buffalo commons," a large park with the return of huge bison herds. They argued that the region should never have been turned into agricultural production and that this experiment had resulted in social, economic, and ecological catastrophes. At a time when the Sagebrush Rebellion (a revolt against federal ownership of lands primarily in the desert grasslands and mountain ranges of the West) was in full steam in the West, the Poppers' ideas, which applied primarily to privately owned lands throughout the grasslands, hit a solid wall of opposition throughout the region.

Bravely, the two East Coast scholars took their ideas to forums in the grasslands. They received icy to hostile receptions in Bismarck, North Dakota; Denver, Colorado; and Laramie, Wyoming. U.S. senators and governors criticized and ridiculed them by name. Undeterred, they persisted and continued to give numerous presentations from 1988 up to the present. What they found difficult to comprehend was that while the population of the grasslands was certainly declining, the amount of land in one form or another of agricultural production was remaining fairly steady. So in the eyes of those living in the region it seemed foolish to remove the productive economy of the grasslands and to replace it with a federally owned, centrally planned and managed bison-covered park. Of course, the Poppers were addressing serious issues besides rural depopulation, and they took note of the mounting ecological problems confronting those still remaining on the land. In fact, their case became persuasive enough that former governor Mike Hayden of Kansas, a harsh critic of the pair while in office (1987–1991), came to embrace their thinking in a public forum held on the campus of Kansas State University in February 2004. At the time, this certainly won him little favor with friends and relatives who still lived in the northwestern portion of the state.

Regardless of what the future holds for the grasslands, the social, economic, and ecological problems identified by the Poppers cannot be ignored. If the grasslands are not to be a "buffalo commons," politicians and residents of the area will still have to forge new ways of living in the region as they adapt to the shifting economic and ecological realties.

See also Bison; *Future of the Great Plains* (1936); Jackson, Wes; National Tallgrass Prairie Park.

Sources

Matthews, Anne. 1992. *Where the Buffalo Roam: The Storm over the Revolutionary Plan to Restore America's Great Plains.* New York: Grove Press.

Cawley, R. McGreggor. 1993. *Federal Land, Western Anger: The Sagebrush Rebellion and Environmental Politics.* Lawrence: University Press of Kansas.

DeBres, Karen, and Mark Guizio. 1992. "A Daring Proposal for Dealing with an Inevitable Disaster? A Review of the Buffalo Commons Proposal." *Great Plains Research* 2 (August): 165–178.

Popper, Deborah Epstein, and Frank Popper. 1987. "The Great Plains: From Dust to Dust. A Daring Proposal for Dealing with an Inevitable Disaster." *Planning* 53 (December): 12–18.

Umberger, Mary L. 2002. "Casting the Buffalo Commons: A Rhetorical Analysis of Print Media Coverage of the Buffalo Commons Proposal for the Great Plains." *Great Plains Quarterly* 22 (No. 2): 99–114.

SAND HILLS In the western half of Nebraska and to the north in South Dakota is one of the more unique landscapes in the grasslands. Created over 8,000 years ago during the dry, hot climate of the hypsithermal, the Sand Hills form the most extensive stretch of grass-covered, dune sands in North America. Most scholars think that the source of this sand came from ancient alluvial deposits. This is a fragile place, one posing significant difficulties for human occupation. Because of their obvious difficulty to farm, the hills have retained much of their grass cover from the time Euro-Americans first encountered them. Consequently, Americans have operated large ranches, on average 4,000 to 6,000 acres, in the hills. Most of the limited farming is by center-pivot irrigation in alfalfa fields for hay production.

The Sand Hills also retain habitat for a large number of grassland species, both plants and animals. The porous nature of the soils creates a large reservoir of groundwater that feeds creeks and wetlands throughout the entire formation. The hills are a haven for a rich diversity of fish, plants, birds, especially waterfowl, and mammals such as mule and white-tailed deer, pronghorn deer (antelope), bison, and elk. Plants are called "borrowed" because they arrived after the retreat of the last glaciation. Botanists refer to the plant communities as "unique association" properly labeled a "Sand Hills prairie." Remnants from the ancient boreal forest survive along the southeastern portions of the hills near the more recent arrivals such as yucca or soapweeds. About 12,000 years ago the area was covered by "pine parkland, a pine forest interspersed with prairie," but this was largely replaced by grasslands shortly afterward. Certain plant communities occupy different ranges within the Sand Hills—for example, some associations live on the upper portions of the dunes while others have taken up residence along the waterways and wetlands. Some plant communities will migrate around to blowout areas or highly disturbed areas and depart once the dunes regain some degree of stability.

American Indian people subsisted in dunes well into the nineteenth century. Evidence suggests that there was human occupation around the Sand Hills as early as 12,000 years ago, but none has been discovered in the interior of hills during the same time. However, there is more evidence of human activity in the region 2,000 to 1,000 years ago. These people crafted ceramics, practiced agriculture, and seemed to have had rather permanent settlements. Between 500 and 1,000 years ago, humans took up extensive residence along the stream valleys, lakes, and wetlands. Many scholars believe these people were the ancestors of the Pawnees and Arikaras. Soon after, hunting possibilities centering around bison along with the presence of wild horse herds made this area a highly contested region, one fought over by the Pawnees and Omahas to the east with the Sioux and Northern Cheyennes to the north. By the late 1800s, Euro-Americans and their ranching operations had largely displaced all of the Indian peoples and bison herds throughout the hills, and cattle operations have become the mainstay of the region's economy.

Many people have attempted farming the Sand Hills but without much success. Judge Moses P. Kinkaid became alarmed about the inability of farmers to succeed on 160-acre homesteads, and he lobbied Congress to give special attention to the problems associated with farming the Sand Hills. This resulted in the passage of the Kinkaid Act of 1904, which allowed 640-acre homesteads in the Hills. It was on such a farm that Mari Sandoz was raised, and the Sand Hills became the source for her stories about grassland life in such works as *Old Jules*. Despite the limited success Jules had in retaining his farm, most "Kinkaiders" failed quickly at farming within less than a decade, and those few who endured fell victim to the harsh ecological and economic realities during the Great Depression. Nearly all who remained after that had given up farming for ranching or some other nonagricultural line of work.

The Sand Hills of Nebraska are a constant reminder of just how difficult and tenuous a foothold humans have had in the grasslands. They are also revealing of just how unpredictable and mutable the ecosystems of the region are.

See also Endangered Species; Ogallala Aquifer; Scarp Forests; Wildflowers.

Sources

Bleed, Ann, and Charles Flowerday, eds. 1990, 2nd ed. *An Atlas of the Sand Hills.* Lincoln: Conservation and Survey Division, Institute of Agriculture and Natural Resources, University of Nebraska.

McIntosh, Charles Barron. 1996. *The Nebraska Sand Hills: The Human Landscape.* Lincoln: University of Nebraska Press.

SCARP FORESTS   Usually, forests are seldom, if ever, considered a part of the grassland biome. In places, however, several remnant Pleistocene-era forests thrive in the grasslands. These trees grow in rugged places, escarpments, or breaks far removed from watercourses or rivers. For example, there are the National Forests of Ponderosa pines (*Pinus ponderosa*) in the Nebraska Sand Hills, and scattered groupings of Ponderosa pines and junipers in eastern Montana, Wyoming, and western Dakotas, and the wooded Black Mesa in Oklahoma.

Now what accounts for these highly isolated islands of trees in the midst of grassy oceans? What has become apparent to many scholars is that conifers have thrived well when planted on the Great Plains in areas removed from scarps or river valleys. Throughout stretches that have been treeless now for several centuries, paleontologists have discovered in soil cores conifer pollens, hackberry fruit, and fossil landsnails. All of this evidence indicates extensive tracts of woodlands ranging from broadleaf trees to the east giving way to conifers. The researchers took into account wind conditions and paleoclimatological data, and concluded that there was little, if any, reason to believe that climate has played a part in keeping the grasslands free of trees apart from riparian woodlands and the scarps. So what accounts for the more recent, isolated and highly limited growths of woodlands? As far as these scholars are concerned, it was fire.

Lightning has often been presented as the prevailing cause of prairie fires. However, lightning normally occurs with thunderstorms, and grasses burn poorly when drenched. Humans, on the other hand, could and did set fires when burning conditions were opportune—the dry late times of fall or early spring. Grasses have always responded well to fire-burning regimens whereas trees do less so. The decline of woodlands began concurrently with the probable arrival of humans in what would become the broad, treeless expanses of the grasslands. Consequently, only one variable adequately explains the disappearance of the woodlands, and that is fire—and more than that, fire set by human beings. In time, the scarps became the refuge for the diminished woodlands, whereas these rugged stone outcroppings presented poor soil conditions for advancing grasses. In short, the scarp forests in the grasslands are telling reminders that the grasslands themselves evolved largely out of a relationship with the fire management practices of humans.

See also Bison; Flint Hills; Pleistocene Extinctions; Sand Hills; Wildflowers.

Sources

Cook, John G., and Larry L. Irwin. 1992. "Climate-vegetation Relationships between the Great Plains and the Great Basin." *American Midland Naturalist* 127 (April): 316–326.

Pyne, Stephen J. 1982. *Fire in America: A Cultural History of Wildland and Rural Fire.* Princeton, NJ: Princeton University Press.

Stewart, Omer C. 2002. *Forgotten Fires: Native Americans and the Transient Wilderness,* edited by Henry T. Lewis and M. Kat Anderson. Norman: University of Oklahoma Press.

Van Auken, O. W. 2000. "Shrub Invasions of North American Semiarid Grasslands." *Annual Review of Ecology and Systematics* 31: 197–215.

Wells, Philip V. 1965. "Scarp Woodlands, Transported Grassland Soils, and Concept of Grassland Climate in the Great Plains." *Science* 148 (April): 246–249.

Wells, Philip V. 1970. "Postglacial Vegetational History of the Great Plains." *Science* 167 (March): 1574–1582.

STATE ENGINEERS Unheralded and often overlooked, state engineers largely shaped the modern infrastructure of the hydraulic society as it exists today on the Great Plains. The office of state engineer flourished during the heyday of the American Progressive Era. At that time, a great many of the reformers believed in efficiency and rational planning of natural resources for the greatest common good, the cornerstone of the conservation movement. The origins of the office date to the 1870s in Colorado. To resolve growing intrastate water conflicts, the Second General Assembly divided the state into ten water districts with the water rights in each to be determined by the courts. Soon it was apparent that an administrator was needed to regulate the water rights, and this led to the creation of the state engineer's office. Many of the occupants of these offices became extremely powerful in their dealings with water rights of the region. Elwood Mead learned hydraulic engineering in Colorado, and when Wyoming created a state engineer's office in 1890, he took the position and turned the office into a powerful regulatory bureau, which he guided until 1899. Mead not only administered water rights, but he also determined water rights. In many ways, Michael Creed Hinderlider, one of the most powerful state engineers to have held office in Colorado (1923–1954), summarized the creed of his fellow engineers, the "western water buffalos," when in 1931 he said: "Controlled and guided by the will of man, water becomes his never tiring slave, turning the wheels of industry . . . and bearing the burdens of commerce."

The achievements of George Knapp, the chief engineer of Kansas (1927–1957), demonstrate this view of water development in the grasslands. World War II stimulated the rapid population growth of the large cities in Kansas, especially Wichita with its burgeoning aircraft factories. Knapp wanted a different water code to support this rapid growth, and he knew what he wanted as a solution. The mechanistic tradition as articulated by Hinderlinder certainly guided Knapp in devising this new code. He unequivocally judged the riparian doctrine, with its protection of stream flows and quality, as a hindrance to the economic development of the state. Allowing water to flow freely out of the state on to the ocean, he reasoned, resulted in economic waste and loss. Stream flows and underground water required human control and use in economic pursuits. Knapp was largely unconcerned with protecting the ecology of streams and groundwater systems, or at least these considerations went unrecorded in any of his extant writings. In his mind, people had to have some beneficial use in mind before they could expect a right to use water. Beneficial uses, a code word throughout the West, meant using water in some state-sanctioned economic enterprise. His thinking became the basis for the Kansas Water Act of 1945.

By Knapp's own estimation, the passage of the Water Act of 1945 marked the crowning achievement of his career. With this law in hand, he encouraged the federal government to embark on its dam building program through the Water Act of 1944. City planners like those in Wichita began pursuing an aggressive program of building pumps and conveyance systems to supply their rapidly growing populations. Farmers in the western portion of the state speedily set to work pumping the Ogalalla Aquifer, which has underwritten the development of large-scale cattle and pig production around cities such as Garden City, Dodge City, Scott City, and Great Bend. Clearly, by the end of his life in 1964, Knapp's efforts had set into motion the legal framework within which economic growth, based on the exploitation of water, could flourish.

Interestingly, a few state engineers initially felt ill at ease with such an approach to water development. In 1889, as Elwood Mead was piecing together the state engineer's office of Wyoming, Hammond Hall wrote to him that the prior appropriation system was a poor mechanism for regulating water. It "presupposes that [water] is without ownership, like a wild beast of the forest or of the plain; and it has been the curse of irrigation from time immemorial, that water has been treated like it was a beast—to be shot down and dragged out by the first brute that came in sight of it. . . . The principle is wrong." Yet the vast majority of water buffalos shared Hinderlider's viewpoint, and still do to this day.

The state engineer's office, which was duplicated in states throughout the grasslands, gave rise to some of the more politically and economically powerful bureaucrats in the region. Certainly, Stephen Reynolds, state engineer (1955–1990) of New Mexico, stands out as one of the most commanding personalities ever to hold that office. Even governors had a difficult time dealing with him. But not all state engineers were able to command the power of a Reynolds; this was the case with Frederick E. Buck of Montana (1941–1963). In the 1940s he complained bitterly about being unable to administer even in the most basic manner the water rights on the eastern plains of his state. During the Great Depression many of state engineer offices were subsumed by the creation of water offices that coordinated state and federal planning for building dams and irrigation projects. Such offices became exceptionally useful with the passage of the 1944 Water Act, which gave states the ability to review plans for federal water projects. Montana led the way with the creation of a Water Conservation Board in 1934. Colorado followed suit in 1936, and other states throughout the grasslands did likewise in the years afterward. Since then, most state engineer offices, with a few exceptions like Wyoming and New Mexico, have lost power and prestige to these conservation and planning boards.

See also Cheyenne Bottoms; Endangered Species; Interstate Water Compacts; *Kansas v. Colorado* (1907); Ogallala Aquifer; Pick-Sloan Plan; Water Use Doctrines.

Sources

Hays, Samuel P. 1979. *Conservation and the Gospel of Efficiency: The Progressive Conservation Movement, 1890–1920.* Cambridge: Harvard University Press.

Sherow, James E. 1989. "The Chimerical Vision: Michael Creed Hinderlider and Progressive Engineering in Colorado." *Essays and Monographs in Colorado History* (Essays Number 9): 37–59.

Sherow, James E. 2002. "The Art of Water and the Art of Living: Review Essay." *Kansas History* 25 (Spring): 52–71.

Sherow, James E. 2004. "'The Fellow Who Can Talk the Loudest and Has the Best Shotgun Gets the Water': Water Regulation and the Montana State Engineer's Office, 1889–1964." *Montana, The Magazine of Western History* 54 (Spring): 56–69.

WATER USE DOCTRINES   Throughout the grasslands, Euro-Americans have devised a complicated set of water doctrines to regulate the conservation and use of water. Generally, the laws governing water uses are the convergence of two traditions: one emerging from English common law, and the other through communal practices that arose out of ancient Spanish traditions shaped in part by Moorish irrigation customs. English common law gave rise to riparian doctrines, and Spanish ways were developed into the prior appropriation doctrine. The word "riparian" is derived from the Latin *rip(a)*, meaning riverbank. Generally, this doctrine grants to the proprietor of any land abutting a stream the right to use of the water flowing by his property unaffected in terms of its quality or quantity by any applications upstream. The prior appropriation doctrine, on the other hand, allows the first person to divert stream flows for beneficial uses—normally defined in terms of economic production—to have the continued use of that flow before anyone else who developed a beneficial use at some later date, or "first in time, first in right." The riparian doctrine inhibited the development of non-riparian water uses, which is what irrigation systems normally do.

People throughout the semi-arid grasslands certainly understood the potential of the prior appropriation system for economic development. Anyone who dallied in channeling stream flows toward farming, city building, or industrial growth risked being left high and dry as others raced to develop rights to the water first. Two institutional forms of prior appropriation, the Wyoming and Colorado systems, have dominated throughout the grasslands. In the Colorado system the courts assign water rights and the state engineer regulates them; in the Wyoming system the state engineer assigns and regulates water rights.

In those grassland states that are divided into subhumid and semi-arid regions, some combination of both riparian and prior appropriation doctrines still prevail. This is the case in states such as the Dakotas, Nebraska, Texas, and Kansas. As opposed to the riparian doctrine, the prior appropriation system can also be used to regulate groundwater development. New Mexico was the first state to regulate groundwater pumping and to create a legal connection of groundwater to surface water. Other states, especially those above the Ogallala Aquifer, have, with various modifications, followed a similar practice. Included in this increasingly complex regulation of surface and groundwater sources has been the development of groundwater and regional surface water management districts. Further complicating the legal water uses are the interstate river compacts, which take precedence over state and local regulations.

The post–World War II growth of environmentalism has challenged the prevailing view that water throughout the grasslands should be used solely for economic development. With the depletion of hundreds of miles of stream flow through the Great Plains, the rapid reduction of the Ogallala Aquifer, the destruction of wetlands, and the pollution of rivers and creeks, environmentalists have demanded, with few significant results in either legislation or litigation, for new legal approaches toward the use of water. Little wonder that water lawyers are kept busy as people throughout the grasslands work to create new institutional controls for the ecological and economical uses of water.

> *See also* Cheyenne Bottoms; Interstate Water Compacts; *Kansas v. Colorado* (1907); Ogallala Aquifer; Pick-Sloan Plan; Playa Lakes; Webb, Walter Prescott.
>
> *Sources*
>
> Lee, Lawrence B. 1980. *Reclaiming the American West: An Historiography and Guide.* Santa Barbara, CA: ABC-Clio Press.
>
> Pisani, Donald J. 1996. *Water, Land and Law in the West: The Limits of Public Policy, 1850–1920.* Lawrence: University Press of Kansas.
>
> Reisner, Marc. 1987. *Cadillac Desert: The American West and Its Disappearing Water.* New York: Penguin Books.
>
> Worster, Donald. 1985. *Rivers of Empire: Water, Aridity and the Growth of the American West.* New York: Pantheon Books.

WEBB, WALTER PRESCOTT (1888–1963)   Although best known as a historian of the American West, Walter Prescott Webb was also a writer, teacher, and conservationist. In 1975 W. Eugene Hollon called Webb's first book, *The Great Plains*, "the most original and significant idea about the American West that has appeared since Fredrick Jackson Turner delivered his famous essay" in 1893. Webb, drawing from his own experience growing up on a farm in north-central Texas, suggested that geography, and specifically the semi-arid to arid climate of much of the Great Plains, led to the development of a uniquely western culture.

Walter Prescott Webb was born in 1888 in Panola County in far east Texas, but in 1892 his family moved north and west to Stephens County, the southeastern edge of the Great Plains. Years later, when asked when he started doing research for *The Great Plains*, Webb's standard reply was "When I was four." He worked his way through high school, teaching for a year in a one-room schoolhouse between his junior and senior high school years to pay for his education. He also received advice, books, and several loans from William E. Hinds, a New York businessman who had seen a letter Webb wrote to the magazine *The Sunny South* in 1904. Web had sought advice on acquiring an education and becoming an author, and Hinds found his entreaty intriguing. As a result of Hinds's support, Webb began taking classes at the University of Texas in 1909, and again had to take several years off to teach high school and earn funds. He received a B.A. in 1915, and after teaching at San Marcos Normal School, then trying his hand at other businesses, Webb was hired by the University of Texas to teach courses on secondary education, and he remained at UT Austin for forty years.

Webb's first major historical work, *The Great Plains*, describes how the geography west of the Mississippi River shaped the ways Native Americans, the

Spanish, and Anglo-Americans responded to the region, and how they shaped the land. In some ways the book is an answer to Fredrick Jackson Turner's "Frontier Thesis," which described the importance of the frontier in shaping American culture. Webb argued that the western environment forced this change, and much of the work highlights Anglo-American cultural adaptations to the semi-arid grasslands. As the book's introduction concludes, "east of the Mississippi, civilization stood on three legs—land, water and timber," but west of the river, only land remained, and civilization "toppled over in temporary failure." Webb focused on the development of ranching, barbed wire, and water laws and how they shaped a developing western culture. Scholars still turn to Webb's geographic determinism, arguing for and against it but acknowledging the importance of his ideas.

Walter Prescott Webb was to return to the idea of frontier and geography in several of his later works. His second book, *Divided We Stand: The Crisis of a Frontierless Democracy*, came out in 1937 and addressed the problem of northern corporations spreading into the West and South, taking the regions' resources and intensifying sectionalist tensions in the country, in part because there was no longer a frontier for people to move to. This work led to *The Great Frontier*, Webb's final attempt at placing the western hemisphere and its frontiers into world context. *The Great Frontier* puts forth Webb's "boom hypothesis" that the great prosperity and rising civilization Europe experienced between 1500 and 1900 came from the resources and population outlet of the western hemisphere's frontier. He speculated on the result of the loss of that frontier after 1900, predicting great changes of some form were in store as society adjusted to this new condition. More of a synthesis than an original history, the work remains less popular than *The Great Plains* or *The Texas Rangers*, a narrative of the Texas Rangers' history from 1834 to 1934.

Webb's background as a farm boy growing up during a drought influenced his concern for the land he saw being ruined by poor farming and wasteful usage. He wrote *Flat Top: A Story of Modern Ranching* about Charles Pettitt's attempts at land restoration and repair on a ranch in central Texas in 1960. Webb would follow a similar course with his smaller Friday Mountain Ranch in the Hill Country near Austin, bringing in cotton gin waste as fertilizer and working to restore native grasses and plants. He also predicted the current problems with Texas's water supplies in the 1954 work *More Water for Texas: The Problem and the Plan*, in which he suggested that either conservation or water importation was Texas farmers' only long-term hope to keep raising eastern crops in a western climate.

Walter Prescott Webb married Jane Oliphant in 1916. They had one daughter, Mildred, who graduated from the University of Texas as a member of Phi

Beta Kappa. Jane died in 1960 after a short illness, and Webb remarried Terrell Maverick, the widow of Maury Maverick, Sr., in December 1961. Webb died in a one-car accident March 8, 1963, while driving back late one evening from a speaking engagement.

See also Malin, James; Popper, Deborah and Frank.

Sources

Butler, Annie M., and Richard A. Baker. 1985. "Walter Prescott Webb: The Legacy." In *Essays on Walter Prescott Webb and the Teaching of History*, edited by Dennis Reinhartz and Stephen E. Maizlish. College Station: Texas A&M University Press.

Furman, Necah Stewart. 1976. *Walter Prescott Webb: His Life and Impact.* Albuquerque: University of New Mexico Press.

WILDFLOWERS   Although the North American grasslands are best known for their native and domestic grasses (big bluestem, gramas, maize, wheat), wildflowers and other forbs make up a large percentage of the plants of the vast and complex biome. Aesthetically pleasing to humans, wildflowers also performed vital services to the grassland botanical community. Most wildflowers of the grasslands are forbs, herbaceous plants with non-woody stems that are not grasses. Some, like blooming cacti and Junegrass (*Koelaria pyramidata*), are succulents or grasses, even though they are sometimes marketed as wildflowers because of their blooms. Grassland wildflowers range in size from the minute, ground-hugging prairie cat's foot to blazing star and rattlesnake master, plants that can reach five feet in height in the tallgrass prairie regions. Yucca and agave tower over the short grasses of the western plains and Great Basin. Grassland wildflowers come in many colors, and different varieties bloom from early spring through late autumn, adding color and texture to the sweeps of grass that once stretched across the Great Basin and central grasslands.

Wildflowers add more than just color to the grasslands. Most of the nitrogen-fixing plants that helped support the grasses were wildflower-bearing legumes like wild indigo, leadplant, and purple prairie clover. These plants return nitrogen to the soil, replacing that used by other plants and ensuring the continued fertility of the soil. Wildflowers also contributed to the variety of insect species present in the grasslands. Grasses pollinate via wind, but wildflowers require insects to do the work. Growing interest in native plants and climate-appropriate landscaping has led to the increasing popularity of wildflowers for gardens and roadside plantings. While some plants, like rattlesnake master, do not propagate or transplant well due to their extended root systems, flowers like columbines, poppies, coneflowers, prickly-pear cactus, and other native plants can be found at garden shops or specialty plant stores. Some states, of which Texas is the best known, encourage wildflowers along roadsides because they require less mowing and water, as well as attracting tourists.

Wildflowers played an important part in life on the North American grasslands. Flowering forbs provided early season ground cover, added nitrogen to the soil, and provided food for insects, birds, and animals. The grasslands are best known for their namesake grasses, but wildflowers were crucial to the grassland ecosystems of the plains and Great Basin.

See also Endangered Species; Johnson, Claudia Alta Taylor "Lady Bird"; National Tallgrass Prairie; Ordway, Katherine.

Sources

Brown, Lauren. 1989. *The Audubon Society Nature Guides: Grasslands.* New York: Alfred A Knopf.

Madson, John. 1995. *Where the Sky Began: Land of the Tallgrass Prairie* (rev. ed.). Ames: Iowa State University Press.

Paulson, Annie, ed. 1989. *The National Wildflower Research Center's Wildflower Handbook.* Austin: Texas Monthly Press, 1989.

WINTER WHEAT Winter wheat has largely been hailed as an example of human adaptation to the "subhumid" environment of the North American grasslands, from Saskatchewan to Oklahoma. Indeed, it was the successful adoption of winter wheat that gave states like Kansas and Nebraska their reputations as the "breadbasket" of the nation.

The origin of winter wheat (*Triticum aestivum L.* and *T. trugidum L. var. durum*), especially in Kansas, has been shrouded in romance and myth. Specifically, hard red winter wheat, the hybrids and descendants of which are grown today, was originally developed for the harsh winters of the Russian steppes. As settlers arrived in Kansas in the 1860s and 1870s to take up homesteads in the grasslands, they planted those crops they were most familiar with, especially corn. Some settlers, however, notably German-speaking Mennonites recently immigrated from Russia under Catherine the Great, planted wheat of different varieties, along with other basic crops for subsistence. After the introduction of hard red winter wheat to the Plains, it was rumored that the original settlers from Russia had brought the miraculous crop with them from overseas, sewing quantities of wheat seed into their clothing for safe passage.

As romantic and heartwarming a story as this is, it is likely not the origin of hard red winter wheat in the midwestern great plains. James Malin, a historian of Kansas and the Great Plains in the 1930s and 1940s, made an exhaustive search for the origins of winter wheat to demonstrate the adaptation of human beings to a new, untried environment. He found that the Mennonite communities adopted red wheat along with most other communities at the same time, and that local newspapers and other contemporary sources were silent on the importation and use of this wonderful new crop.

Malin preferred to believe that the adoption of hard red winter wheat to overcome the environmental obstacles of prairie agriculture (drought, pests) was a spontaneous move on the part of the "average" farmer. Norman Saul, however, found that there were particular individuals who actively worked to convince farmers of the need to switch from soft wheat varieties to hard red winter. The widespread adoption of hard red winter wheat was part of a well-organized and deliberate attempt by scientists to find new crops in the newly established Experiment Station system of the United States Department of Agriculture located on the campuses of land-grant schools in Kansas and throughout the Midwest. In particular, Saul highlighted the efforts of farmer Bernhard Warkentin and wheat scientists W. T. Krehbeil, Edward Shelton, and Mark Carleton at Kansas State Agricultural College in Manhattan Kansas, and at its several substations, particularly the Fort Hayes experiment station.

Mark Carleton stands out in the introduction and successful adoption of both hard red "turkey" winter wheat in the lower Midwest and the successful adoption of durum winter wheat in the northern Great Plains states of Minnesota and the Dakotas. Raised in Kansas, Carleton attended Kansas State Agricultural College (KSAC) and graduated with a master's degree in agronomy in 1893. In December 1907, Carleton was elected the first president of the American Society of Agronomy. He continued to work as a wheat scientist at KSAC from 1894 until 1919. The primary problems associated with wheat farming in Kansas in the late nineteenth century were drought and marketing. Soft wheat varieties needed a minimum amount of summer rain to thrive but were easily transported and milled into flour for a ready market. Hard red winter wheat was better suited to the soils and conditions of the West, but it was difficult to mill and therefore returned far less to the farmer.

Carleton never credited himself with the "discovery" of hard red "turkey" winter wheat. He likely collected it along with dozens of other varieties from local farmers (perhaps Mennonites near McPherson where he had his first teaching position) and then generated experiments to test its viability as a commercial crop. His selection of Turkey Red was aided by a severe freeze in 1890 that definitively selected turkey as the hardiest and best yielding of the winter wheats in Kansas. By 1919, turkey wheat was being planted in 83 percent of Kansas wheat fields and remained the most popular variety in the United States until 1944. Carleton added to the winter wheat gene pool by traveling to Siberia and Russia and bringing back new cultivars for experimentation including "Kharkof" and "Crimean." It was the introduction of "Kubanka" durum wheat, however, that made wheat farming possible in Nebraska and Montana, where spring wheat was grown well into the twentieth century.

Carleton's success at popularizing these new varieties was based on his ability to establish joint research projects with other agronomists across the plains and then publishing the results of their findings in popular newspapers and agricultural journals. Carleton was also especially helpful in the dissemination of winter wheat through his research on leaf rust. His investigations into rust and his development of effective antifungal agents positively affected farmers' willingness to take on this new crop. While Carleton's contributions were clearly recognized (the 1914 *Year Book of Agriculture* listed these), he was dismissed from the USDA for allegedly mishandling funds. He died alone and forgotten while doing research in Peru in 1925. Nevertheless, winter wheat was an established crop on the Great Plains, enabling thousands of farmers to take advantage of the grasslands' seemingly endless fertility.

While steel rollers overcame the problems of milling hard red winter wheat, its popularity has waned in recent years due to the introduction of new hard white varieties. These new varieties combine the environmental advantages of winter wheat with higher yields and a whiter, therefore more desirable, flour. The transition from red to white has been difficult for farmers, as volunteer red wheat invades new white wheat fields, lowering the value of their transitional crops considerably. There is also a reluctance on the part of farmers to make the change given that it takes two to three growing seasons to establish high grade white wheat fields where red wheat had been previously planted. As other countries adopt white wheat and receive a premium price for it, the farmers of the Great Plains will find it easier to incorporate the new variety into their own farming practices.

See also Bonanza Farms; Carleton, Mark; *The Future of the Great Plains* (1936); Jackson, Wes; Malin, James; Popper, Deborah and Frank; Zybach, Frank.

Sources

Isern, Thomas. 2000. "Wheat Explorer the World Over: Mark Carleton of Kansas." *Kansas History* 23 (Spring–Summer): 12–25.

Kansas State University Agricultural Experiment Station and Cooperative Extension Service. 1998. 1998 Kansas Performance Tests with Winter Wheat Varieties. Report of Progress. 186 (July).

Lynn-Sherow, Bonnie. 2000. "Beyond Winter Wheat: The USDA Extension Service and Kansas Wheat Production in the Twentieth Century." *Kansas History* 23 (Spring–Summer): 100–111.

Saul, Norman. 1989. "Myth and History: Turkey Red Wheat and the 'Kansas Miracle.'" *Heritage of the Great Plains* 22 (Summer): 1–13.

YELLOW WOLF (ca. 1779–1864)   Yellow Wolf (*O-Cum-Who-Wast*), sometimes translated as Yellow Coyote, had a renowned reputation as a war leader and later as a peace chief among the Southern Cheyennes. His life illustrates how, like so many others, it was shaped by the horse-borne, bison-hunting culture, and how he came to recognize serious difficulties in sustaining that lifestyle by the mid-1840s. While he greatly profited personally from horses and bison hunting, and in leading warfare expeditions for either horse stealing or punishing enemies, he had correctly observed the great diminution of bison throughout the Central Grasslands. In response, he sought assistance from the United States government to create an alternative way of living in the region. Yellow Wolf had observed farming, and he thought if done properly, the Southern Cheyennes could make a transition from a horse-borne hunting culture to agriculture. On more than one occasion, he besought Indian agents such as Thomas Fitzpatrick and William Bent, and army officers such as Lieutenant Abert to send word to federal officials of his willingness to lead his followers toward a settled, farming lifestyle. He clearly understood that a sizable portion of the Southern Cheyennes would object, but he saw this shift in natural resource use as the only viable way to survive in the grasslands.

With the signing of the Fort Wise Treaty in February 1861, achieving Yellow Wolf's long desired goal seemed at hand. The federal government promised to supply the necessary materials to build an agricultural settlement for the Cheyennes and Arapahos along the banks of the Arkansas River in Colorado Territory. Unfortunately, the beginning of the American Civil War doomed whatever prospects this ecological, social, and economic experiment had. Funding for the project dried up, tensions between Indian peoples and the great influx of Euro-Americans into the area mounted, and before long armed conflict between Indian peoples and Euro-Americans became frequent despite the best efforts of peace chiefs such as Yellow Wolf or Black Kettle to stop it. Tragically, this effort to wean the Cheyennes off bison hunting came to complete disaster when Colonel Chivington's Colorado volunteers massacred the peaceful Cheyennes camped along Sand Creek in November 1864. There the efforts of the elderly Yellow Wolf to bridge the differences between his people and Euro-Americans, and to provide the Southern Cheyennes with a different means of adapting to the grassland biome, came to an end with his death at the hands of the Coloradans. It later fell to his son, Red Moon, to guide a sizable portion of the Southern Cheyennes as they adapted to reservation life, farming, and land allotment in present-day western Oklahoma.

See also Bison; Lone Wolf.

Sources

Berthrong, Donald J. 1963. *The Southern Cheyennes*. Norman: University of Oklahoma Press.

Grinnell, George Bird. 1972, 1923. *The Cheyenne Indians: Their History and Ways of Life*, 2 vols. Lincoln: University of Nebraska Press.

ZYBACH, FRANK (1894–1980)   More than any one person, Frank Zybach transformed the economics and geography of farming the grasslands. Frank, the son of Swiss immigrants living in Nebraska, never really took to farming. More to his liking was tinkering and inventing. Some of his experiments took on rather bizarre forms, such as his self-propelled tractor that neighbors often turned off fearing the operator, Frank, had fallen off and hurt himself. His most notable achievement, however, was his design of the center-pivot irrigation system. This device contributed to the expansion of irrigation throughout the grasslands and now accounts for the irrigation of over 90 percent of all irrigated cropland on the Great Plains—and over one-third of all irrigated cropland in the United States.

Prior to Zyback's work, irrigators applied water to their crops by flooding the ground. At first this was done by digging ditches from streams, then directing the flow by building or tearing down low dirt banks to channel and contain the water. Groundwater irrigation led to the use of pipes to carry the water from the pump to where it was needed. Both methods required very intensive labor and almost perfectly flat ground. Rolling or sloping terrain was impossible to irrigate—the water either pooled in low spots or ran off fast enough to erode the land without watering the crops. Frank Zyback changed all that.

Zybach once watched a demonstration of pipe irrigation and noted the deficiencies. After several experiments with different designs, he had developed a working center-pivot prototype by 1948. He attached irrigation pipe to towers mounted on wheels. The water pressure rising in the well pipe turned cog-driven rods attached to the wheels and slowly drove the entire unit in a sweeping circle. The water flowing in the irrigation pipes powered a mechanism designed to keep the towers in alignment before exiting through the spray nozzles. In July 1952, Zybach had perfected his design to the point that he took out a patent on his "Self-propelled Sprinkler Apparatus." Zybach, and his partner A. E. Trowbridge, lacked the means to capitalize on their creation, and in 1954 they negotiated a deal with Robert Daugherty who headed the Valley Manufacturing Company. Daugherty found marketing this new form of irrigation slow going at first, but his persistence paid off. Now, Valmont Industries, formerly the Valley Manufacturing Company, is the largest manufacturer of center-pivot irrigation systems in the world.

The attraction of center-pivot irrigation was that it opened new opportunities for irrigation as it solved several problems at once. It attaches directly to the well or water pipe at one end. Water flows up and into a series of metal pipes that extend as far as necessary to water a field—up to a half-mile in some cases. Semi-rigid hoses with sprinkler nozzles on them hang from the main horizontal pipe. Water flows out of these nozzles and onto the crops. Modern center pivots often have adjustable hoses so that as the crop grows, the sprinkler head can be kept close to the plants for more efficient water application, and now computers and geographical information systems (GIS) often monitor the rate of water flows. The entire assembly travels on rubber tires mounted on triangular frames that attach to the horizontal pipe at regular intervals. These wheels allow the sprinkler to follow rolling terrain previously impossible to irrigate. Natural gas, electricity, and gasoline engines are the three most common ways of powering the water pumps of the system.

Frank Zyback devised a method of reducing the labor and increasing the acreage of irrigation on the Great Plains. If one looks down on the plains from above, much of the land resembles a quilt, with each square containing a circle of green. While his invention has greatly increased crop production throughout the grasslands, it may have also created its own future obsolescence. The increased efficiency of pumping the fossil water out of the Ogallala Aquifer limits the future use of center pivot on the vast majority of the Great Plains; it will draw down the aquifer to the level that pumping water from deep depths is economically prohibitive, or the groundwater itself will be pumped dry in the more shallow regions.

See also Cheyenne Bottoms; Jackson, Wes; Ogallala Aquifer; Playa Lakes; Popper, Deborah and Frank; Sand Hills; Water Use Doctrines.

Sources

Sheffield, Leslie F. 1981. "Story of 'Farm Shop Inventor.'" *Irrigation Age* (January): 22–23, 36.

Splinter, William E. 1976. "Center Pivot Irrigation." *Scientific American* 234 (June): 90–99.

CHRONOLOGY

ca. 10,000–8,000 BP Depending on the source, this span of years marks the end of the Pleistocene Epoch, a time when the last major North American glaciation ended.

ca. 10,000 BP Archaeologists refer to the people living in the emergence of the North American grasslands as the Clovis culture. These people had developed sophisticated hunting techniques for bagging megafauna such as the wholly mammoth.

ca. 10,000 BP By this time the Pleistocene Extinctions were in full swing. This was a period of dramatic climate change during which the megafauna of North America largely disappeared from the scene. Scientists have had a continuing, unresolved debate over the causation of the extinctions—were they caused by humans, climate, or a combination of the two?

10,000–8,500 BP The period of time scientists call the Postglacial or Boreal, which is marked by a rapid retreat of glaciers and shifts in boundaries marking grasslands and forests.

ca. 8,000 BP The Folsom culture flourishes in the North American grasslands. These people are skillful giant bison hunters.

8,500–3,000 BP A period of time contemporary climatologists call either the altithermal or the hypsithermal, a span of centuries marked by atmospheric temperatures considerably higher than experienced in the previous three millennia.

8,000 BP–around 500 BCE The people whom scholars label the Archaic culture lived in the grasslands. These people lacked agriculture and ceramics, but they probably practiced controlled fire burning, according to the archaeological evidence.

ca. 2000 BP People begin practicing agriculture throughout the Southwest and desert grasslands. Two of the larger groups, generally referred to as the Hohokams and the Anasazis, attain sophisticated civilizations that reached their apex around 1000 BP.

ca. 1200 The Pueblo cultures begin to take form throughout the Rio Grande River Valley.

ca. 1350 Around this time the Little Ice Age begins, a period of time marked by considerably cooler temperatures than in the previous centuries or in the decades following its end, roughly around 1870.

1400 By this time, the *Diné* have established their presence to the west and east of the Pueblos; the people to the west become known as the Navajos, and the people to the east the Apaches.

1530 Some scholars have shown evidence indicating that European diseases are beginning to take a toll among peoples living in the grasslands.

1541 Francisco Vásquez de Coronado makes his trek through the Rio Grande River Valley and then into the Central Grasslands. Most authorities believe he explores to a point somewhere in the central portion of present-day Kansas.

1610 The people living in the Rio Grande Pueblos and those in the Central Grasslands of present-day Kansas encounter Juan de Oñate and his troupe. His expedition begins the permanent establishment of Spanish colonies in the Rio Grande River Valley.

1612 Spaniards establish the outpost of *Santa Fe Nuevo Méjico.*

1680 The Pueblo Revolt begins, and Spanish colonizers flee from the Upper Rio Grande Valley south to El Paso.

1680 The Pueblo peoples begin freely trading horses to the Utes who live to the north of their villages. The Comanches acquire horses from the Utes and begin a migration to the Southern Grasslands.

1700 The Comanches begin their occupation of the Southern Grasslands.

1700 The Spanish conclude their "reconquest" of the Upper Rio Grande Valley.

1700 By this date French Canadians have established a gun trade with the Pawnees.

1700 The people of *Nuevo Méjico* tend a flourishing sheep economy.

1706 General Juan de Ulibarri of *Nuevo Méjico* leads an expedition to retrieve some Picurís who had fled Taos during a revolt in the 1660s. They were living in a place the Spanish called El Cuartelejo, which was located near present-day Scott City, Kansas.

1763 Pierre Laclede establishes an Indian trading post near the confluence of the Missouri and Mississippi rivers that in time becomes the City of St. Louis. Laclede is the stepfather of René Auguste Chouteau and father of Jean Pierre Chouteau, both of whom are exceptionally influential in creating and shaping the fur trade throughout the grasslands. Jean Pierre's sons, Auguste Pierre and Pierre, continue the family's businesses, and in time Pierre becomes a leading citizen of St. Louis.

1779 Don Juan Bautista de Anza leads an expeditionary force of Spaniards and Pueblos that attacks and breaks apart the military strength of the northern

Comanches. The Comanche leader Cuerno Verde (Green Horn) is killed during this engagement. This marks a Spanish and Comanche rapprochement.

1787 *Nuevo Méjico* governor Fernando de la Concha launches a colonization effort intended for the Jupé Comanches who follow Paruanarimuco. The goal is to transform those bison hunters into irrigation farmers. This social experiment utterly fails.

1804 Meriwether Lewis and William Clark and their Corps of Discovery embark upon their trek to the Pacific Coast.

1806 Zebulon Pike and his military contingent cross the Central Grasslands following the trade routes of Indian peoples and Spaniards.

1810 Miguel Hidalgo y Costilla instigates the Mexican Revolution.

1811 At St. Louis, Wilson Prince Hunt begins an overland expedition for John Jacob Astor's Pacific Fur Company. The intention is to establish a trading post in the Pacific Northwest, which becomes the short-lived Fort Astoria.

1812 Robert Stuart leaves Fort Astoria on the Columbia River, and with a small troupe of six others, heads east to report to John J. Astor. In October, Stuart crosses South Pass, which he describes once arriving in St. Louis in April 1813. The St. Louis newspapers quickly report this discovery and its importance for the initiation of a wagon route to the Pacific Coast; in short, the first promotion of the Oregon Trail.

1819 Stephen Long leads an army exploration expedition through the Platte and Canadian River Valleys. He labels portions of the grasslands the "American Desert" on his 1821 map.

1821 The Mexican Revolution comes to a conclusion with the defeat of the remaining Spanish forces and with the ascendance to power of Agustin de Iturbide. Trade with the United States is then opened.

1821 John Jacob Astor and the Chouteau family of St. Louis join together in the American Fur Company and come to exert primary control over the bison robe trade throughout the grasslands.

1821 William Becknell, a resident of Franklin, Missouri, loads pack animals and makes the first successful trading expedition to Santa Fe, New Mexico.

1825 George Champlain Sibley leads the first official mapping expedition of the Santa Fe Trail that is undertaken by the United States.

1825 A growing trade develops between merchants in the United States and those not only in *Nuevo Méjico* but also far to the south in the cities of Chihuahua, Durango, and Zacatecas. These caravans began altering the grassland ecosystems along the Santa Fe Trail.

1832 Nathaniel Wyeth, leading a company of twelve, departs St. Louis. Bound for the Pacific Northwest. Wyeth's party is the first American expedition to trace the route of the Oregon Trail.

1832 or 1833 Charles and William Bent, along with Ceran St. Vrain, establish an adobe trading post along the Arkansas River just downstream from the mouth of the Purgatory River. Their operation dominates trade throughout the Central Grasslands well into the late 1850s.

1834 Chiefs Clarmont of the Osages and Dohason of the Kiowas negotiate with the Army Dragoons who are led by General Henry Leavenworth until he dies in June, and later by Colonel Henry Dodge who assumes command upon Leavenworth's death. These negotiations lead to the Camp Holmes Treaty of 1835 and the Fort Gibson Treaty of 1837, bringing peace between the Osages and Kiowas. It was on this expedition that artist George Catlin begins thinking about the need for a "nation's park" in the grasslands.

1834 John Jacob Astor withdraws from the American Fur Company, and complete control rests with the Chouteau family.

1834 or 1835 Charles Bent marries Maria Ignacia Jaamillo and thereby cements his trading ties to the *rico* class of *Nuevo Méjicanos.*

1835 William Bent marries Owl Woman, a Southern Cheyenne woman of a prominent family; her father is the Sacred Arrow Keeper. This relationship binds the economic interests of Bent's Fort with the Southern Cheyenne nation and consequently leads to significant alterations of the ecosystems in the Central Grasslands.

1836 Marcus Whitman, Henry Spalding, and others take the Oregon Trail with the intention of doing Protestant missionary work among the Indian peoples living in the grasslands of the Columbia River Basin. Whitman's and Spalding's reports and letters become a magnet attracting Euro-American colonizers to the Oregon Territory.

1837 A smallpox epidemic breaks out among the Indian peoples who live in agricultural villages along the upper reaches of the Missouri River. In the Mandan nation, nearly 90 percent of its population is killed by this disease by the spring of 1838. This epidemic follows trade routes and works its way south decimating Indian nations along the way.

1838 Pierre Chouteau reorganizes the American Fur Company as the Pierre Chouteau, Jr., and Company, and its trading interest covers the entire grasslands from the Front Range of the Rocky Mountains to the east.

1838 Colonel John James Abert organizes the Army's Corps of Topographical Engineers, who produced some of the best professional mapping of the grasslands and Rocky Mountains in the two decades to follow.

1839 The first emigrants bound for the Pacific Northwest cross on the Oregon Trail.

1840 At a location on the Arkansas River near Bent's Fort, the Southern Cheyennes and Arapahos enter into a treaty agreement with the Northern Comanches, Kiowas, and Plains Apaches to the south. The agreement permanently ends hostilities between them, creates trade opportunities for the southern nations with Americans, and opens hunting opportunities in the southern grasslands for the northern nations.

1841 John Bidwell leads the first emigrant wagon train to Oregon.

1843 Large-scale emigration in wagon trains along the Oregon Trail begins and continues well into the 1850s. This movement of people through the central and intermountain grasslands disrupts their ecosystems and the horse-borne bison hunting economy throughout both regions.

1844 George Catlin's observations and accounts of High Plains Indian peoples is published. In his *Letters and Notes on the Manners, Customs, and Conditions of North American Indians,* Catlin recommends the creation of a "nation's" park to preserve the wild grasslands.

1846 Diplomats from Great Britain and the United States agree by treaty to establish the boundary line in the Northwest along the 49th Parallel. This ends the Hudson Bay Company's dominance of the fur trade in the intermountain grasslands.

1846 The United States Congress declares war on the Republic of Mexico, thereby beginning the Mexican-American War.

1847 Yellow Wolf, a Southern Cheyenne chief, discusses with Agent Thomas Fitzpatrick the possibility of creating an agricultural sanctuary for his people. Yellow Wolf is concerned about the rapid disappearance of game animals and sought an alternative means of economics for his people.

1847 Mormons, led by Brigham Young, begin their colonizing work based on irrigated agriculture on the western base of the Wasatch Mountain Range.

1848 Eighty-two merchants create and organize the Chicago Board of Trade with the mission of standardizing the selling and buying practices of agricultural commodities.

1851 Writing from Bent's Fort, Indian agent Thomas Fitzpatrick notes that Indian peoples around him were in a "starving state" given the rapid depletion of the bison herds throughout the region.

1854 Congress passes the Kansas-Nebraska Act, opening the Central Grasslands to Euro-American settlement.

1860 This marks the waning of the Little Ice. Some scholars mark this as the end of the Little Ice Age, while a few argue that it persists until nearly 1900.

1862 The United States Congress passes the Morrill Act, thereby creating the land-grant colleges with the declared purpose of serving and educating the "mechanical and agricultural classes" of American citizens.

1862 The state legislature of Iowa accepts the provisions of the Morrill Act, and opens the doors of Iowa State Agricultural College (today Iowa State University) to students in 1869.

1862 The United States Congress passes the Homestead Act, which goes into effect in January 1863.

1863 Formerly Bluemont Central College, Kansas State Agricultural College (today Kansas State University) becomes the first land-grant school in the nation to function under the provisions of the Morrill Act.

1863 Kit Carson's assault on the Navajos results in a forced march of these people away from their homeland along with the destruction of their pastoral sheep practices.

1864 Colonel Chivington attacks the peaceful village of Southern Cheyennes led by Black Kettle. This attack takes the form of a massacre rather than a military engagement, and it results in the expulsion of the Southern Cheyennes and Southern Arapahos from the Central Grasslands.

1864 George Perkins Marsh publishes *Man and Nature*, which marks a major milestone on the path toward the conservation of natural resources.

1865 The Chicago Board of Trade defines and regularizes the selling and buying of "futures" contracts.

1865 The United States negotiates a treaty with the Ute nation, and this agreement places the Utes on a reservation where they were to be taught Euro-American agriculture.

1867 General Hancock leads the Seventh Cavalry in an attack on a Southern Cheyenne Dog Soldier and Oglala Sioux village. He takes an inventory of the possessions left behind by the occupants who flee the area, and this list gives excellent insights to the material culture of these people.

1867 Alexander Gardner, a famed Civil War photographer, begins documenting the route of the Union Pacific Railroad across the central and southwest grasslands of the United States. These photographs give some of the only visual accounts of the grasslands as managed by Indian peoples.

1867 This year marks the beginning of longhorn cattle drives from Texas to railheads throughout the grasslands. Under the guidance of Joseph McCoy, Abilene, Kansas, is the first railhead developed explicitly for this trade.

1867 In October, the United States government negotiates the Medicine Lodge Treaty with the Southern Cheyenne, Southern Arapaho, Kiowa, and Comanche nations, and thereby effectively ends any occupational rights these people have in the grasslands of Kansas and Colorado.

1869 In May, at Promontory Summit, the railroad lines of the Central and Union Pacific companies are joined to complete the first transcontinental railroad.

1869 In May, John Wesley Powell launches his first expeditionary descent of the Colorado River and concludes his exploration in August.

1870 Navajos return to their desert grassland homelands and resume their sheepherding practices.

1870 Edward Walter Maunder, an astronomer at the Royal Greenwich Observatory, begins recording annual numbers of sunspots. His observations, which show a low number of spots occurring during the fifteenth through the seventeenth centuries, lead later scientists to propose that these reduced numbers relate to the onset and duration of the Little Ice Age.

1870 Importers bring the salt cedar (*Tamarix aphylla*) into the port at Galveston, Texas. People grow it as an ornamental shrub.

1871 Early blizzards in November destroy over 70 percent of the Texas cattle being grazed on the mixed, and shortgrass prairies. This results in massive economic losses for Texas cattlemen and is a warning about the perils of open-range cattle grazing on the grasslands during winter months.

1872 Congress passes the Hardrock Mining Act, which provides free and open access to anyone wanting to mine "valuable mineral deposits in lands belonging to the United States."

1872 Congress sets aside Yellowstone National Park.

1873 Theodore C. Henry begins successfully experimenting with growing soft winter wheat and soon beomes known as the "wheat king" of Kansas.

1873 Congress passes the Timber Culture Act, which allows a farmer to acquire 160 acres of public domain provided 40 acres are planted in trees.

1874 E. M. Shelton begins his work as professor of agriculture and superintendent of the farm at Kansas State Agricultural College. He studies hundreds of varieties of wheat and in his bulletin, *Experiments with Wheat* (1888), promotes hard winter wheats as the varieties best suited for grassland climates, based on his results.

1874 Exodusters, former African-American slaves living throughout the southern portions of the Mississippi River Valley, begin their treks to the Central Grasslands in search of havens where they can begin farming and live free from oppressive and virulent racism.

1874 Russian Mennonites arrive in Kansas and begin their wheat farming practices. The story of the Mennonites introducing Turkey Red, a hard winter wheat variety, is most likely apocryphal.

1875 Working with the management of the Northern Pacific Railroad Company, Oliver Dalrymple begins managing large-scale wheat growing operations in the Red River Valley of North Dakota. Undertakings such as these quickly become known as "bonanza farms."

1876 The apex of Sioux power over the northern grasslands occurs with their victory over the Seventh Cavalry at the Battle of Greasy Grass, otherwise known as the Battle of the Little Big Horn.

1877 Theodore C. Henry begins experimenting with growing Red Russian, or Turkey Red, hard winter wheat.

1877 Congress passes the Desert Land Act, which allows a person to claim 640 acres if the land is irrigated.

1878 In Minneapolis, Minnesota, Edmund N. La Croix begins operating the first flour mill outfitted with steel rollers capable of grinding hard winter wheat.

1878 John Wesley Powell publishes his *Report on the Lands of the Arid Regions,* which recommends a planned, controlled settlement of the West based on the relative ecological and economic productive capacity of various landforms to support ranching and farming.

1879 Walt Whitman's record of traversing the grasslands is published. In this work he asserts that the grasslands is America's "characteristic landscape."

1879 Ranchers form the Wyoming Stock Growers' Association with the avowed purpose of creating cooperation among themselves for grazing their herds of cattle on the public domain.

1880 Oliver Dalrymple manages over 100,000 acres of wheat production in the Red River Valley.

1880 George Washington Swink, of Rocky Ford, Colorado, successfully develops a high-value cantaloupe, one that depends on irrigated agriculture. By 1888, rail connections have created markets for his cantaloupes as far away as St. Louis, Chicago, and New York City.

1881 The Sioux led by Sitting Bull returned to United States territory and surrender. This ends virtually all resistance of Indian peoples to Euro-American occupation of the grasslands and the efforts of Euro-Americans to transform both Indian peoples and the wild grasslands into a preconceived notion of "civilization."

1886 Beginning in the late fall and well into the winter of 1887, blizzards and frigid temperatures destroy open-range cattle grazing operations throughout the central and northern grasslands.

1887 The United States Congress passes the Hatch Act, which funds a system of agricultural experiment stations under the auspices of land-grant colleges.

1887 The United States Congress passes the General Allotment Act, otherwise known as the Dawes Act after its sponsor, Senator Henry Dawes of Massachusetts. This act, purportedly a reform for improving the lives of Indian peoples living on reservations, forcibly allots a set amount of acreage per person on a reservation with the intention of turning Indian peoples into Euro-American–style farmers. The act gives the Bureau of Indian Affairs the responsibility for selling the unassigned lands on the open market and placing the proceeds into trust accounts for tribal uses.

1888 Louis G. Carpenter becomes a professor at Colorado State Agricultural College and begins a career in the study and teaching of irrigation engineering.

1888 William Smythe, a newspaper man from Omaha, Nebraska, travels the southwest grasslands and observes how well the irrigated crops of Pueblo Indian peoples and those of New Mexicans flourish despite the harsh drought conditions of that year.

1889 Through the energetic promotional work of Hiram Hadley, New Mexico Agricultural College opens its doors for its first class of students.

1889 Beginning in the late fall and extending into the winter of 1890, severe, cold weather decimates the open-range cattle operations throughout the intermountain grasslands.

1890 In the city of Logan, the faculty open the doors to Utah State Agricultural College, a place for the "education of rural western democracy—a rural college for the masses."

1890 Professor A. E. Blount begins his work on improving irrigated agriculture in New Mexico.

1890 Congress passes the Second Morrill Act, which provides additional federal funding for land-grant schools as long as these schools do not discriminate against African-Americans. However, the act allows for the creation of separate institutions for African-Americans.

1890 From this time through the 1930s, many Euro-Americans make several efforts to establish dryland farming in the Southwest and intermountain grasslands, and on the Great Plains. By the time of the droughts in the 1930s, most of these efforts have been abandoned.

1890 Ranchers begin a combination of raising hay for winter feeding and grazing cattle on the open range during the spring and summer months.

1891 William Smythe organizes the National Irrigation Congress, and its membership holds its first conference in Salt Lake City, Utah.

1891 In the Southwest grasslands, an extended drought begins and lasts through 1893. Open-range cattle grazing throughout the region is grievously hurt, and their owners suffer crushing economic losses.

1892 Washington State Agricultural College and the University of Idaho, both land-grant schools, open their doors for their first classes of students.

1893 Severe lack of rainfall destroys wheat and corn crops over wide portions of the central grasslands.

1893 Frederick Jackson Turner, a history professor at the University of Wisconsin, delivers his address, "The Significance of the Frontier in American History," at the Columbia Exposition in Chicago, Illinois.

1894 George Washington Swink of Rocky Ford, Colorado, successfully demonstrates sugar beet potential to the sugar magnate Henry T. Oxnard. This leads to the rapid spread of irrigated sugar beet fields and processing plants throughout the valleys of the High Plains.

1896 Thomas Cooper, with the financial support of James Hill's Northern Pacific Railroad Company, acquires irrigation systems in the Yakima River Valley of Washington and places them on a functioning foundation.

1897 Langston Colored Agricultural and Normal University opens in Langston, Oklahoma. Authorized by the 1890 Land Grant Act, this school and others like it provide segregated educational institutions for African-Americans.

1899 William Smythe publishes *The Conquest of Arid America*, a work that celebrates the purported successes of irrigated agriculture throughout the semi-arid and arid grasslands.

1900 Botantists from the Nevada experiment station take note of the destruction of desert grassland ecosystems as a result of overgrazing cattle and recommend reseeding the area in native grasses.

1900 Mormon colonization efforts through irrigated agriculture result in 10,000 farms encompassing over 250,000 acres.

1900 Irrigators plant salt cedar along irrigation and stream banks for erosion control. Soon the plant becomes uncontrollable and starts consuming large amounts of water, by some estimates up to 200 gallons of water per plant per day.

1900 The Apaches' livestock operations on the San Carlos and White Mountain Reservations are showing fine returns and create a viable economic base for both nations.

1901 The Kansas attorney general files suit in the U.S. Supreme Court charging the state of Colorado with destroying the flows of the Arkansas River into Kansas and thereby harming the Kansas economy.

1902 The United States Congress passes the Newlands Act, which provides federal support for the creation of irrigated agriculture throughout the semi-arid and arid grasslands.

1902 President Theodore Roosevelt charges the Country Life Commission, chaired by famed horticulturalist Liberty Hyde Bailey, to study the conditions of rural life and to make recommendations for its improvement.

1903 The United States Supreme Court decides the case *Lone Wolf v. Hitchcock*, in which the Kiowas lose their challenge to the constitutionality of the General Allotment Act.

1904 The United States attorney general intervenes in the ongoing United States Supreme Court case *Kansas v. Colorado* to protect the interests of the Reclamation Service.

1905 With Gifford Pinchot as its first head, the Forest Service is created in the Department of Agriculture.

1906 The American Stock Growers Association and the National Livestock Association merge to create the American National Livestock Association, the leading organization in the nation representing the interests of ranchers.

1907 The United States Supreme Court delivers its decision on the *Kansas v. Colorado* suit. In this decision written by Justice David Brewer, the Court institutes the doctrine of equity as the principle for resolving this case, and equity became the legal precedent most often used in deciding subsequent interstate water cases in the United States.

1908 The United States Supreme Court hands down its decision on *Winters v. the United States.* The justices decide that the 1888 treaty with the Gros Ventres and the Assiniboines had created a "reserved" water right for tapping the Milk River for irrigated farming on the Fort Belknap Reservation. This decision established the legal precedent for securing Indian peoples' water rights on their reservations.

1911 Construction of the Theodore Roosevelt Dam on the Salt River of Arizona is completed. One of the earliest projects of the Reclamation Service, this project shows the economic potential of hydroelectric production coupled with irrigation development.

1914 The United States Congress passes the Smith-Lever Act, which creates and funds the extension service. The mission of extension is to provide people not in attendance at a college or university with information on how to improve their agricultural or home economic practices.

1914 Irrigators first take note of the appearance of salt cedar in the Arkansas River Valley. This plant has rapidly spread throughout the valley and become a severe ecological hindrance to irrigated farming practices.

1919 Custer State Park in South Dakota is created and becomes one of the earliest sites for the preservation and increase of bison.

1920 Farmers begin investing in dryland farming techniques over wide portions of the Great Plains from Montana south well into the Panhandle of Texas.

1921 Wheat prices collapse with the resumption of pre–World War I international crop production. This leads to a grievous economic downturn throughout the grasslands and ruins thousands of farmers and small bankers.

1922 Representatives from seven western states meet in Santa Fe, New Mexico, and under the chairmanship of Secretary of Commerce Herbert Hoover, negotiate the terms creating the Colorado River Compact.

1927 The state legislature of New Mexico is the first to bring groundwater under public control.

1932 Near the small town of Clovis, New Mexico, archaeologists E. B. Howard and John Cotter unearth stone points made around 10,000 BP. These long, fluted points were used to hunt megafauna; the people who crafted these points become known as the Clovis culture and their handiwork known as Clovis points.

1933 The lack of rainfall coupled with the farming techniques across the Great Plains sets the stage for the first dust storms to sweep the nation.

1933 Dr. V. E. Shelford, a nationally prominent grassland ecologist at the University of Illinois, makes the first succinct case for the creation of a grassland national park in the scientific journal *Ecology*.

1933 New Dealers, under the direction of John Collier, director of the Bureau of Indian Affairs, and Hugh Hammond Bennett, later the director of the Soil Conservation Service, begin addressing the overgrazed rangelands on the Navajo Reservation. Overgrazing creates a myriad of social, political, and ecological problems.

1934 A massive dust storm originating in Montana and Wyoming carries an estimated 350 million tons of dust into the air, blanketing cities far to the east and even ships in the Atlantic Ocean.

1934 The United States Congress passes the Taylor Grazing Act, creating the United States Grazing Service, which is administered within the Department of Interior. The act effectively ends homesteading and provides a mechanism for conserving and regulating grazing on the public domain.

1935 A. S. Hitchcock publishes his *Manual of Grasses of the United States.* This work is the first to classify and describe the wild grasses of North America. Much of the research for these volumes was done while Hitchcock served as a professor at Kansas State Agricultural College.

1935 One of the worst, if not the worst, dust storms on record occurs on April 14. Known as "Black Sunday," the severity of the storm leads Congress to take strong action to set into place better farming policies and techniques for the nation through the creation of the Soil Conservation Service with Hugh Hammond Bennett as its first director.

1936 The first ski runs created in a National Forest are built in Sun Valley, Idaho.

1936 New Dealers publish *The Future of the Great Plains: Report of the Great Plains Committee.* This report sets forth policy recommendations for achieving better conservation of natural resources and for stimulating greater economic efficiencies in farming. This report establishes the agenda for federal farm policy into the present.

1944 Congress passes the Flood Control Act, which establishes the outline of what becomes known as the Pick-Sloan Plan. This act authorizes the Army Corps of Engineers and the Bureau of Reclamation to build dams throughout the Missouri River Basin.

1946 President Truman's efforts to streamline his administration lead to the consolidation of the General Land Office and the Grazing Service to form the Bureau of Land Management.

1947 Roy Bedechek publishes *Adventures with a Texas Naturalist.*

1947 James Malin publishes *The Grassland of North America: Prolegomena to Its History,* a groundbreaking work combining the most recent scholarship in grassland ecology with traditional history to gain an understanding of the relationship between humans and their environment. In some respects, this work is an early precursor of the methodology incorporated into modern environmental history.

1948 The *Journal of Soil and Water Conservation* reprints an earlier address by Aldo Leopold in which he articulated the ethics of an "ecological conscience."

1949 Frank Zyback patents the first center-pivot irrigation system, an invention that will revolutionize farming practices throughout the High Plains.

1954 Ezra Taft Benson, secretary of agriculture, gives his famous exhortation to farmers to "get big or get out."

1954 Elmer T. Peterson publishes his *Big Dam Foolishness,* in which he questions the wisdom of the massive dam projects proposed and built by the Bureau of Reclamation and the Army Corps of Engineers.

1957 Walter Prescot Webb publishes "The American West, Perpetual Mirage," in *Harper's Magazine.* This criticism of Euro-American ways of life in the American West draws a barrage of rebuke in letters to the editor in weeks following.

1959 The Bureau of Land Management publishes reports indicating the grossly overgrazed condition of the public range.

1961 The fiasco involving rancher Carl Bellinger and Secretary of Interior Stewart Udall closes any possibility of creating a grassland land national park near Manhattan, Kansas.

1963 Rachel Carson's *Silent Spring* is published and leads to a reconsideration of petrochemical uses in American agriculture.

1964 Congress enacts the Wilderness Act.

1968 W. J. Downton and E. Trgunna, both of whom were biologists, are the first to describe the different mechanisms by which warm- and cool-season grasses combine carbon dioxide and visible light to form carbohydrates.

1971 Peter Bogdanovich directs the movie *The Last Picture Show.*

1973 Congress passes the Endangered Species Act.

1974 Wes Jackson returns to Salina, Kansas, from California to launch the work of the Land Institute.

1976 An article by J. D. Hayes, John Imbrie, and N. J. Shackleton in the journal *Science* is the first mathematical confirmation of Milutin Milankovitch's theory that cyclical variations in the orbit of the Earth around the sun produce the waxing and waning of glaciation. Milankovitch, a Soviet scientist, had formulated most of his theoretical work before 1930.

1977 Kathrine Ordway, heiress to a family fortune, provides the funding for purchasing the land that eventually becomes Konza Biological Research Station, the first long-term ecological research site devoted to the study and understanding of tallgrass prairie ecosystems.

1977 In South Dakota, the United Family Farmers defeat the plans of the Bureau of Reclamation to build the Oahe Unit, an irrigation project that was a part of the Pick-Sloan Plan.

1978 The "sagebrush rebellion" gains strength in its opposition to federal land use policies and in its support for transferring federal lands to the ownership and control of the states.

1978 Jim Hightower, the former agricultural commissioner of Texas, publishes *Hard Tomatoes, Hard Times,* a hard-hitting indictment of land-grant institutions' embrace of mechanistic and corporate agriculture.

1980 Garth Youngberg and Charles Benbrook, two economists working in the Department of Agriculture (DOA), and who have important responsibility for publishing the *Report and Recommendations on Organic Farming,* guide the first report by the DOA treating the subject in a positive light.

1982 The Shortgrass Steppe Long-Term Ecological Research site is created in northwest Colorado.

1983 Frank and Debbie Popper begin publishing their work advocating the creation of a "buffalo commons" on the High Plains.

1986 Congress authorizes the Conservation Reserve Program to be administered through the Department of Agriculture.

1987 Ron Arnold publishes *Ecology Wars,* which sets forth an agenda for the "wise use" movement centered in its version of "economic growth, technological progress and a market economy."

1988 Cheyenne Bottoms is designated as a "Wetland of International Importance" under the provisions of the intergovernmental treaty negotiated at the Convention on Wetlands and signed at Ramsar, Iran, in 1971. Currently there are 147 governmental participants and 1,524 denoted wetlands worldwide.

1988 In a conference at Reno, Nevada, attendees, representing various corporations and grassroots organizations, put together an agenda for attaining the goals of the "wise use movement."

1989 Ian Frazier publishes the *Great Plains*.

1989 The Nature Conservancy purchases the Barnard Ranch in Osage County, Oklahoma, thereby creating a tallgrass prairie preserve and reintroducing bison in it.

1990 Kevin Costner directs and stars in the movie *Dances with Wolves*.

1990 Washington State University creates the Center for Sustaining Agriculture and Nature Resources.

1991 William Least Heat-Moon publishes *PrairyErth*.

1992 David Pope, chief engineer for the state of Kansas, curtails pump irrigation for farming to protect the water rights supplying Cheyenne Bottoms.

1993 Published reports by scientists such as Aaron Blair raise concerns over the connection of petrochemical uses in farming to the increasing incidents of cancer among American farmers.

1994 The Soil Conservation Service is renamed the Natural Resources Conservation Service.

1995 The U.S. government approves the planting of genetically modified corn.

1995 Jon Margolis, writing for the *New York Times Magazine*, asks if it is worth keeping North Dakota as a state.

1996 Scholars Fred B. Samson and Fritz L. Knopf's work, *Prairie Conservation*, warns that the grasslands are the most endangered ecosystem on the planet.

1996 Over 150 years after George Catlin recommended it, Congress passes the Tallgrass Prairie National Preserve Act, thereby creating the first national grassland park.

1997 Courtney White organizes the Quivira Coalition, a group of environmentalists and ranchers working collaboratively to protect and restore desert grasslands while maintaining the practice of ranching.

1997 Iowa Beef Processors build the largest meat processing plant in the nation, one capable of slaughtering and dressing 6,000 animals per day.

1999 The Nature Conservancy purchases a 16,800-acre ranch in Logan County, Kansas, and creates a preserve of the shortgrass ecosystem.

2000 The proliferation of ranchettes starts becoming a concern to local governing officials and environmentalists.

2003 On Christmas Eve, the Department of Agriculture announces the confirmation of the first case of bovine mad cow disease in the United States.

SELECTED ANNOTATED BIBLIOGRAPHY

BOOKS AND ARTICLES

Ackerman, Frank A., Timothy A. Wise, Kevin P. Gallagher, Luke Ney, and Regina Flores. 2003. *Free Trade, Corn, and the Environment: Environmental Impacts of US—Mexico Corn Trade under NAFTA*. Medford, MA: Tufts University, Global Development and Environmental Institute, Working Paper No. 03-06.

In 1994, the United States, Canada, and Mexico entered into an agreement to end trade barriers among the countries. In regard to corn production in Mexico and the United States, Americans increased their exports to Mexico and by 2003 accounted for nearly one-fourth of all the corn consumed in Mexico. The authors contend that this has led to harmful environmental effects in the United States such as high chemical use, stream pollution, soil erosion, and biodiversity losses, and has reduced agro-biodiversity in Mexico.

Anderson, Gary. 1999. *The Indian Southwest, 1580–1830: Ethnogenesis and Reinvention*. Norman: University of Oklahoma Press.

The Indian peoples of the Southwest lived in highly fluid societies. New cultures came into existence and existing ones reinterpreted for themselves their own values and beliefs in order to meet changing demographic, ecological, economic, and diplomatic conditions in the region.

Axelrod, Daniel. 1950. *Studies in Late Tertiary Paleobotany*. Washington, DC: Carnegie Institution of Washington, Publication 590.

Axelrod's chapter on the evolution of desert vegetation and how it formed the main components of the shortgrass plains is of interest.

Axelrod, Daniel. 1985. "Rise of the Grassland Biome, Central North America." *The Botanical Review* 51 (April–June): 163–201.

Axelrod argues persuasively that the grasslands encountered by the first Europeans to depict them arose as a result of conscious management practices by the people who had occupied the grasslands in the centuries before. Fire was the main tool used by pre-contact peoples.

Barrell, Joseph. 1975. *The Red Hills of Kansas: Crossroads of Plant Migrations.* Rockford, IL: Natural Land Institute.

Barrell offers one of the few studies of the origin of grassland plants and their dispersal patterns in the postglacial period.

Binnema, Theodore. 2001. *Common and Contested Ground: A Human and Environmental History of the Northwestern Plains.* Norman: University of Oklahoma Press.

This excellent portrayal of Indian peoples living in the northern grasslands describes the ecological relationships shaping their lives. Binnema analyzes the intertwined relationships of bison-hunting economics, horse-tending practices, cultural beliefs, climate, and natural resources and how these together shaped the grassland ecosystems and well-being of the people who occupied them.

Blair, Aaron, and Shelia Hoar Zahm. 1993. "Patterns of Pesticide Use among Farmers: Implications for Epidemiologic Research." *Epidemiology* 4 (No. 1): 55–62.

Blair and Zahm offer a statistical analysis of cancer rates among farmers that has a reasonable connection to applications of chemical pest controls and fertilizers. Their conclusions indicate a likely connection between chemical uses and incidents of melanoma.

Bleed, Ann, and Charles Flowerday, eds. 1989. *An Atlas of the Sand Hills.* Lincoln: University of Nebraska, Conservation and Survey Division.

The Sand Hills of Nebraska cover nearly the entire western half of the state. They are a unique grassland ecosystem, and this atlas is a thorough treatment of the area's contemporary geographical setting, both physical and social.

Blouet, Brian W., and Frederick C. Luebke, eds. 1979. *The Great Plains: Environment and Culture.* Lincoln: University of Nebraska Press.

This collection of essays by a group of notable Great Plains scholars examines the relationships of various cultures to their environments. Subjects range from paleoindian adaptations in the Republican River Valley, the Populist response to the ecological difficulties of farming in the late nineteenth century, to irrigation developments and the growth of an urban complex.

Brown, Lauren. 1985. *Grasslands: A Comprehensive Field Guide, Fully Illustrated with Color Photographs, to the Birds, Wildflowers, Trees, Grasses, Insects, and Other Natural Wonders of North America's Prairies, Fields, and Meadows.* New York: Alfred A. Knopf.

This volume is one of a series titled The Audubon Society Nature Guides. It is a superb, nicely illustrated introduction to the plants, animals, climate, and physical geography of the grasslands—an especially valuable guide when exploring the various ecological realms of this region.

Carrels, Peter. 1999. *Uphill against Water: The Great Dakota Water War.* Lincoln: University of Nebraska Press.

This is a story of farm people who successfully resisted the creation and building of an irrigation district, the Oahe Unit, that was intended to be a part of the Oahe dam and reservoir, a Bureau of Reclamation project that was a part of the Pick-Sloan Plan along the Missouri River in South Dakota.

Clark, Ira G. 1987. *Water in New Mexico: A History of Its Management and Use.* Albuquerque: University of New Mexico Press.

Clark is one of the most knowledgeable historians of western water resources of the Southwest grasslands. This work is the most comprehensive history yet of water use, policy, and institutional developments in New Mexico.

Cline, Gloria Griffen. 1963, 1988. *Exploring the Great Basin,* with a new foreword by Michael J. Brodhead. Norman: University of Oklahoma Press; Reno and Las Vegas: University of Nevada Press.

Beginning with a description of the historical geography of the Great Basin itself, Cline gives an account of Euro-American explorations throughout the region. He describes the results of the Spanish expeditions starting with Fathers Garces and Escalante, along with the later ones of Juan Bautista de Anza, ending with those of John C. Frémont.

Crown, Patricia L., and W. James Judge, eds. 1991. *Chaco and Hohokam: Prehistoric Regional Systems in the American Southwest.* Santa Fe, NM: School of American Research Press.

Crown and Judge edited the contributions of eleven scholars, which form the thirteen chapters in this collection. The authors discuss the rise, functioning, and fall of these two distinct cultures.

Cunifer, Geoff. 2005. *On the Great Plains: Agriculture and Environment.* College Station: Texas A&M Press.

The methodology of Cunifer's work blends some of the newest advances in geographic information systems technology with agricultural census data and reaches some provocative conclusions. For some time, many scholars have observed and charted environmental problems associated with Euro-American farming techniques. Cunifer's data and interpretations suggest that Euro-American practices have resulted in a sustainable "equilibrium of land use that [has] varied only slightly" since the end of World War II.

Dort, Wakefield, Jr., and J. Knox Jones, Jr., eds. 1970. *Pleistocene and Recent Environments of the Central Great Plains.* Lawrence: University Press of Kansas.

The editors divided this volume into four sections, one covering soils and climate, another pre-European contact cultures, the third an analysis of plant changes over 10,000 years, and the last a history of insect, bird, and mammalian changes during the same time period. Thirty authors contributed to this volume.

Fagan, Brian. 2000. *The Little Ice Age: How Climate Made History, 1300–1850.* New York: Basic Books.

Fagan's focus is primarily on western European cultures. However, his work is an excellent analysis of the "complex state of flux" in the dynamic relationships among climate, cultures, politics, and economics. He shows how the Little Ice Age became a central force in shaping the stage on which social and economic change occurred.

Fiege, Mark. 1999. *Irrigated Eden: The Making of an Agricultural Landscape in the American West.* Seattle: University of Washington Press.

This innovative work details how irrigation systems throughout the desert grasslands of Idaho became in essence functioning ecosystems. In other words, the culture and technology of irrigation did not remove people out of nature; rather, their enterprise simply entangled them in the ecological relationships shaping the landscape itself.

Flora of North America Editorial Committee, eds. *Flora of North America: North of Mexico.* 9+ vols. New York: Oxford University Press, 1993–.

For this work in progress, the contributors are providing one of the most, if not the most, comprehensive guide to plants in North America. The coverage takes into consideration climate, geography, and human uses as they relate to the history of plants in North America during the Quaternary period.

Flores, Dan. 1991. "Bison Ecology and Bison Diplomacy." *Journal of American History* 78 (September): 465–485.

Flores's interpretation of the destruction of the great bison herds of North America offers a new interpretation, one that implicates Indian peoples' role in the robe trade as one important factor contributing to the decline of the herds.

Fountain, Steve. 2004. "A Horse Is a Horse of Course: Multiple Horse Cultures and Ethnocultural Change in the North American West." Paper presented at the Forty-Fourth Annual Conference of the Western History Association.

Indian peoples used horses for a number of purposes besides hunting large game. Fountain demonstrates how the North Paiutes and Yokuts of the Great Basin found horses useful not only for enhancing their ability to wage war but also as a food source.

Gumerman, George J., ed. 1991. *Exploring the Hohokam: Prehistoric Desert Peoples of the American Southwest.* Albuquerque: University of New Mexico Press.

Gumerman conducted a symposium on the Hohokam culture in 1988. Out of that meeting came nine papers by fourteen participants discussing various aspects of Hohokam culture from social organization to environmental exploitation.

Harper, Kimball T., et al. 1994. *Natural History of the Colorado Plateau and Great Basin.* Niwot: University Press of Colorado.

Twenty authors contributed to this volume. Most of the chapters deal with various aspects of the Great Basin. The Colorado Plateau covers the northern halves of New Mexico and Arizona, the eastern half of Utah, and the western half of Colorado. The Great Basin is defined as the area south of the Snake River Valley, west of the Sierra Nevada Mountains, north of the Sonoran Desert, and west of the Colorado Plateau. The topics range from climatological overviews, the Pleistocene extinctions and prehistory, and the ecological effects of introduced animals and cultures.

Haynes, Gary. 2002. *The Early Settlement of North America: The Clovis Era.* New York: Cambridge University Press.

Haynes offers a thorough discussion of Clovis culture in North America. He addresses the many differing theories about dispersal, hunting techniques, and material culture. He offers a reasoned argument upholding the "megafauna extinction" theory.

Heaton, John W. 2005. *The Shoshone-Bannocks: Culture and Commerce at Fort Hall, 1870–1940.* Lawrence: University Press of Kansas.

Even though stripped of their traditional material culture, the Shoshones and Bannocks were able to reproduce certain elements of their traditional cultures even within the context of the Fort Hall Reservation and U.S. policy designed to turn them all into farmers shaped by an American agrarian ideal. Their ranching practices, especially, gave them some power to guide allotment policies but not enough to prevent internal divisions along lines of environmental adaptation strategies.

Hurt, R. Douglas. 1994. *American Agriculture: A Brief History.* Ames: Iowa State University Press.

Hurt's work is one of the few historical surveys of American agriculture, and it provides a good overview of the main issues and trends shaping its history.

Hurt, R. Douglas, ed. 1998. *The Rural West since World War II.* Lawrence: University Press of Kansas.

This collection of original essays details many subjects pertinent to the environmental history of the grasslands. Farm policy, environmentalism and farmers, and water development are just a few of the relevant chapters.

Iverson, Peter. 1994. *When Indians Became Cowboys: Native Peoples and Cattle Ranching in the American West.* Norman: University of Oklahoma Press.

Normally, Indian peoples are not associated with successful ranching operations. However, Iverson, by gauging the successes of the Apaches and Tohono O'odham in Arizona, shows how adaptable those people were and how well they understood the economic, ecological, and social problems confronting them.

Jordan, Terry G. 1993. *North American Cattle-Ranching Frontiers: Origins, Diffusion, and Differentiation.* Albuquerque: University of New Mexico Press.

Jordan provides a well-researched overview of cattle ranching tracing its origins to pre-contact pastoral practices in Spain, Britain, and Africa; he describes how these were transplanted to North America and modified over time as they spread from one part of the grasslands to another. Probably his most controversial conclusion is that practices derived from British colonizers played a more important role in developing American ranching than did ones from Spain.

Knight, Richard L., Wendell C. Gilgert, and Ed Marston, eds. 2002. *Ranching West of the 100th Meridian: Culture, Ecology, and Economics.* Washington, DC: Island Press.

The editors have divided the essays in this collection into five parts. The first describes ecosystems of rangelands; the second the cultures of ranching; the third the current ecological concerns in ranching; part fours covers the economics; and the fifth presents a summary conclusion spelling out the direction ranching must take in order to succeed in the twenty-first century.

Lekson, Stephen H. *The Chaco Meridian: Centers of Political Power in the Ancient Southwest.* Walnut Creek, CA: AltaMira Press, 1999.

Lekson contends that Chaco centers of power shifted from Chaco Canyon north to Aztec, then in a straight line to the south at Paquime. He believes that in these transitions, Chaco culture maintained itself, thereby influencing the development of Pueblo culture well beyond the collapse of the culture in Chaco canyon before 1100 CE.

Leopold, Aldo. 1948. "The Ecological Conscience." *Journal of Soil and Water Conservation* 3 (July): 109–112.

This was an early version of Leopold's land ethic. He crafted this version with particular attention to farmers and what he thought they needed to do if they wanted to continue living on the land.

Lewis, David Rich. 1994. *Neither Wolf nor Dog: American Indians, Environment, and Agrarian Change.* New York: Oxford University Press.

On first glance, the Hupas, Utes, and Tohono O'odham seem to share little in common. However, Lewis shows how a similar U.S. policy toward these very different peoples—that is, to Christianize them and transform them into farmers—achieved different results depending on the cultural and ecological practices of each. The chapters on the Tohono O'odham and Utes are pertinent in how they related to the desert grasslands in the Southwest and Great Basin.

Lowitt, Richard. 1993. *The New Deal and the West.* Norman: University of Oklahoma Press.

Lowitt's is the best survey of how the policies of the New Deal addressed the ecological, political, economic, and social problems plaguing the American West during the Great Depression. Particular attention is given to agricultural and ranching policies.

Malin, James C. 1984. *History and Ecology: Studies of the Grassland.* Edited by Robert Swierenga. Lincoln: University of Nebraska Press.

Swierenga has collected and edited nineteen chapter-length writings by Malin. These essays range from Malin's interest in applying the science and findings of ecology to an understanding of grassland history, to his quantitative historical analyses of population demographics and economic trends in the grasslands. These works, written from the mid-1930s to the early 1960s, were important precursors of contemporary environmental history.

Martin, Paul S. 2005. *Twilight of the Mammoths: Ice Age Extinctions and the Rewilding of America.* Berkeley: University of California Press.

In this work Martin further elaborates his argument that the during the last 50,000 years over one-half of the world's 200 genera of large animals has become extinct. In North America, over thirty genera went extinct, and the cause of those extinctions, so Martin argues, lies directly with human beings. He further contends that wherever and whenever possible, the health of North American ecosystems lies in "rewilding" them, or in other worlds, returning species such as elephants to the grasslands.

May, Dean L. 1994. *Three Frontiers: Family, Land, and Society in the American West, 1850–1900.* New York: Cambridge University Press.

This is a study and comparison of three distinct communities in the intermountain West. Settled primarily by southerners, Sublimity, Oregon, was marked by tight kinship and community bonds. Alpine, Utah, colonized by Mormons, developed tight-knit theocratic communal bonds. Enterprising, market-oriented Euro-Americans established Middleton, Idaho. Day carefully illustrates how the social orientation of each city led to different types of economies and relationships to the land itself.

McClaran, Mitchel P., and Thomas R. Van Devender. 1995. *The Desert Grasslands.* Tucson: University of Arizona Press.

Twelve authors contributed to this volume. The topics range in chronology from the end of the last ice age to the present. While the chapters focus on various ecological aspects of this biome, the authors include human agency as a contributing force. The role of fire management is especially noted in how it shaped plant and animal communities. The area analyzed covers the southwestern portion of Texas, the southern halves of Arizona and New Mexico, and the central region of Mexico.

McGinnis, Anthony. 1990. *Counting Coup and Cutting Horses: Intertribal Warfare on the Northern Plains, 1738–1889.* Evergreen, CO: Cordillera Press.

McGinnis provides a sweeping analysis of how the warrior cultures and economies of Northern Plains Indian peoples shaped the history of the region. His study is particularly astute on the horse cultures of those peoples, and it consequently touches on some of the geographical and ecological constraints of their material cultures.

Meinig, D. W. 1971. *Southwest: Three Peoples in Geographical Change, 1600–1970.* New York: Oxford University Press.

This succinct study of the Southwest traces the historical geography of the region as it was shaped by three generalized cultures: "Indian, Hispano, and Anglo." Meinig begins his story in 1598 with the Spanish colonization effort in the Rio Grande River Valley and ends it with Anglo dominance and geographic influence.

O'Neill, Robert V. 2001. "Is It Time to Bury the Ecosystem Concept? (with Full Military Honors, of Course!)" *Ecology* 82 (No. 12): 3275–3284.

The liabilities of ecosystem theory are aptly discussed in this piece. O'Neill takes pains to demonstrate that an understanding of any ecosystem can occur only when all elements forming it are understood, and this includes the human role in shaping them. To underscore this point, O'Neill argues that human beings should be considered a "keystone" species of any ecosystem.

Opie, John. 1993. *Ogallala: Water for a Dry Land: A Historical Study in the Possibilities for American Sustainable Agriculture.* Lincoln: University of Nebraska Press.

Opie combines the research of the hard sciences with traditional history to create a hard-hitting look at pump irrigation on the Great Plains. This award-winning book asks if American farming techniques have much longer to exist before becoming unsustainable either economically or ecologically.

Ostler, Jeffrey. 2004. *The Plains Sioux and U.S. Colonialism from Lewis and Clark to Wounded Knee.* New York: Cambridge University Press.

Within the context of colonialism, Ostler offers a new interpretation of Sioux and Euro-American relationships. He depicts three levels in which American colonialism and Sioux agency manifested themselves. On the first he considers the various ways in which Euro-Americans approached the policy of colonialism; on the second he illustrates the interrelationships of Euro-Americans and Sioux peoples as the former pursued subjection of the Sioux and the Sioux worked to maintain their horse-borne culture; and on the third he shows the diversity within Sioux society itself and how these many ways interacted to shape Sioux agency. Environmental change is an important actor in Ostler's work.

Parman, Donald L. 1976. *The Navajos and the New Deal.* New Haven: Yale University Press.

Parman's study offers keen insights into the origins and consequences of the conservation and reservation policies on the Navajo reservation that were implemented by New Deal reformers such as John Collier.

Reichman, O. J. 1987. *Konza Prairie: A Tallgrass Natural History.* Lawrence: University Press of Kansas.

One of the former directors of what is now the Konza Biological Research Station, Reichman gives a thorough introduction to the natural history and ecological complexities of the tallgrass prairies.

Reisner, Marc. 1987. *Cadillac Desert: The American West and Its Disappearing Water.* New York: Penguin Books.

A highly readable history of the Bureau of Reclamation and the Army Corps of Engineers in the West with good coverage of developments throughout the grasslands, Reisner's work offers a highly critical yet well-researched analysis of the ecological, economic, political, and social problems these projects have brought to the region.

Ritchie, Mark, and Kevin Ristau. 1986. *Political History of U.S. Farm Policy.* St. Paul: Minnesota Agriculture Commission.

Family farming has endured its ecological and economic ups and downs and has retreated from the American scene. Ritchie and Ristau take a hard look at farm crises since the late 1800s up to the passage of the 1985 farm bill, which in their view, adds an additional serious threat to the viability of the family farm as an American institution.

Robbins, William G. 1993. "Landscape and Environment: Ecological Change in the Intermontane Northwest." *Pacific Northwest Quarterly* 84 (October): 140–149.

Robbins provides a summary overview of how Euro-American culture, especially its economic aspirations, has altered the waterscape, forests, and grasslands of the region. He raises critical issues of sustainability.

Ronda, James P. 1984. *Lewis and Clark among the Indians.* Lincoln: University of Nebraska Press.

This is an intensive analysis of Meriwether Lewis and William Clark's experiences with Indian peoples. The economic, diplomatic, and ecological practices of Indian peoples is noted along with a description of the northern grassland environment along the Missouri River and Columbia River basins.

Samson, Fred B., and Fritz L. Knopf, eds. 1996. *Prairie Conservation: Preserving North America's Most Endangered Ecosystem.* Washington, DC: Island Press.

Are grasslands one of if not the most misunderstood and threatened ecosystems in North America? Samson and Knopf certainly answer this question with a resounding "yes!" They emphasize their concerns with a collection of essays mainly by biologists and governmental officials who are currently researching grassland ecology and working on providing policies for preserving what is left of the once vast, wild grasslands.

Schneiders, Robert Kelley. 1999. *Unruly River: Two Centuries of Change along the Missouri.* Lawrence: University Press of Kansas.

This work chronicles the cooperative efforts of Euro-Americans to tame the Missouri River for purposes of economic development by regularizing the flows of the "Muddy Mo" primarily for river traffic and flood control. The environmental harm that dam development has caused to the river below Sioux City, Iowa, is carefully related. Furthermore, the author suggests that perhaps instead of trying to regulate the river through "self-interest," using principles of "ecosystem management" might produce a healthier river for both humans and other life.

Smith, Karen L. 1986. *The Magnificent Experiment: Building the Salt River Reclamation Project, 1890–1917.* Tucson: University of Arizona Press.

This is the most comprehensive look at the initiation and completion of the first Reclamation Service project, the Roosevelt Dam on the Salt River of Arizona.

Smith, Sherry L., ed. 2003. *The Future of the Southern Plains.* Norman: University of Oklahoma Press.

This collection contains writings of some of the more prominent and promising scholars of the American Southwest. John Morris offers an interesting look at the transformations of family farms in the southern Great Plains from immediate family operations to corporate farms governed by an extended, nonresidential family. The piece by Dan Flores on why national park formation in the region has failed to materialize also demands attention.

Stefferud, Alfred. 1948. *Grass: The Yearbook of Agriculture, 1948.* Washington, DC: Government Printing Office.

While some of the material is dated, this yearbook contains a wealth of information about grasses and grasslands. It is still a valuable reference tool.

Stewart, Omer C. 2002. *Forgotten Fires: Native Americans and the Transient Wilderness,* edited by Henry T. Lewis and M. Kate Anderson. Norman: University of Oklahoma Press.

Initially completed in 1954, Stewart's manuscript contains a wealth of information depicting how Indian peoples purposefully managed their environments through burning practices. The chapters covering the grasslands are replete with solid evidence supporting his contention. However, editors uniformly refused to publish Stewart's groundbreaking research, and this remained the case until Lewis and Anderson edited his original work and published it in 2002.

Stuart, David E. 2000. *Anasazi America: Seventeen Centuries on the Road from Center Place.* Albuquerque: University of New Mexico Press.

Stuart presents a controversial thesis about Chacoan culture as a class-ridden system that overexploited its agricultural resources. As this happened, Chacoans became a highly aggressive people who waged extensive warfare on neighboring peoples.

Svobida, Lawrence. 1940, 1986. *Farming the Dust Bowl: A First-Hand Account from Kansas.* Foreword by R. Douglas Hurt. Lawrence: University Press of Kansas.

Svobida took up farming in southwestern Kansas just before the advent of the great dust storms of the 1930s. His is perhaps the best firsthand account of the trials and tribulations of living through those difficult years.

Switzer, Jacqueline Vaughn. 1997. *Green Backlash: The History and Politics of Environmental Opposition in the U.S.* Boulder, CO: Lynne Rienner Publishers.

Switzer traces the current anti-environmental movements to antecedents dating to the time when the United States came into being. She shows how movements such as "wise use" and the "sagebrush rebellion" have much in common with historical attitudes toward federal control over land policy and are movements encompassing highly diversified interest groups. The chapters pertaining to ranchers and farmers are very illuminating.

Unruh, John D. 1979. *The Plains Across: The Overland Emigrants and the Trans-Mississippi West, 1840–60.* Urbana: University of Illinois Press.

In what is considered one of the best accounts of the overland experience along the Oregon Trail, Unruh provides a graphic picture of how Americans fared along the route. The environmental difficulties of travel by draft animals through the grasslands is especially revealing.

Wali, Mohan K. ed. 1975. *Prairie: A Multiple View.* Grand Forks: University of North Dakota Press.

This limited edition, annotated bibliography was the result of the Midwest Prairie Conference at the University of North Dakota in 1974. The volume con tains over 7,000 entries relating to the history of, and research on, the grasslands of North America. Of course, much research has been completed since this publication, but it remains one of the best starting places for anyone seeking references on the grasslands.

Webb, Walter Prescott. 1931. *The Great Plains.* New York: Ginn and Company.

In this classic, Webb depicts the adaptation Americans made in adjusting to the ecological realities of farming the shortgrass regions of North America. In a highly memorable metaphor, Web says that American civilization in the eastern forests rested on a three-legged stool: forests, land, and water. On the shortgrass regions, only land remained, and this forced Americans into a series of novel technological and legal adaptations such as the windmill and prior appropriation water doctrine. Although Webb is often criticized for his research, his theme of adjusting culture to environment still resonates with scholars.

Wells, Philip V. 1965. "Scarp Woodlands, Transported Grassland Soils, and Concept of Grassland Climate in the Great Plains Region." *Science* 148 (April): 246–249.

What accounts for the remnant conifer forests scattered throughout portions of the grasslands? Wells offered an important explanation that gave considerable importance to the role of Indian peoples' fire management for creating open grasslands dotted by pockets of conifer forests that survived beyond the retreat of the last major continental glaciation over 10,000 years ago.

West, Elliott. 1998. *The Contested Plains: Indians, Goldseekers, and the Rush to Colorado.* Lawrence: University Press of Kansas.

Indian peoples and Euro-Americans had very different views of the grasslands in the two decades following the end of the Mexican-American War. West depicts clearly how their perceptions of proper land uses led to serious confrontations along with ecological changes throughout the Central Grasslands.

White, Richard. 1983. *The Roots of Dependency: Subsistence, Environment, and Social Change among the Choctaws, Pawnees, and Navajos.* Lincoln: University of Nebraska Press.

This work is a notable contribution to the fields of environmental history and ethnohistory. The studies of the Pawnees and their ecological role in terms of their hunting and horse culture, and that of the Navajos and their pastoral practices demonstrate clearly that Indian peoples were active agents in shaping their environments.

White, Stephen, and David Kromm, eds. 1992. *Groundwater Exploitation in the High Plains.* Lawrence: University Press of Kansas.

The White and Kromm collection includes the work of scholars covering a broad range of issues and subjects related to the Ogallala Aquifer. Rather than being historical, the works address geography and contemporary questions.

Wood, Judith Hebbring. 2000. "The Origin of Public Bison Herds in the United States." *Wicazo Sa Review* 15 (Spring): 157–182.

This is a thorough historical overview of the people who worked to preserve bison. Wood details the efforts of many whose names are often overlooked in this story. She includes the relatively unknown labors of Mary Goodnight, Samuel Walking Coyote, and Sarah Larabee, a Cheyenne woman, along with the work of other, more commonly recognized people such as William Hornaday.

Worster, Donald. 1979. *Dust Bowl: The Southern Plains in the 1930s.* New York: Oxford University Press.

Worster's account pins the cause of the Dust Bowl squarely on an American market culture, one that empowered people with a God-given right to exploit and transform the grasslands for profit. Government had the obligation to protect and enhance this right. In so doing, much to their own detriment and to that of the land as well, people ignored the ecological realities and constraints of farming the Great Plains.

Worster, Donald. 1985. *Rivers of Empire: Water, Aridity, and the Growth of the American West.* New York: Pantheon Books.

Based on an application of the Frankfort School of Criticism, Worster portrays the history of water development throughout the grasslands as one that has led to the degradation of water systems, the centralization of power in a few bureaucracies and corporate entities, and social impoverishment.

Young, James A., and B. Abbott Sparks. 1985. *Cattle in the Cold Desert.* Logan: Utah State University Press.

At best, ranching has always been a difficult proposition throughout the Great Basin. Young and Abbott clearly portray the development of ranching from the arrival of Euro-American ranchers in the 1860s to the decade of the 1890s when harsh economics and environment devastated the industry.

Zimmerman, John L. 1990. *Cheyenne Bottoms: Wetland in Jeopardy.* Lawrence: University Press of Kansas.

Mostly a natural history of this wetland of international importance, the work also tells how conservationists, hunters, entrepreneurs, and environmentalists have worked to preserve and enhance Cheyenne Bottoms.

SELECTED PERIODICALS

Agricultural History, 1924 to the present.

This journal is a rich source of articles pertaining to the history of agricultural policies at both the state and national levels. Also, articles cover the historical changes in farm and ranch economics, technology, conservation, and rural society.

Environmental History, 1996 to the present.

This is the journal of record for the American Society of Environmental History, which was formed in 1976. In the same year John Opie began as editor of the *Environmental Review,* which became the *Environmental History Review* in 1990, and was renamed *Environmental History* in 1996.

Great Plains Quarterly, 1981 to the present.

This scholarly journal is published on a quarterly basis by the Center for the Studies of the Great Plains located at the University of Nebraska-Lincoln. The journal publishes refereed, scholarly articles on the history, literature, and culture of the Great Plains and is an excellent resource on the history of the grasslands.

Great Plains Research, 1991 to the present.

This journal is also published by the Center for the Studies of the Great Plains. Its focus is more on social science research from such disciplines as geography and economics than on the history and culture emphasized in the *Great Plains Quarterly.*

High Country News, 1970 to present.

Wyoming rancher and environmentalist Tom Bell founded and managed this paper in Lander for its first years. He was concerned with environmental issues regarding ranching, federal land policies, water, conservation, wilderness protection, and community development. The paper normally does not cover issues arising in the states comprising the central grasslands—those in the corridor from North Dakota to Texas; however, it does address grassland issues in the states to the west. Relocated in 1983 to Paonia, Colorado, this paper continues to provide excellent bimonthly coverage of these issues.

Plains Anthropologist, 1947 to the present.

This scholarly journal is published by the Plains Anthropological Society and is the best resource for anthropological research pertaining to anthropological topics associated with the grasslands.

Western Historical Quarterly, 1970 to the present.

The journal of record for historical research of the American West, its articles cover a broad range of topics crucial to the understanding of ranching, farming, and environmental politics throughout the grasslands.

There are several state, local, and specialized journals that are very useful for exploring the environmental history of the grasslands. The following list is by no means exhaustive, but it will give the reader some idea of the periodicals publishing articles on the environmental history of the grasslands:

American Indian Quarterly, 1874 to the present; *Annals of Wyoming,* 1928 to the present; *Chronicles of Oklahoma,* 1921 to the present; *Ecological Monographs,* 1931 to the present; *Ecology,* 1920 to the present; *Ethnohistory,* 1954 to the present; *Journal of Arizona History,* 1959 to the present; *Journal of the Southwest,* 1959 to the present; *Kansas Historical Quarterly,* 1931 to 1977; *Kansas History,* 1978 to the present; *Montana, the Magazine of Western History,* 1955 to the present; *Nebraska History,* 1919 to the present; *Nevada Historical Society Quarterly,* 1957 to the present; *New Mexico Historical Review,* 1926 to the present; *North Dakota History,* 1945 to the present; *Pacific Histori-*

cal Review, 1932 to the present; *South Dakota History,* 1970 to the present; *Southwestern Historical Quarterly,* 1896 to the present; and the *Utah Historical Quarterly,* 1928 to the present.

INDEX